Europe's Space Programme
To *Ariane* and Beyond

T0220251

Springer
London
Berlin
Heidelberg
New York
Barcelona
Hong Kong
Milan
Paris
Santa Clara
Singapore
Tokyo

Brian Harvey

Europe's Space Programme

To *Ariane* and Beyond

Springer

Published in association with
Praxis Publishing
Chichester, UK

Brian Harvey, M.A., H.D.E., F.B.I.S.
2 Rathdown Crescent
Terenure
Dublin 6W
Ireland

SPRINGER–PRAXIS BOOKS IN ASTRONOMY AND SPACE SCIENCES
SUBJECT *ADVISORY EDITOR*: John Mason B.Sc., M.Sc., Ph.D.

ISBN 1-85233-722-2 Springer-Verlag Berlin Heidelberg New York

British Library Cataloguing in Publication Data
Harvey, Brian, 1953–
  Europe's space programme : to *Ariane* and beyond. –
  (Springer-Praxis books in astronomy and space sciences)
  1. Astronautics – Europe
  I. Title
  387.8'094
  ISBN 1-85233-722-2

Library of Congress Cataloging-in-Publication Data
A catalogue record for this book is available from the Library of Congress

Printed by MPG Books Ltd, Bodmin, Cornwall, UK

Cover design: Jim Wilkie
Cover images: courtesy of the European Space Agency
Project management by Originator, Gt Yarmouth, Norfolk

Printed on acid-free paper supplied by Precision Publishing Papers Ltd, UK

Contents

Preface

Rockets were invented in China in the 13th century. Spaceships were designed in Russia in the 19th century. But the first modern rocket was built and flown by Germany from a base in the Baltic Sea on 3 October 1942. It was the A-4, V-2 rocket, later to terrorize wartime Britain and Belgium. The V-2 flew 100 km high, 280 km down-range. It had all the characteristics of the modern rocket – liquid fuel engines, guidance systems, a payload. When it disappeared into the distance, the programme director General von Dornberger turned to designer Wernher von Braun and exclaimed: 'Do you realize that today the rocket was born?' 'Yes', he replied, 'it's a shame it landed on the wrong planet!' But the V-2 came, in the course of happily more peaceful times, to lay the foundation of all the post-war rocket programmes. Although later than the Soviet Union and the United States, Europe came in the course of time to develop a highly successful space programme in its own right, and this book will mark its 60th anniversary.

By the end of 2002, 60 years after the V-2 made its first flight, Europe had become the third space 'nation' of the world, with over 150 launches to its credit – far behind the Soviet Union and the United States, but far ahead of China, India and Japan. This was not a natural or easy process. The development of Europe's space programmes was full of false starts, cancelled projects, misadventures and might-have-beens. In the period after the war, Britain, France and Italy tried to build national space programmes. Weakened by the war and lacking the same political and military imperatives of the two superpowers, their efforts were slow, limited and competed poorly with other projects for post-war social and industrial reconstruction.

In the late 1950s, the path for European space development seemed fairly clear: it would be led by Britain. So thought an American assessment at this time, which perceived Britain to have the most advanced engineering and scientific capability of all the European countries. Few would have disagreed. These analysts were vindicated when Britain took the lead role in creating and sustaining the new European Launch Development Organization (ELDO) and the European Space Research Organization (ESRO), formed over the period 1960–4. Britain offered its *Blue*

Streak rocket as the basis of the European launcher, the *Europa*. Britain was the driver of the satellites developed by ESRO. Britain also had its own small launcher *Black Arrow* to supplement these cooperative efforts. Led by Britain, Europe would emerge in the 1960s as a respectable third space power.

But how different things turned out! France had been working quietly on a small national launcher, *Diamant* and had built a base in the Algerian desert. When General de Gaulle formed a national space agency, Centre National d'Etudes Spatiales (CNES), in 1962, critics presumed that this was just another example of Gaullist vainglorious dirigisme, but they underestimated his seriousness of purpose. De Gaulle had grasped the lesson that a modern space programme required a cohesive and highly structured management. CNES has built on that strength ever since. France became the third space-faring nation in December 1965. The Germans were welcomed back into the space community and began to build a broad-based industrial and engineering infrastructure in the 1960s, one that was to become the basis of an understated, but huge contribution to European space activities. Italian engineers fast-tracked their way into the space industry through some extraordinary lateral thinking, working with the United States to fire small *Scout* rockets from a converted oil rig off the Kenyan coast. More of the Italians in a moment.

The *Europa 1* launcher became, in an industry always a split second from disaster, a byword for heart-wrenching failure. Eleven times the *Europa* launcher was fired. Successful suborbit flights paved the way for what should have been a triumphant entry to orbit. Yet, every time, one of the upper stages failed and satellite debris rained into the oceans. The country that had led ELDO and ESRO now lost interest, tired of sharing the mounting bills. Its patience was doubly tried by the fact that the British first stage *Blue Streak* worked perfectly when upper stages went wrong. For the Wilson government of 1964–70, *Blue Streak* and *Europa* were just two of many Conservative projects with untrustworthy European partners that had to be cancelled in difficult economic times. So Britain withdrew from ELDO, reducing the national *Black Arrow* programme to the lowest level of operation short of cancellation. If the Wilson government strangled the project, the subsequent Heath government acted as undertakers, cancelling the project before Britain eventually became the sixth country in space a year after Japan and China. Fresh efforts to rebuild a British space programme in the mid-1980s ended in tears.

The failure of ELDO and the *Europa* launcher forced European designers, engineers and politicians through a painful reappraisal. They grasped that the real problem was managerial rather than technological. The reason that European rockets crashed had less to do with their engineering, which was basically sound, but more to do with poor coordination, management and leadership. Into the vacuum created by the British exit stepped France. When the eleventh *Europa* fell into the ocean, France decided, at the worst possible time, to put forward plans for a new European launcher, the L3S. This would be coordinated by a new space agency, one that would fuse ESRO and ELDO and provide strong management and direction. Complex institutional and financial formulae, echoing the many compromises that characterized the politics of the European Community and European Union, were devised to keep as many countries on board as possible, all with slightly different, if not downright diverging interests. France said it would manage the L3S effort and, never afraid to put its money where its mouth was, undertook to pay for the project. A huge launching site was carved out of the South American jungle near the notorious penal colony that once was home to Papillon.

Few people gave the L3S project much of a chance. Britain did not participate, preferring to have its satellites put into orbit by cheaper American launchers. Germany was more interested in joining the Americans in building *Spacelab*, which was to fly into orbit in the payload bay of the shuttle. As the French cleared the jungle, the sceptics gathered and the pessimists moved in for the kill. The L3S was renamed *Ariane*, an ominous choice, for in classical mythology Ariane died from wounded love when she was abandoned.

Ariane flew on Christmas Eve 1979. Far from being unloved, or abandoned, it became one of the most successful, reliable and moneymaking rockets in history. Not that *Ariane* did not suffer the occasional setback, failure or disaster: like all good rockets, it did. But each time it was patiently rebuilt. When the shuttle exploded in 1986, *Ariane* was the only Western rocket available for the long queue of commercial satellite companies trying to get their stranded payloads off the ground. Not much has been heard from the sceptics since 1979. *Ariane*'s main problem has been trying to keep up with its long and lengthening order book. Critics of the L3S underestimated French leadership, persistence, management and determination. They did not understand that *Ariane* was engineered to conservative principles. It was not designed for high performance, but to work effectively, lift reliably and operate smoothly.

The care put into the institutional and financial formulae that underpinned the European Space Agency (ESA) showed that the lessons from ELDO and ESRO had been well learned. ESA came under strain many times subsequently. It suffered badly in the mid-1980s from the policies of Britain's Kenneth Clark and was put under great stress from German financial retrenchment following reunification in 1990. In both cases it managed to pull through, battered but still in business. *Ariane*'s early success gave Europe the confidence to do something inconceivable when it was established: to send a mission to a comet. This was no ordinary comet, but Halley's comet, the most famous of them all. Unfazed by America pulling out of a comet flyby mission, ESA devised a small probe to fly only 596 km from the nucleus. Dispatched by *Ariane*, the *Giotto* probe scored a bull's eye in one of the great moments of space exploration. On the experience of *Giotto*, Europe built a fleet of adventurous space missions: *Ulysses* to fly over the solar poles, the infrared solar observatory and the *XMM/Newton* telescope. Others are in the pipeline: *Rosetta*, to circle a comet in the icy wastes of Jupiter's distant orbit; *Huygens* to land on the mysterious moon Titan; *Mars Express* to explore the red planet. The science programme for the next 15 years will attack some of the most fundamental questions of science, like terrestrial planets around stars, gravitational lines and the nature of the big bang.

Much of European space development has been about more mundane things of great, though sometimes unappreciated everyday importance. The space applications programme, originally introduced in the days of ESRO, was developed by ESA for experimental communications satellites like *Olympus* and *Artemis*, weather satellites like *Meteosat*, observation satellites like *ERS* and *Envisat*, Maritime communications systems like *Marecs* and the forthcoming navigation satellite system *Galileo*. ESA designed and flew experimental spacecraft that paved the way for operational European, national and commercial systems. Nowadays, although they probably don't realize it, most Europeans have some connexion, however indirect, with the work pioneered by ESA through receiving satellite television, watching the weather forecast or making a long-distance telephone call.

ESA's growth was paralleled by the development of national space programmes. Most European countries ran national space programmes in addition to the contribution they made to the efforts of ESA (though some, especially the smaller countries, put all their resources into ESA). Britain developed many successful scientific satellites, flown into orbit by the Americans. Italy pioneered a tethered satellite. The Netherlands built their first satellite and went on to build a stunningly successful infrared observatory. Sweden built satellites to explore the magnetic phenomena of high latitudes, with the evocative nordic names of *Viking, Freja* and *Odin*. In Britain, Surrey Satellite Technologies led the world's development of small satellites, microsatellites, picosatellites and nanosatellites. Spain developed a communications satellite *Hispasat*. An Austrian cosmonaut was flown by the Soviet Union up to the *Mir* space station.

This brings us to the question of the first Europeans in space. The first opportunity to do so came in the early 1970s, when the United States invited Europe to join them in the post-*Apollo* programme in general and the space shuttle in particular. Led by Germany, ESA built the scientific module flown in the shuttle's cargo bay *Spacelab*. This was an expensive project, and there was an assumption that many Europeans would fly into orbit for week-long experimental missions on *Spacelab*. An astronaut corps was even formed for the purpose, and one lucky member flew aboard the first *Spacelab* in 1983. In the event, *Spacelab* flew only 22 times and few Europeans got to fly onboard. Instead, many of the Europeans who made it into space in the 1980s and 1990s did so as guests of the Soviet Union on the *Salyut* and *Mir* space stations. They flew as part of national, bilateral missions involving France, Germany and Austria. The high point of this experience was the two ESA *Euromir* missions in the mid-1990s. Here, Europeans flew to *Mir* for missions of 35 and 188 days at a fraction of the cost and of much longer duration than *Spacelab*.

Europe's experience of *Spacelab* and *Mir* posed the final question for Europe: should it have its own independent manned space programme? For the French, this was not a question deserving much reflection. Plans for a European, manned space programme were already on French drawing boards before the first *Ariane* flew. The French proposed, in the mid-1980s, a three-legged approach to European manned spaceflight. First, Europe would build a space laboratory *Columbus* to be attached to the American-led international space station. Second, Europe would build a new booster rocket *Ariane 5*, a logical development of the earlier *Ariane 1, 2, 3* and *4*. Third, Europe would build its own spaceplane *Hermes*, crewed by European astronauts, to ride into orbit atop the *Ariane 5* and land back in France. Fourth, though this evolved slightly later in the sequence, Europe would build a large, cargo-carrying spaceship to bring supplies up to the space station. Originally, this was functionally called the Automated Transfer Vehicle (ATV), but later was more poetically named the *Jules Verne*.

Ariane 5 may have been portrayed as a logical successor of previous versions of *Ariane*, but this was where the similarities ended. It was a huge, hydrogen-fuelled monster, with two gigantic, strap-on, solid engines and was able to put two large comsats into 24-hr orbit. The space centre in French Guiana saw its third great expansion. The sceptics and critics had an unexpected day out when *Ariane 5* exploded spectacularly on its first mission. But within a few years *Ariane 5* was hauling cargoes into orbit in a routine, unnoticed, profitable manner, taking with it half the world commercial satellite market.

Although *Ariane 5* was a success and *Columbus* and *Jules Verne* were well in train, ESA suffered one of its rare defeats in *Hermes*. The project may have been one step too far, too soon. Many American and Russian design bureaux were littered with the designs of failed spaceplane projects, from *Dynasoar* to *Spiral*, and the Japanese were soon to add *HOPE* to the sorry list. *Hermes* was over-cost, over-weight and over-designed and fell victim to the German budget axe of 1992. Its one legacy was the European astronaut corps. In Cologne, 17 European astronauts now train together in a standard blue European overall. Who could have imagined that when *Europa* went down in flames?

The European space industry now employs 200,000 people. Its rockets and spacecraft are recognized for their excellence. The European space budget is second only to the United States. An entire city, Toulouse, has been regenerated around the space industry. The space centre of Kourou in French Guiana has joined the list of the world's great space centres, alongside Baikonour (Kazakhstan), Cape Canaveral, Plesetsk (Russia) and Jiuquan (China). Its deep-space missions combine ambition with adventure. None of this has been achieved easily, but through the persistence of engineers, scientists, thinkers and bureaucrats who struggled to overcome immense difficulty, heartbreak and multiple setbacks. This is the story.

NOTE ON CURRENCIES

From the 1970s the ESA used the notional European currency, called the Monetary Accounting Unit (MAU), also referred to as the European Accounting Unit (EAU). This became the European Currency Unit (ECU), sometimes called the 'ecu' in the 1980s and subsequently the euro (€). For the sake of convenience this will be the unit most commonly used and the € sign will be used retrospectively to refer to the MAU and ECU. Reference will be made to national currencies where this is more appropriate. Exchange rates given will be those operating at the time of the reference, though readers should be warned that exchange rate changes over the period of the study have ranged from the substantial to the volatile.

ACKNOWLEDGEMENTS

Many people made this book possible. First, I wish to thank Clive Horwood of Praxis Publishing who conceived the idea of a book on the European space programme. Second, I wish to thank my friends and colleagues who follow contemporary space programmes and who provided ideas, information, sources and much more, principally Rex Hall, Phil Clark and Dave Shayler. Third, I wish to thank the many space agencies, companies, institutes, other bodies and individuals who provided information, photographs and other material for use in this history. They are: Laurent Zimmermann, Alcatel; Michaela Gitsch, Austrian Space Agency; Audrey Nice, Surrey Satellite Technologies Ltd; Helen Westman, Swedish Space Corporation; British National Space Centre; Helga Schib, Contraves; Daimler-Chrysler; JenaOptronik; Astrid Foadi, DLR, Deutsches Zentrum für Luft & Raumfahrt; MAN Group, Germany; Christine Windmejer, NIVR, The Netherlands; B.J. Meier, NLR, The Netherlands; J-L. Lacroix, Belgospace; Viviana Panaccia, Alenia Spazio; Augusto Trivulzio, Galileo Avionica; FIAR; Snecma; CNES; Martha Davidovics, Hungarian Space Office; Gisela Eibisch, JenaOptronik; Sandra Cicchitti, Agenzia Spaziale Italiana; Patrice de Lanversin, EADS Launch Vehicles. I am grateful to them all.

Abbreviations

AATSR	Advanced Along-Track Scanning Radiometer
ACRV	Assured Crew Return Vehicle
ADM	Atmosphere Dynamics Mission
AGI	Année Géophysique Internationale
ALTEC	Advanced Logistic & Technological Engineering Centre
AMPTE	Active Magnetosphere Particle Trace Explorer
ARCAD	Arctic Auroral Density
ARD	Atmospheric Re-entry Demonstrator
ARTISS	Agricultural Real Time Imaging Satellite System
Artemis	Advanced Relay Technology Mission
ASAP	*Ariane* Structure for Auxiliary Payloads
ASAR	Advanced Synthetic Aperture Radar
ASI	Italian Space Agency
ASTRA	Applied Systems and Technologies for Future Space Transport
ATLAS	Atmospheric Laboratory for Applications and Science
ATSR	Along-Track Scanning Radiometer
ATV	Automatic Transfer Vehicle
AU	astronomical units (the mean distance from the centre of the Earth to the centre of the Sun)
BDLI	German Aerospace Industries Association
BIS	British Interplanetary Society
BNSC	British National Space Centre
CalTech	California Institute of Technology
CAT	Capsule Ariane Technologique
CDTI	Centre for the Development of Industrial Technology
Cerise	Caractérisation de l'Environnement Radioélectriques par un Instrument Spatiale Embarque
CERN	European Centre for Nuclear Research
CHAMP	CHallenging Minisatellite Payload
CIPE	Interministerial Committee for Economic Planning
CNES	Centre National d'Etudes Spatiales

COBE	COsmic Background Explorer
COF	*Columbus* Orbital Facility
COPERS	Preliminary European Commission on Space Research
COROT	COnvection, ROtation & planetary Transients
COSPAS	COsmocheskaya Sistema Poiska Avarinykh Sudov
CRISTA	Cryogenic Infrared Spectrum Telescopes for the Atmosphere
CRV	Crewed Return Vehicle
CSG	Centre Spatial Guyanais
CST	Centre Spatiale de Toulouse
DARA	Deutsche Agentur für Raumschifffahrtangelegenheiten
DASA	Deutsche Aerospace
DERA	Defence Evaluation and Research Agency
DFS	Deutsche FermeldeSatellit
DLR	Deutsches Zentrum für Luft & Raumfahrt
DIAS	Dublin Institute for Advanced Studies
DMA	Délégation Ministerielle pour les Armaments
DMC	Disaster Monitoring Constellation
DORIS	Doppler Orbitography and Radio positioning Integrated by Satellite
DSB	Danish Space Board
EAU	European Accounting Unit
EADS	European Aerospace and Defence Industry
ECS	European Communications Satellite
ECU	European Currency Unit
EGPM	Earth Global Precipitation Mission
ELA	Ensemble de Lancement Ariane
ELDO	European Launch Development Organization
EPONA	Energy Particles ONset Administrator
ERA	European Robotic Arm
ERS	Earth Resources Satellite
ESA	European Space Agency
ESC	Etage Supérieure Cryogénique
ESDAC	European Space DAta Centre
ESLAB	European Space LABoratory
ESOC	European Mission Control (European Space Operations Centre)
ESRANGE	European Sounding rocket RANGE
ESRIN	European Space Research INstitute
ESRO	European Space Research Organization
ESTEC	European Space TEchnology Centre
Eureca	European Retrievable Carrier
FCB	Functional Control Block [FGB in Russian]
FESTIP	Future European Space Transportation Investigation Programme
FIRST	Far InfraRed Submillimetre Telescope
FLEVO	Facility for Liquid Experimentation and Verification in Orbit

FLTP	Future Launchers Technology Programme
FPS	Forschungsinstitut für Physik der Strahlantriebe [Research Institute for the Study of the Physics of Jet Propulsion]
FP4	Fourth framework programme for research
FP5	Fifth framework programme for research
GAUSS	Gruppo di Astrodinamica dell'Università degli Studi la Sapienza
GCHQ	General Communications Head Quarters
GDR	German Democratic Republic
GFZ	GeoForschungsZentrum (Geological Research Centre)
GLONASS	Globalnaya Navigatsionnaya Sputnikova Sisteme
GMES	Global Monitoring for the Environment and Security
GOCE	Gravity Ocean Circulation Evolution
GOMOS	Global Ozone Monitoring by Occultation of the Stars
GPS	Global Positioning System
GROC	Geophysics & Space Research Committee
GSOC	German Space Operations Centre
HEOS	High Eccentric Orbit Satellites
Hipparcos	High Precision Parallax Collecting Satellite
HOTOL	HOrizontal Take-Off & Landing
HOPE	H-II Orbiting PlanE
HTP	High-Test Peroxide
Hyper	High precision cold atom interferometry in space
IASTPP	Inter-Agency Solar Terrestrial Physics Programme
ICE	International Cometary Explorer
IKI	Institute for Space Research in Moscow
IML	International Microgravity Laboratory
IMSL	International Microgravity Sciences Laboratory
Inmarsat	INternational MARitime SATellite Agency
IRAS	Infra Red Astronomy Satellite
IRF	Institut for RymdFysik [Institute for Space Physics]
ISAS	Institute for Space and Astronautical Research
ISEE	International Sun Earth Explorer
ISO	Infrared Space Observatory
ISPM	International Solar Polar Mission
ISS	International Space Station
ISU	International Space University
IUE	International Ultraviolet Explorer
IWACKS	Imaging Wide Angle Camera for X-ray Sources
JPL	Jet Propulsion Laboratory
LAS	Large Astronomical Satellite
Lisa	Laser Interferometer Space Antenna
LMS	Life and Microgravity Sciences
LRBA	Laboratoire de Recherches Ballistiques et Aérodynamiques
MAROTS	MARitime Orbital Test Satellite
Master	Mars and Asteroid

MAU	Monetary Accounting Unit
MBB	Messerschmitt Bölkow Blohm
MERIS	MEdium Resolution Imaging Spectrometer
Metop	METeorological Operational Polar Satellite
MIPAS	Mitchelson Interferometric Passive Atmosphere Sounder
MIRAS	*Mir* Infra Red Atmosphere Spectrometer
MIRKA	microgravity re-entry capsule
MMH	monomethyl hydrazine
MOP	Meteosat Operational Satellite
MSG	Meteosat Second Generation
MSL	Microgravity Sciences Laboratory
MWR	Micro Wave Radiometer
NASA	National Asronautics and Space Administration
NGST	Next Generation Space Telescope
NISO	Netherlands Industrial Space Organization
NIV	Institute for Aeroplane Development
NIVR	Netherlands Agency for Aerospace Programmes
NLR	Nationaal Lucht en Ruimtevaartlaboratorium [National Aerospace Laboratory]
OAO 1	Orbiting Astronomical Observatory
ONERA	Office National d'Etudes et de Recherches Aéronautiques
OSCAR	Orbiting Satellites Carrying Amateur Radio
PoSAT	Portuguese imaging SATellite
PROBA	PRoject for On Board Autonomy
PRODEX	PROgrammes de Développement d'EXpériences Scientifiques [Programme for the Development of Scientific Experiments]
PSN	National Space Programme
RA-2	Radar Altimeter 2
RAE	Royal Aircraft Establishment
RAM	Research and Applications Modules
RKA	Russian Space Agency
RLV	Reusable Launch Vehicles
ROSAT	ROentgenSATellit (X-ray satellite)
RPE	Rocket Propulsion Establishment
SARSAT	Search and Rescue Satellite Aided Tracking
SAX	Satellite Astronomica raggi-X
SCA	Système de Contrôle d'Attitude
Sciamachy	Scanning Imaging Absorption Spectromeer for Atmospheric Cartography
SCOT	Service de Consultation en Observation de la Terre (Consulting Services for Earth Observation)
SEP	Société Européenne de Propulsion
SES	Sociéte Européene des Satellites
Sicral	Satellite Italiana di Communicazioni Riservate ed Alarmi
SLED	Solar Low Emission Detector

SLS	Space Life Sciences
SMART	Small Missions for Advanced Research and Technology
SMOS	Soil Moisture and Ocean Salinity
SNAP	Surrey Nanosatellite Applications Programme
SNPE	Société Nationale des Poudres et Explosives (Company for Solid Fuels and Explosives)
SOHO	SOlar Heliosphere Observatory
SPAS	Shuttle PAllet Satellite
SPELDA	Structure Porteuse pour Lancements Double Ariane
SPELTRA	Structure Porteuse Externe de Lancements TRiples Arianes
SPOT	Satellite Pour l'Observation de la Terre
SRET	Satellites de Recherche et d'Etudes Technologiques
SRON	Space Research Organization Netherlands
SRTM	Shuttle Radar Topography Mission
SSC	Swedish Space Corporation
SSME	Space Shuttle Main Engine
SSTL	Surrey Satellite Technology Limited
STENTOR	Satellite de Télécommunications pour Expérimenter les Nouvelles Technologies en ORbite
SYLDA	SYstème de Lancement Double Ariane
TD	Thor Delta
TDF	TéléDiffusion de France
TEKES	Space Technology Development Centre
UDMH	Unsymmetrical Dimethyl Methyl Hydrazine
ULV	Unmanned Launch Vehicle
UoSAT	University of Surrey SATellite
UPM	University Polytechnic of Madrid
USML	United States Microgravity Laboratory
Vega	VEnus and HAlley
VfR	Verein für Raumschifffahrt
XEUS	X-ray Evolving Universe Spectroscopy
XMM	X-ray Multi Mirror

Figures

Tables

Chapter 4

Chapter 5

Chapter 6

Chapter 7

1

The first real rocket

Europe's experience of rocketry got off to a bad start. When British forces under Arthur Wellesley invaded Mysore in India in 1799, they came under rocket attack from Tipoo Sultan at Seringapatam [1]. The rocket was a weapon known to the Indians for some time and to the Chinese since the 13th century.

Suitably impressed by their experience in India, the British set about building their own rockets. The person who did most to develop early British rocketry was William Congreve (1772–1828), son of the Comptroller of the Royal Arsenal in Woolwich [2]. He was an inventor and entrepreneur, interested in such things as fire alarm sprinklers, canal locks and colour printing.

In Woolwich, Congreve became familiar with the rockets brought back from India. He adapted and improved them to serve the British army during the war against Napoleon Bonaparte. Congreve first developed a 15-kg powder rocket able to fly 2,000 m. During the Napoleonic wars, he had iron-cased powder rockets 1.07 m long, 10 cm diameter developed at Woolwich Arsenal. They were used devastatingly by the Royal Navy against Boulogne in 1806 (some overshot and set the town on fire) and in the attack on the French fleet in Copenhagen, Denmark in 1807. Some of his rockets were quite heavy, over 100 kg in weight.

During the Anglo-American war of 1812–14, the British used Congreve rockets in the attack on Baltimore, Maryland, an event noted in song in the anthem *The Star-spangled Banner*. Recording the American resistance, it referred to 'the rockets' red glare'. To Congreve goes the honour of writing the first rocket manual, in 1807. Later, he set up a rocket export factory and built small rockets for carrying lifelines to ships in danger of sinking at sea. His four generations of rockets have now been documented and some examples have even been recovered from old battlefields [3]. Over 100,000 rockets of subsequent inventors were manufactured for the British army in the 19th century, the last batch as late as 1901.

1.1   THEORISTS AND PIONEERS

Even as practical progress was being made, Europe had its fair share of writers on space travel. Cyrano de Bergerac (1619–1655) wrote accounts of imaginary journeys to the Moon and the Sun, called *Comic History of the States and Empires of the Moon* (1657) and *Comic History of the State and Empires of the Sun* (1662). Of the pre-20th century writers, the best known was Jules Verne (1828–1905) who wrote *From the Earth to the Moon* (1865) and *A Trip around the Moon* (1870). Verne's astronauts were launched by canon, a method that in reality would have killed them instantly. Prophetically, though, they were launched from a hillside not far from the present location of the Kennedy Space Centre. After circling the Moon, they splashed down in the Pacific Ocean where they were picked up by an American naval vessel. H.G. Wells described a Martian invasion in 1898 (*War of the Worlds*) and a landing on the Moon three years later (*First Men in the Moon*).

During the 1920s and 1930s, the first amateur experiments were made with liquid fuel rockets. The best known are those of Robert Goddard (first flight, 1926) and Sergei Korolev (first flight, 1933). However, European rocket societies were also active. Before we look at them, it may be important to consider the theorists who paved the way for the practical experiments.

1.2   HERMANN OBERTH AND HERMANN POTOCNIK

Perhaps Europe's greatest rocket theoretician was Hermann Oberth (1894–1990). He was born in Hermannstadt, now Sibiu in northern Transylvania, then part of the Austro-Hungarian empire and now part of Romania. As a young boy he remembered the arrival of electricity, then the telephone and then the car. At age 11, his mother gave him Jules Verne's *From the Earth to the Moon* and he was transfixed, reading it time after time. He spent many of his school hours, which he hated, dreaming of air and rocket flight and even did the mathematical calculations to see how they could be achieved. He fought in the war, was wounded in 1915 and spent the remaining years as an ambulance sergeant, where his talents may have been under-used. During the war he offered to the government the idea of an alcohol-fuelled rocket missile, guided by gyroscope and vanes.

He wrote his doctorate in 1922 for the University of Heidelberg. A version taking up its ideas was later published as *Die Rakete zu den Planeträumen*, (Rockets into Interplanetary Space) (1923). In this he set down the essentials of rocketry, spaceflight and interplanetary travel, much as Tsiolkovsky had done in Russia many years earlier. He wrote about fuels, tanks, pressures, coolants and guidance and even designed a model rocket to explore the upper atmosphere. In 1923, Oberth was awarded the title of professor in Mediasch, Romania. In 1924 he moved to Germany where a banker financed him for six months to make rocket experiments. In 1929 he received the recognition denied him by the university, winning the Esnault-Pelterie prize. He expanded on his earlier book with a new, much more developed version *Wege zur Raumschifffahrt* (Ways to Space Travel).

Hermann Oberth earned his living as a secondary school teacher in Transylvania. His real interest was rocketry and he had to get leave to engage in his intermittent experiments and writing. In 1927 film director Fritz Lang invited him to be technical consultant for his film *Frau im Mond* ('Woman in the Moon'), which involved him in the construction of a real liquid fuel rocket. He planned to build a rocket able to fly up to 20 km high from an island in the Baltic Sea propelled by methane or alcohol mixed with liquid oxygen. Oberth was injured during an explosion testing the rocket. In the event the film was made without a real rocket and models were constructed instead. The project broke up and Oberth returned to Romania for 1928–30. One of those who tried to assist Oberth at the time was a student engineer at a Berlin locomotive shop, a tall, blue-eyed boy called Wernher von Braun. In 1938 Oberth was offered a chair at the Technical University of Vienna, but moved back to Germany (Dresden) soon after the outbreak of the war. There, he was involved in designing fuel pumps for what became the A-4.

Even small countries had their pioneers and theorists, many of whom have not achieved the recognition they deserve. An example is the Slovene pioneer Herman Potocnik (he wrote under the *nom de plume* of Hermann Noordung). Born in Pula, Croatia, in 1892, he graduated from Vienna University in Mechanical Engineering, and in 1922 wrote *The Problem of Space Travel*. This was published in Berlin and quickly translated into Russian. In it he described a space station in geosynchronous orbit and made over a hundred drawings of how a space station might be constructed. Sadly he died of tuberculosis in Vienna in 1929, aged only 37. His life was eventually commemorated by a summer camp held in Maribor, Slovenia, on 9–10 September 1999.

1.3 ROBERT ESNAULT-PELTERIE

Returning to practical experiments, the pioneer of French astronautics was engineer Robert Esnault-Pelterie (1881–1957). He combined a mixture of theoretical and practical work and is well known as the inventor of the control stick of the aeroplane, but his interests were wide-ranging. He developed his interest in rockets in 1912 and in 1920 built liquid-fuel rockets for the French army. From 1927 to 1931 he worked on propulsion systems for the artillery with a co-worker, Jean Jacques Barré. He began giving lectures to the French Astronomical Society in 1927 and the following year he got together with a banker to offer a 5,000 franc prize to be awarded annually to the person who did most to advance space travel (Hermann Oberth was the first winner). That same year, he wrote *The Exploration by Rockets of Great Heights and the Possibility of Interplanetary Voyages*. In the 1930s he began writing for the Société Astronautique de France on the exploration by rockets of the atmosphere. In 1930 he published *L'Astronautique*, which outlined the key steps involved in the launching of an Earth satellite, interplanetary travel and rocket trajectories.

Robert Esnault-Pelterie began building his own rocket engine early the following year. In 1932 a rocket engine experiment went horribly wrong when the tetranitromethane fuel with which he was working exploded and he lost four fingers. In 1936 he built an engine that developed a thrust of 125 kg for 60 sec. Jean Jacques Barré built a nitrogen peroxide motor in 1937, which led to the launch of the first French liquid-fuel rocket, the EA-41, in secret near Lyons in 1941. Its first formal, recorded launching was near Toulon in March 1945. The main popularizer of French astronautics was Alexandre Ananoff who wrote and lectured widely about astronautics. In the 1937 universal exhibition he staged an 'Astronautics Hall' and his moon rocket became the basis of Tin Tin's cartoon moon rocket.

1.4   GERMANY: HOME OF EUROPEAN ROCKETRY

The home of European rocketry was Germany. Here, amateur societies played the key role in paving the way for the technology of rockets and satellites. The principal amateur society promoting an interest in rocketry was the VfR, or the Verein für Raumschifffahrt (Association for Orbital Flight). The VfR was founded at 6.30 p.m. on 5 July 1927 in an alehouse called the Golden Sceptre (*Goldenes Zepter*) at 22 Schmiedebrücke in Breslau [4]. The court in Breslau initially refused to register the association on the basis that the term 'Raumschifffahrt' did not then exist in the German language, but was eventually persuaded.

VfR quickly attracted over a thousand members, including Herman Oberth. The rest were a roll-call of the early days of German rocketry: Max Valier, Johannes Winkler, Eugen Saenger, Rudolph Nebel, Klaus Riedel and Willy Ley. Three years later a young 18-year-old student called Wernher von Braun joined. Many later became part of the Peenemünde team. VfR had its own magazine, appropriately called *Die Rakete* (The Rocket), which published for three years and carried articles on how to build rockets and explore the solar system. Although *Die Rakete* did not last, it is important not to underestimate its value. Many of the ideas that subsequently inspired pre-war and post-war rocketry and engineering were published there. For example, Baron Guido von Pirquet wrote a series called *Fahrrouten* (ways of travelling) outlining how space stations might be built.

VfR was interested not just in theory but in practice as well. The late 1920s and early 1930s saw a number of individuals and groups working on rocketry in Germany, sometimes connected to the VfR, sometimes not, some working in collaboration, some in an individualistic way. The first VfR rocket was called the *Kegeldüse* or cone motor, a long, thin rocket. In July 1930 they fired the *Kegeldüse* for 90 seconds in the Reich Chemical Technical Standards Office where it delivered a constant 7- kg thrust, consuming 6 kg of liquid oxygen and 1 kg of petrol. Oberth, back in Germany, persuaded the government to note it as an official record.

Next the VfR moved on to develop what it called a minimum rocket, or *Mirak* (which was tested on a farm in Bernstadt, Saxony) and then a bazooka-shaped rocket called the *Repulsor*. They got permission to make tests in the grounds of an old, disused, overgrown army munitions dump near Berlin, which they pretentiously called the *Raketenflugplatz* (rocket flight field). The platz had the advantage of having bunkers and blast walls, where rockets could be let off with impunity. They begged materials from anyone – like Shell or Siemens – who might be prepared to give them. Much of the work was done by unemployed craftsmen who had time on their hands. The first *Repulsor* reached an altitude of 61 m before it crashed into a tree, but a later version was more successful and got to 1,000 m before coming down on a parachute. *Mirak* was a thin, pencil-like rocket, which was launched several times from a rail, reaching 300 m before a parachute opened and it could drift back down. Over 1930–1, VfR fired three versions of *Mirak* and four *Repulsors*. In 1931 von Braun temporarily left the group to go to study in the Institute of Technology in Zurich, Switzerland. Here with another student he built a centrifuge to simulate the effects of rocket take-offs and acceleration on mice.

1.5 VALIER, VON OPEL AND WINKLER

Max Valier was one of the leaders of the early days and was a German of French Huguenot descent, born in Bozun in 1895 [5]. He was already studying physics, astronomy and mathematics in Innsbruck when the war broke out, and after his army service he resumed his studies. Originally, he tested solid and liquid rocket engines on the back of cars rather than on flying machines. Energized by von Oberth's book, he tried to find backers for practical rocket experiments.

Car-maker, racing driver and speedboat ace Fritz von Opel backed him with his resources. With von Opel, Max Valier experimented with rocket-propelled cars and aircraft, even developing an early liquid-fuel engine running on nitrogen tetroxide and benzol. He made a rocket-powered car and a rocket-powered glider called *Die Ente* (the duck). On 11 June 1928 Fritz Stamer made the first-ever flight of a rocket-propelled glider when he flew *Die Ente* for 60 seconds over the Wasserkuppe mountains. *Die Rakete* hailed this as the 'first-ever manned rocket aeroplane' on the cover of its subsequent 15 July issue [6]. On 30 September 1929, Fritz von Opel flew a rocket-powered glider at Rebstock, Frankfurt. He attached 16 solid propellant rockets, each of 22.7-kg thrust. With the help of the rockets, the glider flew 1,525 m for 75 seconds and reached 153 km/hr.

Valier developed powder rockets to help get Junkers aeroplanes airborne. However, he was soon won over to the power of liquid-fuel engines. He believed in testing the engines in man-operated machines first and got the cooperation of Paul Heylandt, who owned a factory making industrial gases. In April 1930 Max Valier fitted an engine making 30-kg thrust and demonstrated it on a car racing around Tempelhof aerodrome in Berlin. The car looked like a conventional racing car of the period, but lifting back the top cover revealed two cylinders along its length – one for fuel, one for oxidizer. His activities attracted much publicity and media excitement in a Weimar republic hankering for escape from the humiliation of the war defeat and the economic depression at the time. With his colleagues he planned to demonstrate the engine at an aviation show in Berlin the following month. Not long before the event, there was a terrible explosion in the workshop where Valier was testing his engine and he was killed (17 May 1930). The factory continued to support the experiments, the fuel injection systems were improved and in April 1931 Valier's colleagues at last ran their rocket-propelled car at the aerodrome. These experiments continued intermittently, interrupted by the economic depression.

At around the same time, Johannes Winkler got a contract from the Junkers aircraft company to investigate powder rockets and in April 1931 tested the idea of a rocket-assisted take-off for a Junkers seaplane. This inspired Winkler to make further tests. What was almost certainly the first recorded, successful liquid fuel rocket in Europe was fired by Johannes Winkler. Burning liquid oxygen and liquid methane, using stabilizing fins, Winkler's rocket reached 305 m near the city of Dessau, Germany on 14 March 1931. Winkler was the President of the VfR and Editor of *Die Rakete*.

Winkler, von Opel and Valier were three rocketeers pursuing independent, connected and parallel experiments. There were others. In neighbouring Austria, in Vienna, Dr Eugen Saenger had made a series of rocket experiments in the university there. Saenger was Austrian, born in Pressnitz in Bohemia in September 1905 and, inspired by Oberth's book, he obtained a doctorate in Aeronautical Engineering. He built an engine in 1931 able to generate pressures of 50 atmospheres and a thrust of 25 kg for half an hour, though it seems that combustion was less than complete. Austrian engineer Friedrich Schmiedle developed a small rocket to deliver mail across Austrian valleys, firing them 3 km from Schnoeckel to Radegund. In April 1931, German Reinhold Tiling showed off a folding-wing rocket with a solid-fuel engine (sadly he died with two assistants when mixing fuels in their laboratory the following year). VfR engineers Klaus Riedel and Rudolf Nebel managed in January 1931 to build a new, lightweight engine. That spring they developed an engine capable of generating 160-kg thrust.

1.6 ENTER THE GERMAN ARMY

The VfR activists paid for these experiments out of their own limited pockets or from whatever they could obtain from philanthropists and benefactors. They realized that as long as they were dependent on themselves and public subscriptions, their progress would be limited to very small rockets. In 1932 the VfR activists from the *Raketenflugplatz* gave a demonstration to the military at Kummersdorf outside Berlin. The military were interested, not least because rocket development was not listed as a prohibited activity under the Treaty of Versailles (the authors of the treaty had probably never imagined such a thing). More positively, several army leaders were genuinely open to the possibility of what the rocket could achieve, notably Lt Col. Karl Emil Becker. The first tests did not go well. *Mirak II* flew only 70 m before crashing and the army accused VfR of publicity-seeking amateurishness.

Despite this, the military opened the possibility of the *Raketenflugplatz* people coming to work for them. Many of the VfR were wary of the military – they would have to work secretly, and be subject to military discipline, which would cramp their style. Von Braun disagreed, arguing that their funding would make the building of real rockets possible at last. Von Braun's colleagues had, for the most part, thought they could get sufficient funding from wealthy sponsors or industrialists – and they grossly underestimated the costs that would be involved. The army, by contrast, was the only way to obtain the large, long-term funding that they needed, an experience mirrored in other countries. In an interesting parallel, the Jet Propulsion Laboratory in California struggled like the VfR – until the military were drawn in and budgets ballooned.

The military appointed Gen. Walter Dornberger to head up a special section of the Army Weapons Department to develop rockets. In November 1932, Dornberger hired von Braun, put him on the army payroll and put him in charge at the precocious age of 20! By January 1933 von Braun had developed an engine capable of generating 140-kg thrust for a minute. He was a civilian employee of the army, a technical officer, given the opportunity to develop rockets and write his PhD at the same time. The combination of von Braun and Dornberger was a critical one for the success of the German rocket effort. Their partnership ran from 1932 to 1945. Von Braun was the designer, the technical expert, the blueprint-maker; Dornberger the administrator, the director, the man who steered the project through the organizational and bureaucratic labyrinths of the Third Reich. Such military–inventor relationships underpinned the rocket efforts of many countries.

Many of the other VfR people moved over to von Braun's group over the next few months. The *Raketenflugplatz*, which the VfR had used, returned to its previous use as an ammunition depot for the German army. The VfR collapsed in the course of 1933 under the weight of internal divisions, the loss of much of its membership when *Die Rakete* ceased publication and unpaid water bills from its use of the *Raketenflugplatz*. The new Nazi government prohibited private rocketry experimentation (indeed, Goebbels even ordered the word 'rocket' not be mentioned in the press).

1.7 VON BRAUN, FATHER OF GERMAN ROCKETRY

Wernher von Braun was the technical architect of German rocketry. Born in 1912, he came from a well-off, aristocratic Prussian family – his full name was Freiherr Wernher von Braun, Baron von Braun. He might well have followed his father into politics, the military or banking, which is what his parents had hoped of him. At the age of 12, his mother gave him a telescope. He read and was captivated by Oberth's book, which turned him overnight into a spaceflight fanatic. To his parents' chagrin, he elected to go to the Technical University of Berlin, an institute that, 70 years later, was in the vanguard of small satellite research. He was awarded his doctorate in 1934 [7].

The Luftwaffe now began to show an interest in rockets. Major Wolfram von Richthofen, a cousin of the ace pilot, now head of technical development in the Luftwaffe, brazenly got von Braun and his colleagues to develop engines for aircraft and fighters, even though they were officially working for the army. Generous Luftwaffe grants smoothed out the bureaucratic hurdles. In 1937 a rocket engine was fitted to a Heinkel 112. Later, wartime Germany was to build and put into production a rocket-propelled fighter, the Messerschmitt 163. By war's end, a vertically launched rocket interceptor had been built, the Bachem Ba 349 Natter, although it crashed on its first flight.

Von Braun's first rocket, the A-1 (Aggregate-1), a finned rocket that used alcohol as fuel and liquid oxygen as oxidizer, was able to generate thrust of 300 kg. Having taken six months to build, in its first attempted flight, the A-1 exploded due to difficulties combining ignition with the precise time at which fuel was fed into the combustion chamber. 'Half a year to build, half a second to blow it up,' von Braun caustically commented.

Instead of continuing with the A-1, von Braun redesigned from scratch, this time building a new finless rocket with an internal stabilizer. The new rocket, the A-2, was a success. Two versions called *Max* and *Moritz*, after cartoon characters of the period, were fired from the North Sea island of Borkum in December 1934, reaching an altitude of 2.5 km before coming down on the beach later.

The launch of *Max* and *Moritz* so impressed the German army that substantial money now became available. His budget was increased from RM80,000 to RM6m. Von Braun and the army relocated to a small island 9 km into the Baltic Sea called Griefswalder Oie. It was a small, virtually uninhabited island with cliffs, woods and a lighthouse. Here he built the larger A-3, with a 1,500-kg thrust engine and many technical improvements such as vanes, a rudder and a three-axis gyro control system, a liquid nitrogen pressurization system and new valves to prevent explosions at the point of ignition. This was the first rocket with its own guidance system, one designed by a naval officer who had spent the war designing gyros (hitherto, rockets had been fired from ramps or rails to steer them in the right direction). It was 6.5 m tall, 70 cm in diameter and 750 kg when ready to fire.

These tests were unsuccessful. The first one had an air of farce about it. The test started ominously during the countdown when the moist sea air, combined with the cold liquid oxygen, melted the dye on the side of the rocket, dye intended to mark the splashdown point in the sea. This trickled down the side, causing electrical systems to short-circuit. Observers out at sea in a boat radioed in their impatience, brought on by a sickness-inducing swell. The first rocket went out of control after 5 seconds and the next two did no better. The guidance system seemed to be the recurrent problem.

1.8   A-5 BREAKTHROUGH

Von Braun once again went back to the drawing board, moving on to a new design, the A-5 (he was saving the A-4 title for a bigger project later. In fact the A-5 and the A-4 looked almost identical, the difference being in size). Several entirely new guidance systems were designed, even radio beams. Just before the outbreak of war, von Braun and his team fired the A-5 into clouds on an overcast day from Griefswalder Oie. Down below, von Braun and his colleagues heard the roar of the rocket continuing to burn for over a minute until, presumably, it ran out of fuel. Silence. Then, five minutes later, through the clouds they could see the A-5 drifting down into the Baltic a mere 200 m offshore. It was in perfect condition and their joy was uncontained.

Over 1939–40, the A-5 was fired 25 times (sometimes the same rocket was fired again if in sufficiently good condition). The main tasks were to test the engines and the different guidance and control systems. Initially, the A-5 was fired to achieve altitude (12 km being the norm) and then range. A-5s were even fired from underneath a Heinkel 111 bomber (the idea of launching rockets from planes was resurrected in the Pegasus project in the 1980s).

Von Braun's military backers continued to provide the financial resources, so he was able to think about setting up a proper rocket base. In 1935 he settled on what became the world's first purpose-built rocket launch centre, Peenemünde, a tongue of land of pinewoods and reedy promontories stretching out from the island of Usedom into the Baltic Sea 97 km north-west of the town of Stettin (Peenemünde means 'the town at the mouth of the river Peene'). It was a land of ducks, deer and white-tailed sea eagles. Indeed, the von Braun family knew of it because some of its members used to go duck-shooting there and his mother recommended the site [8]. Construction concluded two years later. The facilities included a wind tunnel able to generate velocities up to Mach 5 and a flight mechanics computation office to develop guidance systems. Rockets could be launched due east down the Baltic without the risk of crashing onto land (unless they went off-course). The facilities comprised engine test stands, launch pads, observation posts, sheds, factories, workshops, a power station and an airfield. Camouflage netting was installed after the war broke out. It was not far from the old launch site of Griefswalder Oie, which was just off the coast. Von Braun was able to hire all his old colleagues from the *Raketenflugplatz*. How times had changed: from working in sheds for nothing to a RM17m purpose-built rocket centre with an operating budget of RM6m and himself as Head of Works at the age of 25! In 1941 he was joined by Hermann Oberth who had become a German citizen and joined the Peenemünde team, working on the A series and also on anti-aircraft missiles in Reinsdorf.

If von Braun thought that the outbreak of war would lead to increased funding for his rocketry, he was soon disappointed. Some Peenemünde personnel were quickly drafted into army units. The A-5 could not be immediately applied to the conflict that was, in any case, not expected to last more than two years. The rocket project languished in the Reich's bureaucracies. The war economy meant that there was a huge demand on labour, skilled labour, steel and all kinds of production, a more competitive environment in which Peenemünde inevitably fared less well.

1.9 ORIGINS OF THE A-4

Dornberger did explore with the army the potential of an upgraded A-5 as a long-range artillery shell. Large A-5s could be transported by railway to near the front line and then fired up to 300 km at military targets way beyond the range of existing artillery. This led to the A-4, a scaled-up version of the A-5. The A-4 was seen, in

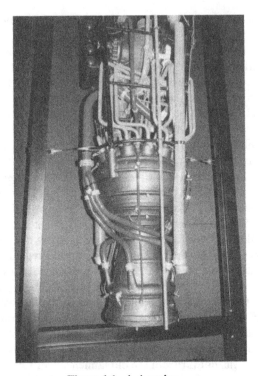

Figure 1.1. A-4 engine.

effect, as a long-range gun ahead of the advancing German army. Some saw it as a means of hitting English factories where the Luftwaffe had failed in 1940.

Scaling up the A-5 to the A-4 was easier said than done. It involved considerable engineering challenges in improving the turbopumps to feed in oxidizer and fuel, injector systems for the thrust chamber, and control and guidance systems. The A-4 was built to develop a thrust of 25 tonnes. Developing the rocket involved mastery of high-temperature turbines running at over 500°C, pressurizing gases, coolants and ignition systems. Building such a large engine took from 1937 to 1941 and was the ultimate achievement of the senior engine designer, Walter Thiel. Von Braun had also to make significant advances in the understanding of supersonic aerodynamics (made possible by the wind tunnels of Peenemünde), guidance, control, fins and stabilization systems.

The A-4 took the better part of five years from conceptualization through to test flight. It was an achievement of technical brilliance, lavish funding, tight organization and Dornberger's insistence that all rocketry research must be gathered together 'under one roof'. Construction of the A-4 took place during an opportune time in German rearmament when virtually any project was able to obtain some funding. Ironically, no-one seemed to have a clear idea of how to use the weapon strategically: it was almost like an updated version of the Paris gun of 1918, able to bombard the

French capital from behind the German lines, except that the A-4 could send a bigger warhead further. Despite this, Dornberger was talking about a large-scale production plant in Peenemünde from as early as 1938.

The first results suggested that they had a long way to go in mastering these problems. When on 13 June 1942 the first A-4 lifted off after its first, flawless countdown, it headed into low clouds at 80 m. There the battery failed, the engine cut out, the rocket began rolling, lost its fins and toppled out of the clouds 600 m into the sea. A gyro failure, soon after lift-off, was later blamed. The armament minister Albert Speer, who was watching, was far from discouraged and gave the rocket the super-priority rating.

The second launch, on 16 August, went much better, reaching the speed of sound after 45 seconds. Just at this point, the A-4 swerved away from its flight path and broke up in a puff of smoke, debris coming down 870 m away. Later analysis suggested that a turbopump had failed, the engine cut, the rocket could no longer be controlled without propulsion and, in the still dense layers of the lower atmosphere, it quickly tore apart.

Would it be third time lucky? This time, they painted a picture of the '*Girl on the Moon*' (*Frau im Mond*) from the Fritz Lang film to wish it well. General Walter Dornberger, whose title was commanding officer of the Peenemünde Army experimental station, recalled that there was a real do-or-die feel about that day and that a continuation of the programme would have been difficult had there been a third failure. He knew that the senior political patrons of the programme, right up to the Führer, would soon lose confidence in the project.

1.10 THE FIRST FLIGHT OF A REAL ROCKET: 3 OCTOBER 1942

The 3rd October 1942 was a warm, autumn day with clear skies. The order was given for the 3-min countdown. The rocket weighed 13.5 tonnes. The tanks pressurized. The ventilators closed. A green flare was fired over the site at T-10 seconds as a final warning to take cover. At zero, sparks came out of the base of the rocket, a flame emerged and steadied to an orange glow, the electrical umbilical fell away. The engine began to really generate thrust now, with 150 litres of fuel pouring in each second and the turbopump revving at 4,000 r.p.m. When thrust reached 25 tonnes, up the A-4 rose, gathering speed above the pines of Peenemünde, leaving behind a cloud of dust for the hopeful watchers.

The rocket kept a steady course, bent gradually over in its climb out over the sea. From the ground, Dornberger and his colleagues saw a steady flame as they were drowned in the rumbling roar of the engine. The A-4 began to tilt over at 4.5 seconds, eventually reaching 45°, the angle needed for maximum range. In the observation tower, a signal relayed the rocket's speed by an acoustic pitch, rising from a gentle hum to a shrill, piping sound. The sound steadily increased. At 20 seconds the A-4 went through the sound barrier – and nothing happened, for it continued its climb completely unperturbed. The controllers, watching from the tower and on television, could see the white rocket, the black and white fins, the flame and the vapour trail. At 36 seconds the A-4 was going twice the speed of sound, the fastest velocity ever reached by a human-made self-propelled object. And the A-4 flew on and on, arcing over the edge of the Pomeranian coast.

At 40 seconds they had a fright, for the white vapour trail suddenly began to zig-zag. It wasn't the rocket veering off course, as they feared, but strong high-altitude wind currents breaking up the vapour trail. Still the A-4 climbed ever higher, followed by binoculars, radar, theodolites and all kinds of measuring devices. The A-4 exhausted its fuel at 63 seconds, reaching an altitude of 85 km and a speed of 5,000 km/hr. The flame cut out, the vapour trail ceased abruptly. It was just a tiny dot in the distance way out into the Baltic.

To the amazement of the team on the ground, they continued to hear its signal for the next five minutes, meaning that the rocket was still high above the atmosphere, curving back to Earth in a gentle, ballistic descent. The acoustic pitch had ceased to rise, but stayed steady. Ground control continued to relay the number of seconds from launch: 100, 150, 200, 250. Would the rocket tumble and break up as it met greater air resistance and then overheat as it fell back? At 296 seconds the acoustic pitch, lower but still quite high, cut out.

Several minutes later, a small spotter plane saw the green dye from the fallen rocket in the sea 190 km downrange. They had done it! Dornberger turned to von Braun and exclaimed: 'Do you realize what we have done today? Today the spaceship was born!' Von Braun remarked, 'Yes, but a pity it landed on the wrong planet!' Hermann Oberth was there, though we don't know what he made of it.

Back at Peenemünde, the scene was a riot. Normally restrained engineers let out years of tension, jumped up and down, hugged one another, poked the launch stand to see how hot it still was, all the while chattering to each other their excited observations of the tremendous events that had just taken place. Dornberger gathered his colleagues around him that evening – a dark, heavy night – and made a short speech. With the war on, the rocket must be perfected as an artillery weapon. But, he said, 3 October 1942 will always be marked as 'the day we first invaded space with our rocket and proved rocket propulsion was possible for space travel.' It was a day to match the inventions of the wheel, the steam engine and the aeroplane.

In the months that followed, von Braun and his colleagues fired more and more A-4s so as to improve reliability, range and accuracy. They didn't always work and the 3 October 1942 success was a lucky event in what turned out to be a run of failures. There were four failures in a row over the new year. Film of the period is full of A-4's toppling over, blowing up in mid-air, shooting off at all kinds of improbable angles, bursting apart, and crashing back down exactly where they started. On one celebrated occasion, the A-4 rose to about 10 m after the turbopump stopped accelerating. The A-4 just hovered there, for several seconds, before gradually gaining speed when consumed fuel made its mass ever lighter (it reached 25 m and then toppled). But over time Von Braun and his colleagues mastered its idiosyncrasies and habits until most could now be fired with certainty of success.

1.11   THE A-4 AS A WEAPON, THE V-2

News of the Baltic successes quickly reached Berlin. By this stage, the war was beginning to go badly wrong: November 1942 saw the Germans rolled back from El Alamein in North Africa and the encirclement of the 6th army at Stalingrad. The A-4 offered the opportunity not only to develop a new artillery weapon, but to bombard London across the English channel. The allies, who had begun air raids on Germany in earnest, could at least be paid back in kind. The A-4 was renamed V-2, 'V' for vengeance.

Hitler turned from A-4 critic to one of its greatest fans. The A-4 was ordered into production, probably prematurely since its testing programme took a full 18 months. Committees were set up to sweep away bureaucratic hurdles. Production plants were established in Peenemünde, Vienna (Wiener Neustadt) and in Friedrichshafen in the old Zeppelin airship sheds. As was the case with most parts of the Reich's economy at this stage, slave labour of Poles, Russians, French and other concentration camp victims was used in production.

1.12   PEENEMÜNDE ATTACKED

Peenemünde had attracted the interest of the Royal Air Force (RAF) aerial reconnaissance teams since early 1943. Allied intelligence had received many reports from diverse sources about secret weapons in preparation there and elsewhere. Intelligence experts rightly figured out that rockets, rocket planes or other forms of secret weapons were being tested and produced there and a massive air raid was ordered up to put Peenemünde out of business. On the summer night of 17 August 1943, Peenemünde was flattened by 600 RAF bombers in a 3-hr raid.

The raid, it turned out, was less than successful. Many of the bombs overshot their target. Although there was much physical damage, the eleven rocket test stands were intact, and neither the wind tunnels nor the guidance and control laboratory had even been hit. Few prominent scientists had been killed and the main causalities – 800 fatalities – were Russian prisoners-of-war on construction work. Another loss – and a heavy one – was engine designer Walter Thiel.

The raid led to the most sinister phase of the development of the A-4. The production facilities for the A-4 were moved underground into a disused oil depot in the Harz mountains. This was the notorious Mittelwerk, where A-4s could be produced far from prying RAF eyes and invulnerable to air attack. Again, slave labour was used and a concentration camp called Dora was established. There they were worked to death in appalling conditions and treated with the utmost savagery by their guards, even a suspected minor insubordination being met by mass executions. From early 1944, A-4s began to emerge from the Mittelwerk at the rate of 300 a month, the production rate later to rise to 900. Von Braun's tests were moved to a new site, Blizna near Krakow in southern Poland, far outside the reach of the allied bombers. Another production plant was established in Zement, 100 km east of Salzburg in Austria. The glory days of Peenemünde were over.

1.13   A-4 AS A MISSILE

Originally, the Germans had planned to use the A-4 against London from huge, cemented launch complexes that combined silos, fuel complexes, testing systems and the means of measuring the impact of the missile. This was von Braun's approach. In late 1943 Hitler ordered construction of the first silo missile site in the world, La Coupole near St Omer in northern France. Slave-labourers carved the huge site out of a mountain, fitting a 5-m-thick dome on top. Inside the bunkers was an entire V-2 servicing, fuelling and launching site. The allies bombed it at will, dropping over 3,000 tonnes of bombs. They were unable to penetrate the concrete, but they did destroy the surrounding railway lines and La Coupole never fired a missile in the end [9].

General Dornberger took exactly the opposite view. The best way to evade the allied air forces was not to bury the A-4s so that they could be pummelled from the air, but to develop mobile launch crews. All that was needed was a truck that could swivel the launcher into position, a command caravan and two fuelling wagons – and a handy bit of forest. The army favoured the use of mobile sites that could be erected and removed quickly in forest clearings. Although relatively unprotected, this method would conceal the V-2 more effectively from an enemy who had command of the air.

Dornberger was right. The allied air forces found ways of shattering the concrete fortifications, be they A-4 silos or U-boat pens. They rarely caught an A-4 in the forest ready to fire, although one lucky RAF pilot did, shot at it as it ascended from the trees and, by a fluke, hit the detonating pin, blowing the A-4 apart. Dornberger's approach eventually became the norm for Soviet missile forces in the Cold War and some other countries in real wars (e.g., the Iraqi Scuds), and von Braun graciously admitted that he had been wrong.

As the A-4 was made ready for use in February 1944, von Braun was summoned to Himmler's headquarters and invited to join his staff. In effect, this was a pitch by the SS to take over the programme. Von Braun refused, preferring to continue to work under Dornberger as he always had. He paid a price for his refusal. Not long after, following a midnight knock on the door, he was hauled off by the Gestapo. He was charged with expressing regret that the A-4 would be devoted to military use, preferring it to be fired into space instead. Von Braun was held in a jail for two weeks until Dornberger told the Führer's headquarters that, unless he were released, there would be no A-4 weapon.

There was a long gap between the first flight of the A-4 and its first use in anger 23 months later. The Peenemünde raid was a serious setback, although, thanks to the SS, the relocation of production into underground factories took only four months. Put simply, the A-4 was in no way ready for production and endless technical and production problems still had to be ironed out. The A-4 still had to compete with the Reich's other top projects, like the V-1 flying bomb and the Wasserfall anti-aircraft missile. Many of the A-4s produced by the Mittelwerk for tests had poor parts and workmanship, an inevitable consequence of the awful conditions under which they were produced (some parts were probably sabotaged as well). During

Figure 1.2. V-2 launching site.

the Polish tests, a new problem called air bursts appeared. This involved A-4s blowing themselves apart a short distance above their target. The search for a solution to this problem was elusive, but a strengthening of the steel panelling eventually did the trick. The builders of the A-4 were not the first to encounter persistent teething troubles, as the builders of the later *Molniyas, Atlases* and *Ariane* were to find out.

1.14   V-2 AS A WEAPON OF WAR

The first V-2 was fired at London on 7 September 1944 from a mobile site. This was the first time a city had been bombarded by a rocket from several hundred kilometres away. Ironically, for Londoners, who were already enduring bombardment from the V-1 flying bomb, the V-2 was less psychologically terrifying. With the V-1 you could hear the drone of its engine, fearing the moment it would cut out and the warhead would begin its plunge to Earth. The arrival of the V-2 came from nowhere: a huge explosion followed by the arrival of the sonic boom – but hearing it meant that you had already survived. In the end 4,320 V-2s were launched between 7 September 1944 and 27 March 1945 of which a quarter (1,120) were directed against London. There were 2,511 fatalities and 6,000 injuries.

A-4/V-2

Dimensions	14.50 m tall, 1.65 m diameter, fins 3.65 m
Weight	12.9 tonnes (incl. 8.9 tonnes propellant)
Altitude	up to 150 km
Propellant	3,810 kg methyl alcohol and 4,900 kg liquid oxygen
Thrust	25 tonnes for 68 seconds
Velocity at cut-off	5,400 km/hr
Range	322 km
Warhead	750 kg

Although the bombardment of London is strongest in popular and historical memory, in reality the Germans fired more V-2s at Antwerp, in an attempt to halt the allies' use of the port for the invasion of Germany. In the winter of 1944–5, the Germans used their bases in still occupied Holland to attack both London and Antwerp. In the event that the launchings might be forced back into the Reich itself, von Braun and his colleagues were put under pressure to develop a longer range version, known alternately as the A-9 and the A-4b (the reason for the double designation was that the A-4 series had been given 'national priority' and the A-4b title ensured the requisitioning of scarce materials). These could hit Britain from the Reich territory itself, or, if still launched from Holland, penetrate further into other British cities.

1.15 WINGED ROCKETS TO AMERICA

To test the A-4b, two large sweepback wings were fitted to two A-4s launched from Blizna. They looked impressive enough on their launch pads, deep in the Polish forest, but in reality little development work had been done to ensure their likely success. The first failed completely on 27 December 1944. In January 1945, when the war had less than five months to run, the wing broke off on the second, though not before it became the first winged guided missile to break the sound barrier and reach Mach 4. The final *fin de guerre* extravagance was Project XII, which was set up to develop a sea container for the V-2, have it towed to the east coast of the United States by U-boat and then launch it at Manhattan. The Vulkan works of Stettin even got a contract to develop Project XII, with a deadline of April 1945. The concept was not entirely impractical (Russia later developed similar seaborne rockets), but as a weapon of war at this stage it was fantasy.

More advanced designs reached the blueprint stage. The two-stage A-4b envisaged, for the upper part of the rocket, the use of gliding trajectories able to deliver payloads or warheads 5,000 km away (the distance of New York). Peenemünde blueprints included a piloted cabin on the A-4b with a tricycle undercarriage so that the cabin could land safely, though von Braun insisted that these blueprints were always kept out of sight whenever army personnel were in the building. The piloted A-4b would, in effect, have sent an astronaut on a suborbital flight. It is unlikely that von Braun's colleagues told the army much about their plans to use the A-4 to explore the upper atmosphere to collect air samples, measure cosmic rays, take the spectra of the Sun or record atmospheric pressures and temperatures. Two A-4s were even prepared to test out the possibilities for such missions. Also under consideration at Peenemünde at war's end were:

- A-10, with a thrust of 200 tonnes, eight times that of the A-4;
- A-11, intended for 'projecting a pilot into a permanent satellite orbit of the Earth'; and
- A-12, using the A-10 and A-11 stages, able to launch a 30-tonne space station into orbit around the Earth.

The Peenemünde team was not the only one with such ambitions. Comparable theoretical studies were done by Austrian Eugen Sänger and his assistant Irene Bredt. Educated in Cologne, she was one of the century's few women rocket-designers [10]. Inspired by von Opel's rocket cars, she was awarded a first-class degree in Physics. From 1929 to 1936 she and Sänger carried out experiments and designed rockets in the Technical University of Vienna. His main text was *Raken-flugtechnik* ('the technique of rocket flight') published in 1933. In the 1930s Sänger designed a number of high-altitude winged rockets called 'Silverbirds', precursors of the later spaceplanes.

Sänger and Bredt moved in 1936 to the Luftwaffe-funded Institute for Rocket Flight at Trauen near Hannover. Here they designed what was called an antipodal bomber, called the *Raketenbomber*. This was a two-stage rocketplane that would take off from a 1,860 m monorail, fly high into space, skip across the upper atmosphere and deliver a warhead 24,000 km distant (New York was the favoured target). The 900-page manuscript was recognized as a breakthrough and only 125 copies were permitted to be published (three survived the war). However, the centre never attracted the same kind of budgets as the army's research centre in Peene-münde. From 1942 to 1945, they designed ramjets for the Luftwaffe, one even operating successfully on a Dornier 217 bomber [11].

As the Russian armies marched westward, the sounds of gunfire to the east began to be heard in Peenemünde. The V-2 programme collapsed into chaos, although the slave-labourers suffered to the very end. Von Braun and his colleagues evacuated Peenemünde for Bavaria, with the hope of falling into the hands of the Americans. In this they were eventually successful. Sänger and Bredt fled to France, with Stalin's agents hot on their heels and a reward offered for their kidnap and exfiltration.

1.16 V-2 POSTSCRIPT

The allies were quick to appreciate the value of the V-2 and the need to obtain both the hardware and personnel involved. All the allied nations recognized that in the rocket the Germans had achieved a breakthrough in technology far ahead of anything that existed on their side (the Americans were especially relieved to see that Germany had made less progress than expected on the atom bomb). The French, British and Americans circulated between themselves the interrogation reports from von Braun, not quite sure what to make of the German's claim that his interest all along had been in sending people into outer space.

Allied specialists combed Germany in the summer of 1945 for A-4 parts, equipment and even complete rockets – and the scientists involved. The Americans were quickest off the mark, being especially sharp in denuding border zones of German technical equipment in the event of these areas having to be handed over to the Russians in the subsequent readjustment of territory. When they duly were readjusted and the Russians asked for 'their' parts back, the Americans duly shipped them tractor parts. The Americans were equally quick to find the mine where the V-2 archive was hidden, spiriting away the treasure trove of blueprints just hours before the British.

Under 'Operation Paperclip' the Americans arranged for German scientists – headed by von Braun – to emigrate to the United States. The horrors of the Mittelwerk were overlooked and blamed on the SS. The official American allocation was 100 scientists, but von Braun's list had 115 names on it. The von Braun group began to arrive in September 1945 and was in White Sands, New Mexico by the following month. The first V-2s were fired from there in April 1946. By 1950, the Germans had been relocated to Huntsville where they became the top rocket development body in the United States. Von Braun, the architect of German rocketry went on to an honoured career in the United States, taking American citizenship in 1955. His *Redstone* rocket was an A-4 derivative and was used to launch America's first satellite in 1958 and America's first astronaut, Alan Shepard, in 1961. He was a leading advocate and popularizer of space exploration, idealized by young Americans interested in rocketry (as the film *October Sky* later illustrated). He did much to bring a sense of purpose to American space exploration in the traumatizing period after *Sputnik*. He masterminded the *Apollo* project and designed the *Saturn V* launcher, arguably the greatest rocket of all time. It was launched 13 times. It succeeded 13 times. He was a decisive figure in the *Apollo* programme, having a key role in deciding the key strategies of the programme, including the method selected for the lunar landing (indeed, a list of the engineers for *Apollo* is a veritable who's who from Peenemünde). Von Braun died of cancer after a long illness in June 1977. Years later, as historians reappraised the Second World War, his own role during the Nazi period came under intense scrutiny and questioning. But nobody ever doubted his technical abilities and it is unlikely that anyone ever will.

What about the great theoretician, Hermann Oberth? After the war he was detained by the allies briefly before returning to a mixture of teaching, experimentation and consultancy, based in Feucht, near Nuremberg. He wrote the prize-winning book *The Way to Sidereal Space*. Not long after he thought he had retired in 1954, he joined von Braun's team in the United States to work on the *Redstone* rocket. A museum was opened in his honour in Feucht in 1971. Oberth lived long enough to be in mission control for the night the first men walked on the Moon.

When the Russians reached Peenemünde, there was little left. The leading

German to join their side was the guidance expert Helmut Gröttrup. The Russians shipped several hundred Germans east to Seliger Lake, whence they eventually returned in the early 1950s. The first Russian V-2s were fired from the Kapustin Yar Volgograd station in October 1947 and were called in the Soviet programme the R-1 (*Raket-1*). The Russians then built their own version, the R-2. Further versions were built in the 1950s, such as the R-5, used for sending animals into the stratosphere in side-mounted cabins. But the shape of the old A-4 is unmistakable. Not until Sergei Korolev built the R-7 did the Russians break free with their own fully indigenous design.

The A-4 cast a long shadow. When in 1956 the Chinese began their space programme and sought Soviet help, the Russians shipped them R-2s as their learning rocket. Like the Americans and the Russians before them, they learned how to reverse-engineer the A-4 and made it the basis of their *Dong Feng* military rockets. This in turn was the building block for the *Long March* rocket, the descendants of which would later put the first Chinese cosmonauts into orbit.

What happened to Peenemünde itself? After most of the facilities were dismantled by the Soviet Union, it became a naval, then air force base for the German Democratic Republic. The military authorities of the reunited Germany left the area in 1996 and it became a Baltic seaside resort once more. Most of the wartime facilities and fortifications are badly overgrown, but a small museum has opened. For visitors, admiration of the world's first rocket port are heavily qualified by the wartime overtones of the Nazi period. Only in recent years has more attention been paid to the ghastly Mittelwerk where the V-2 rockets were made underground [12].

1.17 BACKFIRE

What use did the Europeans make of it? The British took 23 of the Peenemünde personnel. Many of them were those who did not make it onto von Braun's list. Granted the size of the list, this meant that he must have disliked them, regarded them as rivals or just not good enough. The only prominent ones were Walter Riedel of the Zement plant and, most importantly, General Dornberger himself. The British had decided to put him on trial for heading up the rocket bombardment of London, but, fearing that he would highlight the horrors of the allied bombing campaign of German cities, they had second thoughts and let him out in 1947. In the end, he went to work in the United States, living to the fine old age of 94 in 1980.

The idea of Britain firing German V-2s came from a British woman, a commander at Supreme Allied Headquarters, Joan Bernard [13]. Her boss Maj. Gen. Alexander Cameron quickly won approval from the War Office in London. Rockets were key to the future of warfare, and here was an opportunity to gain experience to see how they operated, from the sending end rather than the receiving end for a change. Cameron's teams scoured Germany for anything to do with the V-2 and in weeks had filled 200 trucks and 400 goods-wagons with V-2s and V-2 parts. Helpfully, some were found in perfect condition. The operation soon acquired a name, 'Backfire'. As a launching site, the former Krupp gun testing range near Cuxhaven on the North Sea coast was used. They interviewed and temporarily borrowed the services of some of the Peenemünde Germans who had surrendered to the Americans and enlisted the former commander of the V-2 Group South, Wolfgang Weber. Prisoners-of-war and local labourers were brought in to assist the operation.

The launch campaign for the British V-2s began on 1 October 1945 on this flat, North Sea headland surrounded by tall conifers. After a couple of false starts, the first British V-2 was fired the following day, soaring over the North Sea and impacting five minutes later west of the Danish coastline near Ringkøbing. Britain's wartime allies were invited to watch the tests and the Americans present included Theodor von Karman and Dr William Pickering of the Jet Propulsion Laboratory. Since the Russians were still, officially at least, allies, they were invited too. The Russians sent Colonel Glushko, but another representative Captain Korolev fell foul of paperwork problems at the perimeter gate and had to watch from outside the wire. The only people who objected to the tests were, ironically, the British Interplanetary Society (BIS). The BIS argued that the V-2s should be adapted for technical and scientific tests rather than wasted as artillery shells.

Three British V-2s were fired altogether over a two-week period. Cameron sent off a five-volume lengthy technical report and film to London. In fact, it was the most detailed account of the V-2 up to that point, since the German wartime project was so secret that no manuals had been produced. Twenty Germans were persuaded to work for the British after the war and they went to Farnborough. The rest were already en route to the United States, along with over 100 V-2s that the Americans intended to fire.

A year later, the Ministry of Supply took over the old RAF airbase of Westcott and turned it into the Guided Projective Establishment, later the Rocket Propulsion Department of the Royal Aircraft Establishment. Here, Britain's post-war work on liquid and solid rocket fuel engines was carried out [14]. This involved not only guided missiles, but the *Skylark* sounding rocket and later, the *Black Arrow* and the *Blue Streak*.

1.18   ASSESSMENT AND CONCLUSIONS

Europe was one of three arenas where rockets were first developed, the others being the United States (Goddard) and the Soviet Union (Korolev). The development of rocketry in the three different locations had much in common. All depended on a number of key individuals with the insight, energy and persistence to bring theory to experiment. The role of amateur societies, science fiction writers, theorists and fantasists played an important role. There were a number of false trails and lines of development that led to nothing. Just as the Soviet military had given Korolev the opportunity to make substantial progress, so too did the peculiar circumstances in Germany in the 1930s provide possibilities for von Braun that could never have materialized from voluntary effort and popular subscriptions alone. Goddard, von Braun and Korolev all realized the importance of conquering the problems around the liquid-fuel rocket, one of the most significant aspects of their work, for the development of satellite-lifting solid-fuel rockets lay far in the future. Why European rocketry should find its home in Germany is a question worth asking. France might have seemed a more natural home for rocketry, granted the writings of Jules Verne and the efforts of Esnault-Pelterie, French inter-war aviation and the publicizing efforts of Azanoff. Britain, in circumstances to be examined later, ruled itself out by banning rocketry through the law. Germany had, to its advantage, a solid theoretical base (Oberth), a strong scientific and industrial tradition dating back to the 19th century and a remarkably vibrant civil society to promote amateur effort. The circumstances of early rearmament explain the rest.

The value of the A-4 as a military weapon has been discussed in many analyses of the V-2. The A-4 programme cost Germany something in the order of RM300m in development costs, RM450m for the 6,000 missiles built and probably RM2bn by the time the whole operation was accounted for. This was about $500m in the money of the period, compared with the $2bn of the Manhattan project to build the atomic bomb. However, if adjusted to the scale of the two economies, the A-4 was roughly similar in scope and effort to Germany as was the Manhattan project to the United States [15]. The production of V-2s was very costly for Germany in terms of effort, organization, manpower and resources. The results though were much less impressive. The A-4 made no serious impact on the outcome of the war – except by diverting scarce German resources away from conventional weapons. If anything, it hastened its end in favour of the allies. Had the same effort been directed into building less glamorous, but much needed fighter planes, U-boats or tanks, the war might have turned out differently. The damage inflicted on the people of London and Antwerp, while terrible to those concerned, was much less than the Luftwaffe had done and infinitely less than what the RAF could do on a single night raid over Germany. More slave-labourers died producing the V-2 than the number of people in bombing raids.

The real value of the A-4 was clearly not what it achieved in the short term (very little), but in its enormous long-term impact. It was after all the first modern rocket, able to reach the edge of space – some had reached as high as 176 km. The A-4 was the basis of the post-war Soviet, American (and even Chinese) space programmes. The A-4 did bequeath on post-war civilization the first real rocket, the first machine to reach the edge of space, the first modern rocket engine and the first rocket to use inertial guidance, fins and targeting.

That is its historical significance. It is fair to say that the wartime allies did not underestimate its value at the time either, judging by the efforts to spirit its legacy away to White Sands and Kapustin Yar. Curiously, the Europeans, the people who had suffered most directly from the V-2, had no plans for its future.

2

European cooperation – ELDO and ESRO

There were two reasons the European countries were slow to use the V-2 after the war. Both the leading countries that were candidates to do so, Britain and France, were exhausted by the war. Scientific rockets competed poorly against the urgent tasks of post-war social and industrial reconstruction. Nevertheless, both Britain and France were able to begin work on sounding rockets (*Skylark*, *Véronique*) and Britain constructed the first post-war European missile, the *Blue Streak*. By 1960, 15 years after the V-2, both countries were prepared to bring their two national programmes together in a common European project.

2.1 FRANCE'S EARLY NATIONAL PROGRAMME

France took an important role in the regeneration of astronautics in the years following the Second World War and hosted the first international astronautical conference, held in Paris in 1950, which led to the formation of the International Astronautical Federation that year. Its founding members were the Groupement Astronautique of France, the British Interplanetary Society (BIS) and the Gesselschaft für Weltraumsforschung ('Association for Orbital Research') of Germany.

The French went quickly into Germany in 1945 to inspect the V-2 rocket. Professor Henri Moureu of the armaments led the investigation and recommended the establishment of a rocket study centre for the armed forces. This led to the centre for the study of self-propelled projectiles, the Centre d'Etude des Projectiles Autopropulsés. Separately, Jean Jacques Barré resumed his wartime work on rocketry at the laboratory for ballistics and aerodynamics research, the Laboratoire de Recherches Ballistiques et Aérodynamiques (LRBA) in Vernon. LRBA persuaded the government to fund a sounding rocket called the *Véronique*, able to go up to 70 km into the upper atmosphere. Eleven such sounding rockets were fired over

Figure 2.1. Europe post-war.

1951–3. The French defence research agency ONERA (Office National d'Etudes et de Recherches Aéronautiques) dismantled the Munich Aviation Research Institute's leading wind tunnels from the Otz Valley in the Tyrol, shipping them out of the former Reich and reassembling them wholesale in France.

The two leading German rocketeers captured by the French were Eugen Saenger and Irene Bredt, both of whom came to work in France from 1945 to 1954, first for the Ministry of Aviation (Arsenal de l'Aéronautique) and then for the Matra company. To build up their post-war rocket programme, the French recruited a German engineer, Rolf Engel. Engel, born in 1912, was one of the enthusiasts of German rocketry in the 1920s, having been involved in experiments in the *Raketenflugplatz*. During the war he set up the Grossendorf Jet Propulsion Experimental Establishment, which made anti-aircraft missiles for the Reich. He was recruited by ONERA in 1946 where he put his accumulated knowledge on the record in technical papers on German rocket development and basic satellites. He was permitted to recruit German colleagues, and together with ONERA they formed the basis for French–German collaboration in post-war rocketry.

The International Geophysical Year acted as a stimulus to rocket development in France as it did in many other countries. The government provided sufficient funding for a new version of the *Véronique* to be fired, the *Véronique AGI* (Année Géophysique Internationale). These 1,370-kg sounding rockets were able to carry a 60-kg payload to an altitude of 210 km. *Véronique* was 7.07-m long, 55 cm wide and fired for 45 seconds. Fifteen *Véronique AGI*s were fired to a height of up to 210 km, from a launch site in French Algeria, Hamaguir. In 1963 a number of biological experiments were carried out, carrying Félicitte the cat and Hector the rat.

2.2 ARRIVAL OF DE GAULLE

Up to this point, French rocket research had been carried out by diverse agencies and groups acting in a largely uncoordinated way. This situation changed when General Charles de Gaulle became President in 1958. Attributing the failure of the Fourth Republic to disorganization, lack of coordination and insufficient direction, he resolved to make good its deficits in as short a time as possible.

In January 1959 his government appointed the Space Research Committee, which would report to the Prime Minister. Its role was to map existing French activity in the field, present proposals for space research programmes to the Prime Minister and once approved execute them. By April 1959 the Committee had adopted a six-year plan for scientific observations. Twenty-two *Véronique* sounding rockets were fired over 1959–61.

The Space Research Committee quickly took the view in 1961 that the rapidly expanding rate of space activities in France was too much for a modest committee to handle. Accordingly, with the support of Prime Minister Michel Debré, the Committee presented a plan for the formation of an independent agency to guide France's space programme. This was sent to a presidential meeting in the Elysée Palace in July 1961. Although General de Gaulle only attended part of the meeting and said very little, he needed little persuasion. Uncharacteristically for a ministry of finance, its representative declared the high costs of an independent French space programme to be 'not too expensive' and the matter was settled. 'Just do it,' said de Gaulle as he marched out. A bill was presented to the National Assembly in August 1961 and approved.

2.3 ROBERT AUBINIÈRE, THE FIRST DIRECTOR OF CNES

The agency was called the Centre National d'Etudes Spatiales (CNES), inaugurated 19 December 1961 and formally set up 1 March 1962. General de Gaulle appointed as its first director Robert Aubinière (1912–2001), then aged only 39. Aubinière was a Free French veteran from the war, after which he was sent to Algeria to manage

Figure 2.2. Robert Aubinière.

France's rocket base there. He led CNES for 10 years before he briefly became
director of the European Launch Development Organization (ELDO) (he lived to
89, dying in Paris in December 2001). The early CNES comprised an uneasy mix of
older military men, grisled veterans of sounding rocket campaigns of the 1960s, with
often irreverent, American-trained, enthusiastic young men and women just out of
college – but the results seemed to get the best from both groups [16].

The organization of CNES was significant in-so-far as it was given financial
independence, reported to the Prime Minister (an indication of its importance)
and included, as well as scientific objectives, the building up of a French space
industry. Although geared primarily toward a national French space programme,
it also had a mandate to pursue bilateral cooperation (e.g., with the United
States) and international cooperation within Europe. CNES began with 17 staff.
Construction began of the Brétigny-sur-Orge space centre during the early, snowy
months of 1963. The centre covered 42 ha, the site of an old aerodrome. Several years
later the decision was made to relocate all the main space activities to the Centre

Spatiale de Toulouse (CST), the space centre of Toulouse. Brétigny-sur-Orge was closed in 1974.

CNES was set up on the authority of the Prime Minister and was later made responsible to the Minister for Scientific Research, Atomic and Space Affairs and given an initial budget of €5m. Its tasks were to:

- develop and guide scientific and technical aerospace research;
- prepare programmes and ensure their execution, either directly or through research contracts; and
- cooperate with other countries either on an international or bilateral basis.

CNES was led by a president and director general who were responsible to an 18-member administrative council (Conseil d'Administration), later including six members directly elected by the employees.

CNES was to become the most powerful of the national European space agencies, with the largest budget, the greatest autonomy and the widest range of facilities in its own name. CNES became the dominant body within the European Space Agency, driving it with ideas and resources. Based on the original calculations of participation according to size of gross domestic product, France's contribution to ESA should have been in the order of 17%, but in practice the French financial contribution has been about 30%. French government or political support for CNES has changed little according to governmental or presidential changes and has never wavered.

At the same time, military space activities were made the responsibility of the Délégation Ministerielle pour les Armaments (DMA) under the Defence Ministry. The DMA was given responsibility for the development of a national launcher, the *Diamant*, jointly funded by DMA and CNES. Missiles were tested from a launch site at Biscarosse, south-west of Bordeaux, whence rockets would head south-westward toward the Azores.

2.4 IDEA OF A FRENCH EARTH SATELLITE

It was the military who opened up the possibility of France putting a satellite into orbit. General de Gaulle took an early decision to make France a nuclear power, which required a means of delivering such weapons. In 1959 the government established a ballistic missile research company. One of its early tasks was to test rockets able to lift nuclear weapons and to develop their guidance and delivery systems (much as *Black Knight* did for Britain). Several such rockets were developed and were called the 'precious stones' series, each being named after a precious stone: *Agate*, *Topaze*, *Emeraude*, *Saphir* and *Diamant*.

The last of these, *Diamant*, had the capacity to put a small satellite of about 85 kg into orbit. Such a conclusion was reached by a French military report in December 1960. The government approved the development of the *Diamant* for such a purpose in late 1961 with a view to launch in March 1965. However, it was

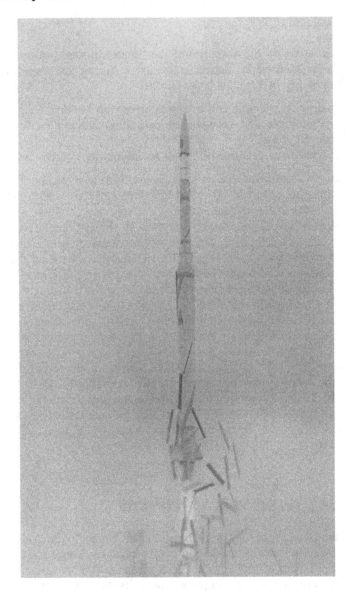

Figure 2.3. *Diamant.*

conscious that this might prove a difficult venture and so, at the same time, approved cooperation with other countries in projects to put satellites into orbit. Specifically, agreement was reached in March 1961 with the United States for the Americans to put a small French-built satellite into orbit and leading French scientists were sent to NASA centres for training.

2.5   PROBLEM OF A LAUNCH SITE

One of the earliest challenges faced by CNES was to find a launching base for France's rockets. France already had a temporary base in North Africa. Following the war a missile and sounding rocket launching site was established in Algeria in the region of Hammada du Guir, shortened to Hamaguir, situated on a desert plateau. It included three launch pads, storage areas and housing for 600 people. As part of the post-colonial settlement, the Evian agreement, France was required to evacuate the site by 1 July 1967.

The search was on for a new launch site. For its tests the military selected Landes, on the French Atlantic coast between Biscarosse and Le Bacares near Perpignan, launching westward out over the Atlantic. As a satellite launching base, Landes had a double disadvantage: it was near inhabited areas and the site meant launching westward. Almost all satellites are launched eastward, where they benefit from the continuous spin of the Earth. Conversely, launching westward requires additional thrust. Only one launch base in the world goes westward: Palmachim, Israel. If it were to launch east, Israel's rockets would overfly Syria and might well be mistaken for a rocket attack. Instead, Israel fires its satellites westward right across the Mediterranean, over Sicily and the straits of Gibraltar and must use extra thrust in doing so.

CNES did not wait for the expiry of the Algerian leases in 1967, and began searching for alternate sites as early as 1962. Four were initially considered: two in France (Perpignan and Narbonne), the Ivory Coast and – with Algerian agreement – continued used of Hamaguir. The criteria were tightened in 1963: political stability, proximity to the equator, low population density, deep sea port, airfield and nearness to Europe. Fourteen were considered:

French launch sites under consideration, 1963

Seychelles		Darwin	Australia
Trinidad		Tricomalee	Ceylon
Nuku Hiva	French Polynesia	Fort Dauphin	Madagascar
Tuamotou	French Polynesia	Mogadishu	Somalia
Desirade	Guadeloupe	Port Etienne	Mauritania
Djibouti	French Somalia	Belem	Amazon delta, Brazil
Kourou	French Guiana		

Kourou (French Guiana) was chosen on 14 April 1964. The others were turned down for a variety of reasons. Many were regarded as politically unstable. Some lacked port or airport facilities, or they could only be provided with great difficulty. Some were at risk of cyclones or too far from Europe (e.g., Darwin). Tuamotou had no freshwater. Kourou required port improvements and the high level of humidity was considered a serious, though surmountable problem. The various sites were graded in this order: Kourou, Darwin, Belem, Tuamotou, Trinidad. CNES

Figure 2.4. Kourou in 1965: jungle.

director Robert Aubinière went out to French Guiana to see for himself. He knew French Guiana had a bad reputation on account of Devil's Island, but wanted to see for himself. He took a helicopter trip along the coast and persuaded himself that it was possible.

In the end it turned out to be a choice between Kourou and ... Rousillon! Rousillon came onto the map as a domestic French contender. Rousillon was much cheaper, €15m to build and €2.3m a year to operate. By contrast, French Guiana would cost €40m to build and €6.9m a year to operate. However, Rousillon went against domestic regional planning policy, was far too distant from the equator (43°N) and was too small for large rockets. Prime Minister Georges Pompidou decided on French Guiana. President de Gaulle travelled there and announced that it would be the site of a great French undertaking, one that would be recognized throughout the world. The centre got a boost when, in 1966 it was decided to move ELDO operations to Kourou, on the basis that ELDO would pay 40% and France 60%. With the collapse of ELDO, it reverted to being a French site. The original cost of the site was €40m.

Kourou had the advantage of being French territory and only 5.14°N (52.46°W) from the equator, giving France a considerable advantage (17% to be precise) over Cape Canaveral (28°) and even more over Baykonur (49°). Rockets could fly from Kourou straight out over the sea, with nothing to fly over but blue water for 3,000 km – either eastward toward equatorial orbit or northward to polar orbit. The only disadvantage was its hot and steamy climate: the wet season is very wet. The site covers 1,000 m², about 1% of the land area of the country, and is a strip of

Figure 2.5. Islands off the coast.

coastal land 29 km wide and 60 km long. Temperatures range from 18°C to 34°C with an average of 26°C. Rainfall is 2.9 m a year. On the positive side, Cayenne and the coastal strip is a windless open plain.

French Guiana was settled between 1763 and 1765 when Louis XV sent Choiseul to lead an expedition there. Around 10,000 settlers perished in the jungles of the mainland, but 2,000 survived in three islands off the coast, called the Iles de Salut ('islands of salvation'), 134 km off the coast. These three islands were Devil's Island, Ile Royale and Ile Saint Joseph. The islands became the home of an artillery regiment from 1800 to 1859 when the decision was taken to turn them into a penal colony, the first convicts arriving in 1852. Ile Royale and Ile Saint Joseph were used for ordinary convicts and deportees, but Devil's Island was for political prisoners, and all of them were used for repeat offenders. Captain Dreyfus was exiled there in 1905. The conditions were so appalling that they developed a literature of their own, from Albert Londres to Henri Charrière, otherwise known as *Papillon*. The penal colony closed in 1947, and when the rocketeers arrived they found nothing but ruins and jungle. The history of Devil's Island will last for ever, immortalized by Dustin Hoffman in the film *Papillon*.

Nowadays, 75,000 people live in French Guiana. The interior is dense jungle and is used for training by the French Foreign Legion where it still has a base. Piranhas and alligators infest its rivers. The native Amazonian people still hunt there in

Figure 2.6. Early Kourou.

traditional ways, but entry into the interior is prohibited to visitors in order to protect them. Before the rocketeers arrived, the colony had been peopled by French settlers, runaway slaves, Chinese and Indians. Indeed, there are many new Chinese there, as the local restaurants indicate.

The great undertaking duly began with land surveys and the acquisition of the site. First construction began in 1965 and the first equipment (the radar station) became operational in 1969. Two 12-m dishes were built to receive telemetry from ascending rockets: at Montagne des Pères and Cayenne. The first sounding rocket was fired from French Guiana on 9 April 1968 (a *Véronique*). The early 1970s saw the launches of *Diamant* and ELDO rockets. There was then a period of contraction while the new *Ariane* rocket was developed. Since 1979, the Centre has always been busy, with waves of expansion for the second launch site (ELA-2, inaugurated 1986) and the third for *Ariane 5* (1996).

More of this later. Meanwhile, what of the other likely post-war space power, Britain?

2.6   BRITAIN: THEORISTS AND PIONEERS

The British Interplanetary Society (BIS) was formed at 7 p.m. on Friday 13 October 1933 by a young structural engineer Phillip E. Cleator in Liverpool [17]. Its purpose was to study interplanetary travel, report on rocketry and carry out practical work. In furtherance of the last, Cleator paid an early visit to the *Raketenflugplatz* in Berlin. The first BIS intention was, like its Continental counterparts, to make amateur rocket experiments. Here they ran foul of the Explosives Substances Act 1875, which prohibited anyone from manufacturing or experimenting with 'gunpowder, nitro-glycerine, dynamite, gun cotton, blasting powders, fulminate of mercury or other metals, coloured fires *or rockets*'. This was called the Guy Fawkes law, after the man who tried to blow up Parliament in the 17th century. Eventually, in 1936 they were told that they could proceed, if they found a suitable site, if they got police permission, if they confined themselves to powder rockets (liquid-fuel rockets were absolutely prohibited) and if the Secretary of State adjudged their design to be safe and sound. The prospects of the latter appeared slim, granted that two years earlier the government had refused to invest in jet propulsion, also adjudging it to be impractical. The BIS met with little sympathy from the academic community either, the journal *Nature* commending the wisdom of the government's scepticism. Incidentally, the law still remains on the statute book.

In the event, they confined themselves to theoretical and practical studies, but not experiments. Even these fell under police surveillance, for Arthur C. Clarke later recalled being stopped by the police carrying out routine checks during the IRA bombing campaign of the late 1930s. His briefcase was full of studies of mixing rocket fuels, which sounded suspicious, and his attempts to convince the police that he was building a Moon rocket were received equally sceptically. In 1939 the BIS completed its first major study – an examination of the techniques that should be considered in organizing a manned flight to the Moon. A key theoretical break-through was that the study established the use of the bottom stage of a lunar module as the lift-off for the top stage. The society devised and built a coleosat, a sighting device so that a lunar spaceship could navigate. When war broke out the Society felt obliged to suspend activities for the duration.

The Society blossomed in the post-war period. Membership rose from about 100 (pre-war) to 3,000 by the time of *Sputnik*. The BIS was to develop many long-range papers over the following years, such as *The Atomic Rocket* (Val Cleaver and L. R. Shepherd, 1948), and many years later the first star ship project *Daedalus* (by A. Bond and A. R. Martin, 1977). Other famous BIS studies examined a lunar spacesuit, the use of an adapted V-2 for manned suborbital flight and a minimal satellite. A week-long seminar was held on satellite launchers in 1951. A journal JBIS (*Journal of the British Interplanetary Society*), still publishing, was started as far back as 1934: initially, it dealt not only with futuristic questions, but in the course of time reported space history and was one of the first journals to publish the technical details of the V-2. A popular magazine *Spaceflight* was begun in 1956 under the editorship of Patrick Moore. The BIS was a major contributor to the founding of an international society for astronautics, the International Astronautical Federation and was a founder member. It also submitted memoranda to the British government that it hoped might inform national space policy. Philip Cleator, born 1908, saw most of his ideas come to fruition. He lived to 1994.

In May 1948 two BIS writers, R. A. Smith and H. E. Ross, outlined how the German A-4 could be adapted as a man-carrying rocket to send a pilot into space for a brief experience of weightlessness. The proposal was submitted to the Ministry of Supply where it was rejected (the mission eventually took place in May 1961 when Alan Shepard flew into space on the *Redstone* rocket).

2.7   ORIGINS OF THE POST-WAR BRITISH SPACE PROGRAMME

Britain's launch of German A-4 rockets from Cuxhaven did not lead, as did the comparable American launches in White Sands and the Soviet launches in Kapustin Yar, into a national rocket programme. There were several reasons for this. First, the enthusiasm of amateurs in the BIS was not matched by official support, still less enthusiasm. Second, Britain's post-war rocket and space effort was divided between many different establishments and locations. The main centres for innovation were the Royal Aircraft Establishment (RAE) in Farnborough and the Rocket Propulsion Establishment (RPE) in Westcott. The Royal Society was the national organization for the promotion of a better understanding of science. The BIS was the home of the enthusiasts, but there was no one centre coordinating their efforts.

The RAE became the main driver of the British post-war space programme. There, Desmond King-Hele put forward over 1953–4 a series of ideas for the development of high-altitude sounding rockets. The King-Hele proposal for a sounding rocket became over time the solid-fuel *Skylark* rocket, developed by the RAE and the Royal Society and attracted a small government grant. Good progress was made and the first flight of the *Skylark* took place on 17 February 1957 from Woomera in Australia. A scientific programme was devised and became the basis of a broad range of experiments conducted over the next few years into the nature of the atmosphere, the ionosphere and solar radiation. *Skylark* was still flying 40 years later, making it one of the most successful early rocket programmes developed. *Skylark* has been through many different evolutions, the *7* and *12* models being the most recent, with launchings having passed the 430 mark. The *Skylark 7*, for example, is 15.3 m long and weighs 1,700 kg in its launch rail. *Skylark* has two solid-fuel rocket stages. It is a long, thin rocket, but able to reach altitudes of 1,000 km and to provide up to six minutes of microgravity.

2.8   IDEA OF A BRITISH EARTH SATELLITE

In the early 1950s BIS experts set to work on a series of studies about how to put a small Earth satellite into orbit with simple instrumentation. The seminal publication was *Minimum Satellites Vehicles* by Kenneth Gatland, A. M. Kunesch and A. E. W. Dixon, published in the *Journal of the British Interplanetary Society* in November 1951. Again, the ideas were not taken up at home, but were of great interest to the American navy, then considering a satellite project. The Royal Society set up an artificial satellite committee in 1956. That same year Desmond King-Hele and

D. Gilmour wrote a paper that showed that there were no insuperable difficulties to putting a 1-tonne satellite into orbit with a liquid-fuel rocket.

It seemed that, after an uncertain start, a British space programme was now in the making. British scientists were well networked internationally and active participants in the International Geophysical Year. In this connexion a group of Russian scientists attended a meeting on rocket technology at Cranfield in July 1957, but no-one paid much attention to reports circulating at the time about the frequencies on which some Russian satellites would shortly be transmitting.

The launching of *Sputnik 1* changed everything. *Sputnik 1, 2* and *3* were tracked extensively from Britain both visually and by radio over 1957–8. The Royal Society set up the British National Committee for Space Research to develop British participation in space activities. A memorandum was sent to the government proposing the orbiting of instruments on a British satellite and the development of a national launcher. It is interesting that in 1958 an American assessment of European space efforts doubted whether Europe could develop a launcher for some time, but that Britain was clearly the most advanced country in the technologies involved.

These proposals were complicated in March 1959 when the United States offered to launch instruments and even full satellites for other countries free of charge on its new small launch vehicle the *Scout*. This prompted a statement by the Prime Minister in May 1959 in which the government expressed its open approval for British instruments or satellites being put into orbit, but the government wished to keep its options for a launcher open, which might be a national project based on a military launcher or a collaborative project with the United States or the Commonwealth. A British team visited the *Scout* launching base in Wallops Island, Virginia and agreement was quickly reached on the launch by the United States of a series of British satellites. This became the highly successful UK series called *Ariel* once they entered orbit.

2.9 A BRITISH ROCKET

During the 1950s Britain developed its own national long-range ballistic missile programme, called *Blue Streak*. This was done in collaboration with the United States, an agreement to this purpose having been signed in 1954. Approved in 1955 Britain intended to install batteries of *Blue Streaks* in 100 reinforced, underground silos in both Britain and the Middle East, ready to hit back at any aggressor (presumably the Soviet Union). To speed up the design, the prime contractor De Havilland entered a cooperation agreement with Convair and General Dynamics in the United States, as did the rocket engine manufacturers Rolls Royce with Rocketdyne. As a result, the *Blue Streak* followed similar design principles to the *Atlas* rocket, then being developed as the United States' intercontinental ballistic missile. *Blue Streak* used two liquid oxygen and kerosene engines to develop a thrust of 123 tonnes. It looked a little like the *Atlas*, but with a stubby, flat top on its otherwise sleek, grey, steel body. The RZ2 first stage engines were developed by Val

Figure 2.7. *Blue Streak* engines.

Cleaver in Rolls Royce with Rocketdyne in the United States, the developers of the *Atlas* engines. The first test firings were made March 1959.

***Blue Streak* details**

Dimensions	21 m tall, 3.05 m diameter
Thrust	117,995 kg
Lift-off weight	93,193 kg
Range	200 km high, 1,100 km downrange

To test the warheads for the *Blue Streak*, the RAE in Britain developed a rocket called *Black Knight*. This was then tested in Australia as Project Dazzle. Armstrong Siddeley Aircraft Co. designed the Gamma engines based on German wartime rockets and made static tests at High Down on the Isle of Wight. All the rocketry, equipment and launching gear then had to be flown or shipped out to Woomera in Australia for their actual missions. On 7 September 1958 they fired the *Black Knight* 564 km high over the Australian desert, then a record for a sub-orbital mission [18]. Teams were sent downrange to watch the re-entries and recover the mock warhead payloads, a task done by night to facilitate tracking. Although many tests had been scheduled, the reliability of *Black Knight* meant that not all the

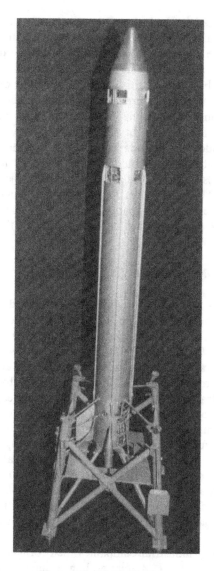

Figure 2.8. *Black Knight.*

rockets were used. The series cost only £5m. *Black Knight* later became the basis of *Black Arrow*.

Black Knight was fired 22 times until the last launch on 25 November 1965. Except for minor anomalies (test BK21 records 'success except lanyard snapped'), these were almost entirely successful. Scientific experiments from Imperial College London and University College London were carried on a number of missions [19].

When *Sputnik 1* was launched, the government was asked about the idea of a British space programme. With its *Blue Streak* rocket, Britain was in a strong position to develop the upper stages necessary to put a satellite into orbit and become the world's third space power. A combination of *Black Knight* and *Blue Streak* provided a firm basis for a British satellite launching capability. But, despite Britain's technical pre-eminence, there was little enthusiasm at senior level for a British space programme. The Prime Minister, though interested in satellites, would not commit to a launcher. In 1959, the Advisory Council on Scientific Policy described the idea of an independent British space programme as 'the greatest folly' and that Britain should confine itself to international projects. Britain in the late 1950s was full of exciting technical projects and world-beating aircraft, like the TSR-2. The Saunders Roe team, who designed the SR 53 rocket interceptor, even sketched plans for spaceplanes in 1958. The problem was that few projects got beyond the drawing board and many of those that did were cancelled.

2.10   CANCELLATION OF *BLUE STREAK* MISSILE

British military experts became more and more concerned that *Blue Streak* was vulnerable and could be wiped out in a nuclear strike long before it could be fuelled up and dispatched. Because it was a liquid-fuel rocket, it could not be kept on the pad ready to go. Preparing it for launch would take some time, in the course of which the rapid arrival of Soviet missiles from eastern Europe would mean that there would be no country left to defend. In February 1960 the chiefs of staff let it be known that they had lost confidence in *Blue Streak* as a weapon. *Blue Streak* was cancelled as a missile on 13 April 1960 and the House of Commons informed accordingly. The government told the House that it would investigate the possible use of the missile as a launcher for space satellites. Britain opted instead to take the *Skybolt* missile from the Americans. This was a weapon that could be dropped off from a British V-bomber off the Russian coast and speed to its target (this too was cancelled in turn, Britain eventually settling on the *Polaris/Trident* system of submarine-launched nuclear weapons). The Russians with their R-7 rocket faced exactly the same problem – but they used the rocket as a satellite launcher instead. It is still flying.

British scientists and industrial companies then formally proposed combining the *Blue Streak* with the *Black Arrow*, adding a small third stage and thereby developing a launcher capable of putting a 790-kg satellite into a 300-km orbit. This was called *Black Prince*. There was nothing new about this. As far back as May 1957 Desmond King-Hele had outlined for the RAE how the two could be combined as a satellite launcher. A meeting of the BIS had discussed the possibilities in 1959. However, one of the problems of the 1960 cancellation was that no governmental department was prepared to invest anything in turning the cancelled missile into a satellite launcher. This left the nagging suspicion that the concept of modifying *Blue Streak* as a satellite launcher was purely a formula designed to save face and dent criticism of the £80m wasted so far [20]. In the period that followed, during summer 1960, the RAE pressed the government to adopt the *Black Prince* proposal while the government sounded out some of the Commonwealth countries as to their interest in a collaborative venture. Saunders Roe suggested an upgraded *Black Prince* with a high-energy upper stage to put payloads into geosynchronous orbit.

The history of the period records that, hoping to share some of the costs, Britain approached France to ask the French if they were interested in joining such a project. At the time Britain was preparing its application for membership of the European Economic Community, and this was one of a number of collaborative projects then under consideration (the others being *Concorde* and the channel tunnel). In reality it was the French who first sounded out the British, although the prime point of French interest was in *Blue Streak* as a military weapon and whether the British were prepared to share that knowledge (they were not, even though it was officially obsolete).

2.11 ORIGINS OF EUROPEAN COLLABORATION

In 1960 both Britain and France had embryonic national space programmes. Both countries were interested in getting satellites into orbit, France more so than Britain where enthusiasts were still doing battle with a sceptical officialdom. Sounding rockets had been built by both countries. Ultimately, these or their descendants could be upgraded to small national satellite launchers. Launching a satellite with American help had also opened up as a possibility. Now a third avenue appeared: European collaboration.

European collaboration may be traced to the efforts of two professors at Europe's nuclear research facility CERN – Italian Eduardo Amaldi and Frenchman Pierre Auger. CERN was set up after the war as the European Centre for Nuclear Research. It seemed to offer a satisfactory model, for CERN enabled work to be done that individual nations could not do, offered economies of scale and had brought Europe up to American standards. There had been much scepticism when it had originally been established, but it had confounded the critics and different nationalities worked harmoniously together there.

Amaldi and Auger invited eight scientists from eight European countries to meet discreetly in Professor Auger's home on 29 February 1960 to consider the idea of a European organization for space research. They came from Britain, France, Italy, Germany, Belgium, the Netherlands, Sweden and Switzerland with an apology from Norway. The British proposed that they put their discussions on a more formal basis and so the Royal Society convened the next meeting for London on 29 April. At this meeting 20 European scientists met with their British colleagues. There was a *tour de table* in which each country reported on national developments. The most promising way forward for a European satellite was to use the British *Blue Streak* missile. It was agreed that the matter be raised with the British government. Thanks to Professor Auger, their work received the blessing of the French government and he was given the go-ahead to convene a small conference in Paris in June 1960 to progress the idea. By coincidence, on the previous day 28 April 1960, the Council of Ministers of the European Economic Community recommended consideration be given to the the concept of a space agency for Europe.

2.12   FOUNDING ESRO

The Paris conference in June duly proposed the convening of a commission to progress European cooperation in satellite and launcher development. It was a sign of changing official attitudes that five countries sent governmental representatives rather than scientists. The Swiss government indicated its willingness to sponsor a conference in Meyrin, Switzerland, from 28 November to 1 December 1960, where 11 countries agreed to form COPERS, or the preliminary European commission on space research. This established the European Preparatory Commission for Space Research, which was mandated to work on setting up a body:

> to provide for and to promote collaboration among European states in space research and technology, exclusively for peaceful purposes.

COPERS was based in Paris, with Pierre Auger as Executive Secretary. Auger and his colleagues drew up a draft convention. The convention was open for signature between February 1961 and March 1964 and became the basis of the European Space Research Organization (ESRO). Twelve European states signed – Belgium, Britain, France, Germany, Italy, the Netherlands, Norway, Spain, Sweden and Switzerland. The convention was formally signed on 14 June 1962, Denmark joining subsequently. It came into operation formally on 20 March 1964. The budget was set at €230m for eight years, France contributing 18.22%.

Members of ESRO	
Belgium	Italy
Britain	Netherlands
Denmark	Spain
France	Sweden
Germany	Switzerland

2.13   FOUNDING ELDO

In September 1960, although it was not a member of the EEC, Britain formally offered the recently cancelled *Blue Streak* missile as a first stage of a European launcher project. There had been some discussion in Britain about what to do with this now-cancelled rocket, and this was felt to be a plausible justification for the high costs already invested. In November 1960 French Prime Minister Michel Debré agreed with the condition that the second stage of the launcher be French and that Britain make available its knowledge of warhead re-entry systems. At a meeting to discuss British entry to the EEC in January 1961, Harold MacMillan and General de Gaulle agreed in principle on cooperation on *Blue Streak* (and on *Concorde*). The project was presented by France and Britain to the other European governments in Strasbourg in February 1961, where they outlined their terms: a British first stage, French second stage, the third stage from other member states, with a launching site in Australia. The cost was set at €196m, of which Britain would pay 33%, France 25% and the others according to their contribution to CERN which was in turn based on gross national product. The project would be run in cooperation with COPERS (subsequently ESRO) set up under the Meyrin agreement.

A second conference was held in Britain (Lancaster House) from 30 October to 3 November 1961 to draw up the terms of an agreement for a launcher development organization. A convention was agreed for the formation of a European Launcher Development Organization (ELDO). The Lancaster House conference also agreed that the third stage would be supplied by Germany, the satellites by Italy, ground guidance Belgium and telemetry systems by the Netherlands. The formal agreement was signed 29 March 1962 and the secretariat moved from London to Paris in June. The agreement came into force on 29 February 1964.

ELDO member countries

Australia	Germany
Belgium	Italy
Britain	The Netherlands
France	

Policy-making was made the responsibility of the European Space Conference. This was a meeting of the ministers of these states. The tradition developed of holding a European space conference every second year in July, though *ad hoc* conferences could also be convened. In between meetings of the conference, there were meetings of what were called 'alternatives' attended by ministerial civil service representatives.

One of the big problems ESRO and ELDO had to face was agreement on where to locate their various facilities. All international organizations face difficulties in making such decisions, trying to reconcile fairness and efficiency. Three facilities had to be agreed: a headquarters, technical centre and tracking centre. ESRO hired Dr O. Dahl to arbitrate the competition. Headquarters were proposed by France, the Benelux countries and Switzerland. For the technical centre, there were six proposals: Brussels, France, Rome, Delft in the Netherlands, Geneva and Bracknell near London. The tracking station was proposed for Darmstadt in Germany, Geneva and Bracknell. From his report it is clear that Dahl received a lot of passionate and contradictory advice. Dahl recommended Darmstadt for the tracking station and that the technical centre be combined with the headquarters in Delft. The Dutch location was the most central, was at a good communications point, offered suitable housing and schooling and offered the lowest costs. In the event, the Delft site proved unsuitable (there was a risk of subsidence) and the Dutch provided, instead, a 40-ha site at Noordwijk 20 km north of The Hague and 30 km from Schipol Airport.

Britain was able to contribute substantial resources to ELDO: its *Blue Streak* rocket, its launch base in Woomera, Australia, up to 38% of the budget and ground facilities in Britain and Australia. Britain was also in the forefront of radio astronomy, with the new radio telescope at Jodrell Bank known worldwide and would have no difficulty establishing a worldwide tracking network through friendly Commonwealth countries [21]. For testing and launching the ELDO rocket, Britain offered the Woomera range in the southern Australian desert. This had been designated a test range as far back as April 1946 when Britain had been searching for a place to test post-war rockets [22]. The nearest city was Adelaide, 500 km distant, though a weapons and research establishment was soon built at the nearer town of Salisbury. The principal advantages of Woomera were its bright weather and the fact that the territory roundabout was relatively lightly inhabited, though the government went to considerable efforts to protect both the sheep-farmers of the region and the aboriginal people. The Woomera 'range' extended 2,000 km across the desert. Rockets out of Woomera were fired westward, tracked by a series of stations and observing posts across the desert until they impacted in the Indian Ocean. The Americans came to install a tracking station there too. Woomera's clear skies, both by day and by night, made it an ideal launch centre.

2.14   *EUROPA I* ROCKET

The first objective of ELDO was the construction of the *Europa I* rocket. This comprised the British first stage, *Blue Streak*; the French second stage, *Coralie*; and the German third stage, *Astrid*. The initial cost of *Europa* was set at £70m, with a first launch set for 1966. Italy was brought into the project as the maker of satellites, and the Netherlands and Belgium to provide tracking, telemetry and ground facilities.

Europa I launcher

Dimensions	31.4 m tall × 3.05 m diameter		
Weight	124 tonnes		
Capability	907 kg into Earth orbit		
	First stage	Second stage	Third stage
	Blue Streak	*Coralie*	*Astrid*
Dimensions	18.4 m × 3.05 m	5.5 m × 2 m	3.81 m × 2 m
Weight	90 tonnes	11.5 tonnes	4.6 tonnes
Fuels	Kerosene, liquid oxygen	UDMH, nitrogen tetroxide	UDMH, nitrogen tetroxide
Thrust	136 tonnes	28 tonnes	2.5 tonnes

Figure 2.9. *Blue Streak* first stage.

So, after a faltering start, more so in Britain than in France, the European countries had now agreed to establish a rocket and satellite programme. There now seemed to be the real prospect that, after the early lead established by the Soviet Union and the United States, Europe would now emerge as the third space power. ELDO and ESRO had the backing of Europe's leading space-faring country Britain and the strong support of France. Structures had been negotiated over three years by governments and scientists and agreed. A well-tested launcher provided a strong base for the new effort, combined with the French experience in smaller stages. The future was bright.

2.15   EARLY ELDO LAUNCHES

In preparation for the launches of *Europa I*, the *Blue Streak* was tested exhaustively at a static stand in Spadeadam, Cumberland. Four hundred ground tests were carried out, the engines being held in a clamp as the level of thrust was raised to ever higher levels. Only when these were complete was the *Blue Streak* taken to its base in Australia. Even as the *Blue Streak* at last made progress, the Treasury continued to snipe at the project, querying whether any savings had been achieved from its development as a collaborative project.

The first suborbital missions were largely successful. Five firings were made from Woomera over 1964–6. Only the *Blue Streak* was fired and dummy upper stages were used. Overall, these were satisfactory and created confidence that, with a reliable first stage, the rest of the programme would be successful. Three of the launches were entirely successful, two partially so.

The first launch F-1 developed problems at 145 seconds into a 154-sec burn, the fuel sloshing inside the tank leading to an early cut-off. The full mission time had almost been accomplished. The first of the 1966 launchers disappeared off the radar screens two minutes after lift-off and was destroyed – but a later look at the data suggests that the radar was in error and the destruction needless. The wreckage was not found until an aerial survey stumbled across it in 1993, 70 km north of Woomera in the Simpson desert. For a first series of launcher tests, this was a record that the early American and Russian rocketeers would happily have settled for. Now they could proceed toward all-up testing.

The first live firing of the second stage was planned for the launches that took place in August 1967 and December 1967. However, on both occasions the second stage failed due to electrical problems. These were carefully investigated and overcome in anticipation of the first all-up launch of three live stages. This time, in November 1968, just as the third stage was due to light, electrical interference triggered its destruct system. ELDO put considerable work into troubleshooting both the technical problems and the question of coordination between the teams of the different countries. There was a similar malfunction in July 1969 when the German third stage did not fire and dropped the 197-kg Italian satellite into the Pacific Ocean near New Guinea 13 minutes later. At the pre-launch press conference, the ELDO team had seemed confident of success, so this failure was a bitter disappointment.

On the F-10 flight, which took place at Woomera on 12 June 1970, it was at first thought that the satellite had entered orbit. When it crashed instead into the ocean, the Australian controllers realized something had gone wrong. It turned out that the third stage had generated insufficient thrust, the nose cone had failed to jettison and as a consequence it did not have enough momentum for orbit.

2.16 CHANGE OF HEART IN BRITAIN

The continued failures of the ELDO launcher exasperated Britain at a time when all the country's foreign, defence, post-colonial and industrial policies were under review. The Wilson government (1964–70) came under extreme economic pressure, which led at its peak to the devaluation crisis of autumn 1967. The government believed that the country's foreign commitments during the previous Conservative governments had become greatly overextended. The Wilson government cancelled many defence projects, such as the TSR-2 and the programme for building aircraft carriers, and tried to pull out of *Concorde*. Relationships with Europe had already soured when General de Gaulle vetoed British entry to the European Economic Community.

We now know that the Wilson government had decided to withdraw from ELDO at an early stage [23]. The new minister responsible for aviation Roy Jenkins commissioned two reports on British space policy, called the Bondi report and the Space Policy Review. Both told him that ELDO was not worth the expense of continuing. There were several related reasons. First, there were fears that costs would continue to escalate. Second, the French had already realized that Europe needed a more powerful launcher to send communication satellites into 24-hr orbit, tentatively called the *Europa B* (also *Europa III*) – and this in turn would lead to even more costs. Third, the reports before Jenkins questioned the underlying assumptions that had led to British participation in ELDO in the first place and estimated the benefits to British industry to be virtually zero. Wilson, in particular, saw ELDO as a means of rescuing and finding a home for the Tories' ill-judged *Blue Streak* missile, but felt no commitment to it.

Accordingly, the British government decided on 16 December 1965 to withdraw from ELDO. This decision was not announced at the time for political reasons. The government feared that to do so would be to send the wrong signal to Britain's allies in Europe, who were then pressing for British entry to the European Economic Community. Roy Jenkins' approach was to try to persuade ELDO that there should be a re-evaluation of the ELDO programme, so that the other countries could persuade themselves of the futility of *Europa I*. Britain spent the next few years trying to find a means of announcing the December 1965 decision in such a way as to minimize the political damage while costs continued to go up. Despite the lack of a formal political announcement, Britain's faltering enthusiasm for *Europa* and ELDO became quickly apparent.

Figure 2.10. *Europa* takes off.

2.17 *EUROPA I* REVISED: *EUROPA II*

The *Europa I* launcher also came under technical pressures in 1966. France began to press for the upgrading of the launcher, calculating that, with the use of a small fourth stage namely the *Diamant B* solid propellant motor, *Europa* could put a 200-kg satellite into geosynchronous orbit. This would enable Europe to launch its own communications satellite, provide European entry into the field and pave the way for operational European systems [24]. The name for the improved version of *Europa I* would be *Europa II*.

The European space conference of July 1966 tried to resolve the contradictions of France wanting to expand *Europa* while the British wanted to reduce their contribution. It was agreed that Britain reduce its financial contribution from 38% to 27%. The French and German contributions were rebalanced at 25% and 27%, respectively. In exchange, *Europa II* was agreed. The cost of the *Europa I* and *II* programme was reset at €626m. The new launcher *Europa II* would fire from the French base in Kourou in French Guiana, then in construction.

2.18 BRITAIN PULLS OUT

Far from resolving the ELDO problem, the July 1966 conference was a warning of worse to come. In advance of the July 1968 meeting, Britain announced on 16 April 1968 that it was ending its financial commitments to ELDO from 1971. Britain would continue to make *Blue Streak* available until 1976, but without a financial contribution. Britain seemed to indicate that it regarded direct competition with the United States in the making of launchers as unjustified. Britain took the view that a satellite launch from the Americans would cost about £3m, compared with an equivalent European rocket for £4m, never mind the development costs that had by this stage totalled £250m for *Europa*. ELDO was portrayed in the British press as having been expensive, ill-judged and mismanaged.

The July 1968 ELDO conference was left with the sorry task of cutting back on the launcher programme. A further conference was convened that November in Bad Godesberg. Positively, this appointed a team of civil servants to draft a convention for a united European space agency, embodying the principle of mandatory and optional programmes. Negatively, ELDO could not agree a budget for 1969 and was close to collapse. The 1969 budget was not agreed until April 1969, well into the year, with Britain and Italy reducing their contributions (France, Germany, Belgium and the Netherlands making up the shortfall) and the *Europa* programme was trimmed still further. In 1969, after the failure of F-9, Britain stopped making payments and Britain's withdrawal from ELDO was completed by end 1969. The French were, however, reluctant to give up the idea of a European launcher. In 1970 the French had persuaded their European colleagues (Britain now absent) to adopt what were called the Bad Godesberg principles, under which European countries should as a matter of course give priority to a European launcher, provided that it was technically suitable and that it was not more than 125% the cost of an equivalent American launcher.

This heralded the end of the line for Woomera. Although it was to see Britain's only satellite launch in 1971, the base fell into disuse for over a decade. Britain formally left in 1980. It was not reopened until a new British and American sounding rocket programme began there in August 1987. Japanese shuttle landing tests took place there in 1996.

2.19   EUROPA II

Britain's partners now proceeded with the *Europa II* without her. Fitted with a kick motor, *Europa II* could send small communications satellites into geostationary orbit. If the test programme was successful, the first payload would be the Franco-German *Symphonie* satellite. This time the new French rocket base in French Guiana, South America was used.

The first launch of *Europa II*, on 5 November 1971, ended after 161 seconds when the rocket veered off course, became overstressed and exploded. Initially, electrical interference in the third stage was blamed again. More detailed analysis of the data showed that at 27 km the inertial guidance in the third stage had failed. The *Blue Streak*, still burning, was no longer getting instructions for its flight path and was now bending over at an angle of 35°. This imposed unacceptable strains on the stack, leading the second stage to explode. What was left of the rocket crashed into the south Atlantic 483 km downrange.

A commission was convened not just to look at the technical aspects of the failure but the underlying reasons as well. A programme review was done with the assistance of American launch vehicle engineers [25]. ELDO's commission blamed the constant failures in the first instance on highly unintegrated project management and an inefficient secretariat lacking authority and political pressures from the member states. There were technical problems as well, such as a lack of homogeneity in the three stages, poor integration of the stages, bad wiring, an unflightworthy guidance computer and insufficient documentation. The report recommended technical improvements, but above all the ELDO secretariat needed to become an effective management tool with a strongly centralized and project-oriented attitude coupled with a direct contractual relationship to the manufacturers. The report was especially critical of how the project for the third stage had been organized. If all these matters were put right and with additional funding, the ELDO commission said that *Europa II* might then have 'the normal probability of correct functioning'. The Commission's report was very frank and specific, making no attempt to blame external circumstances.

Following the release of the commission's report, ELDO took a number of steps to rectify what were broadly accepted as organizational and managerial deficits. The secretariat was reorganized and a number of new high-level appointments made. ELDO assumed direct supervision of launch vehicle integration and checkout. The contracts with the suppliers of the four stages were revised to give ELDO authority over the work of the manufacturers [26]. These were significant interventions, even if they were open to the accusation of being taken after the stable doors had been opened and the horses escaped.

2.20   *EUROPA III* INTRODUCED

The November 1971 failure was most disheartening for ELDO. Despite this, France was still looking ahead, beyond what was hoped to be the first successful flight of *Europa II*. At the July 1970 ELDO conference, France, Germany and Belgium had supported a French proposal for the study of *Europa III*, a launcher capable of putting 700-kg broadcasting and telephone relay communications satellites into 24-hr orbit.

Europa III was defined by October 1971: it would be a three-stage launcher 40 m high with a weight of 190 tonnes. Germany made plain its interest in building a cryogenic upper stage. Things began to look up at last and General Robert Aubinière moved across from CNES to head up ELDO. The most crucial element in *Europa III* was a hydrogen-fuelled upper stage, where the German company MBB and the French SEP made remarkable progress in designing a highly efficient, high-pressure, cryogenic upper stage. They designed the H-20, recycling the gases in the pumps, using copper lining for improved heat conductivity and achieving pressures of 130 atmospheres. Its efficiency level was comparable with anything achieved by the United States at the time [27].

2.21   ELDO: END OF THE ROAD

Despite the progress made in improving the management of *Europa II* and the prospect of *Europa III*, ELDO entered a new dark hour. Europe became badly divided on how best to respond to the American invitation to join the post-*Apollo* programme. The United States invited participation in a project to build a laboratory for the payload bay of the space shuttle, *Spacelab*. Germany favoured participation with the Americans in the *Spacelab* project, France was against and Britain was unconvinced. France, supported by Germany, favoured an independent European launcher, but Britain did not. The July 1972 European Space Conference was called off because the prospect of bridging these different perspectives was adjudged too difficult. The European Space Conference was reconvened for October 1972, but abandoned when it was clear that no agreement could be possible. Eventually it met for a day on 20 December, and the French put forward the idea of a simplified *Europa III* launcher. This was not shot down at once, though it left the future of *Europa II* uncertain [28].

Disheartened engineers worked hard to try and bring this disappointing string of failures to an end. F-12 was prepared for launch. If successful F-13 would fly with the communications satellite *Symphonie* on board. F-12 was scheduled for 1 October 1973, and there was sufficient money in hand to continue preparations for the launch.

On 27 April 1973, as F-12 was en route to Kourou, the German and French finance ministers pulled the plug and the programme was cancelled. ELDO had now spent almost €1bn on *Europa* by this stage. The ministerial decision cancelled *Europa II* and *Europa III*, bringing ELDO to a close on 1 May 1973. The remaining resources allocated to F-12 were used to wind down and close the programme. Three thousand jobs were lost immediately, Rolls Royce's engine factory being the most severely affected. ELDO staff publicly condemned the decision. They pointed out that they had carried out the changes mandated by the commission that followed the F-11 failure, but had never been given a chance to prove themselves. They had been confident of success with F-12. Indeed, Rolls Royce had authorizations in hand for engines up to F-18.

F-13 is now in a museum in Scotland while the other *Blue Streak*s were dispersed to Redu, Belgium (F-14) and Munich (F-15). F-12 reached French Guiana and was sold off as scrap when it arrived. It was towed into the jungle where the shell was eventually used as a chicken coop. It was an ignominious end for what was at the time the largest rocket in the world outside the United States or the Soviet Union [29].

ELDO's flight record

Europa I		
F-1	5 June 1964	Partial success (first stage only)
F-2	19 October 1964	Success (first stage only)
F-3	22 March 1965	Partial success (first stage only)
F-4	24 May 1966	Success (first stage only)
F-5	15 November 1966	Success (first stage only)
F-6	4 August 1967	Second stage failure
F-7	5 December 1967	Second stage failure
F-8	30 November 1968	Third stage malfunction
F-9	3 July 1969	Third stage malfunction
F-10	12 June 1970	Third stage failure
Europa 2		
F-11	5 November 1971	Exploded at 150 seconds
F-12	October 1973	Cancelled

2.22 POST-MORTEM

What went wrong? At a superficial level, ELDO went wrong because its rockets didn't work. High failure rates were not unusual in the early rocket programmes and they dogged the early American and Soviet programmes as well. More perceptive analysts believed that there were deeper organizational reasons behind the ELDO launch failures. The ELDO programme suffered from the inexperience of the different countries, all at an early stage of their rocket development. All rocket projects require high levels of technical ability, rigorous management, the proper sharing of responsibilities and persistence – and these qualities were not universally in evidence.

Both ELDO and ESRO had weak secretariats. Member states were inclined to take sole responsibility for the work assigned to them, and there was no-one to knock heads together to make international programmes work internationally. A major problem with the ELDO project was the fact that the different parts of the launcher were shared out crudely between the member states concerned. Massey and Robins [30] point out that 'there was no nominated prime contractor for the overall launching system and a tenuous management chain in a project of such complexity could not but aggravate the problems of overall system design.'

Following the 1966 crisis over *Europa I* funding, in the following year ELDO asked Jean-Pierre Causse to address the question of a better coordinated European space policy. He recommended a single, strong space organization – but, by the time it was considered in 1968, ELDO was in an even deeper crisis with the British carrying out their threat to withdraw. He recommended Europe increase its space budget by 10% and aim at putting 2-tonne application satellites into 24-hr orbit (something that was eventually done, but 10 years later). He expressed confidence in *Blue Streak*, dismissing British arguments that it was now obsolete. Thirty years later he pointed out that many of its 'obsolete' contemporaries are still flying profitably.

2.23 ESRO

Of the two organizations set up over 1960–4, ELDO attracted the most public attention, debate and controversy. Almost unnoticed, ESRO began its work of establishing an initial European presence in space.

The first thing ESRO did was to devise a satellite programme for the following three years. Its initial efforts were focused on two small satellites (these became *ESRO 1* and *2*), three medium-size projects and three larger ones, including an astronomical satellite project (ultimately, this was not brought to fruition). About 100 scientists submitted proposals, of which 45 were adopted for the programme. The principle of *juste retour* was adopted. This was the principle that work, contracts and responsibilities should be shared out between the member states according to the financial contributions they put into ESRO in the first place. Although the principle of *juste retour* seemed to contain the seeds for further division, in reality it rarely proved to be a problem within ESRO. This was the financial arrangement:

ESRO financial contributions (%)

Belgium	4.42	Netherlands	4.24
Britain	25	Spain	2.66
France	19.14	Sweden	5.17
Germany	22.56	Switzerland	3.43
Italy	11.17		

ESRO's first task was to set up its infrastructure, entailing six main establishments and a tracking system as follows:

ESRO establishments

ESRO headquarters	Paris
European Space Technology Centre (ESTEC)	Noordwijk, Netherlands (for the research and development of spacecraft)
European Space Laboratory (ESLAB)	Noordwijk (for the integration of payloads)
European Space Data Centre (ESDAC)	Darmstadt, Germany
European Space Research Institute (ESRIN)	Frascati, Italy (for advanced research)
European Sounding Rocket Range (ESRANGE)	Kiruna, Sweden
Tracking facilities	Redu (Belgium), Alaska, Spitzbergen and Falkland Islands

2.24   ESRO'S SATELLITES

Although ELDO never had a successful launch, ESRO lasted long enough to see satellites with the *ESRO* name get into orbit. ESRO's first event was the launch of a sounding rocket from the Andoya range in Norway in 1966. Because of the delay in Europe in developing its own launcher capacity, ESRO decided to approach the United States for the launch of its first satellites on the *Scout* rocket. An agreement to this effect was signed in 1966. ESRO's scientific programme made good initial progress.

ESRO satellites

20 May 1967	*ESRO 2A*	Launch failure (fourth stage)
1 May 1968	*ESRO 2B/Iris*	Solar corpuscular radiation
3 October 1968	*ESRO 1A/Aurorae*	Geomagnetic field, ionosphere
5 December 1968	*HEOS 1*	Solar radiation over 11-year cycle
1 October 1969	*ESRO 2/Boreas*	Geomagnetic field, ionosphere
31 January 1972	*HEOS 2*	Magnetic fields
12 March 1972	*TD-1A*	Full sky survey in X, gamma and ultraviolet
22 Nov 1972	*ESRO 4*	Earth polar atmosphere

In ESRO's initial programme, *ESRO 1* was chosen as a satellite to study the polar

Figure 2.11. *ESRO 2* satellite.

ionosphere while *ESRO 2* would be for solar astronomy and X-ray studies. *ESRO 1* had instruments for the study of electrons and protons and auroral photometry, while *ESRO 2* had instruments to study solar X-rays, cosmic rays and high-energy particles. Due to a variety of circumstances, *ESRO 2* came to be launched first, although in the event the first attempt failed. Called *Iris* and *Aurorae*, respectively, they plotted the travel of solar particles into near-Earth space and monitored magnetic storms.

HEOS 1 and *2* were ambitious probes approved in March 1965. *HEOS* (High Eccentric Orbit Satellites), made by the Junkers company, were so-called because their orbits went as far out as 225,000 km. Their aim was to establish the strength and direction of the geomagnetic plasma mantle, a region of rarified, high-temperature ions.

The *TD* series was so named because it required the powerful American *Thor Delta* launcher. They were intended as a standard bus to provide a platform for astronomical, geophysical and solar observations. In the event the *TD* series proved to be overambitious, too complex and expensive, so much so that *TD-2* had to be abandoned (the experiments intended for *TD-2* were transferred to *ESRO 4*). *TD-1A* was a 500-kg observatory with seven major experiments designed to scan the heavens for X-rays, gamma rays and high-energy emissions from an orbit of 528 km, 97°. An ultraviolet telescope was developed by the British Science Research Council and the Institute of Astrophysics in Liege.

TD-1A conducted a full-scale sky survey and catalogued over 15,000 stars. The probe cost €70m with a further €10m launch costs and went into polar orbit at 97°. Although it suffered an early electrical failure, ground control in Darmstadt was able to recommence operations. In late July–early August 1972, a solar storm was studied by *HEOS 1* at 177,000 km, *HEOS 2* at 230,000 km and *TD-1A* at 550 km, successfully triangulating the phenomenon.

The last satellite, *ESRO 4*, was launched on an 18-month programme of research into a polar orbit of 240–1,100 km, 98 mins, to study solar, ionospheric and magnetospheric particles. Weighing 115 kg, it was built by Britain (Hawker Siddeley), Germany and Italy (FIAR) for €11m, with additional experimentation from Sweden and the Netherlands. *ESRO 4* was launched by a *Scout* from Vandenberg and controlled from Darmstadt in Germany. The Italians built the power system, and this mission was a significant breakthrough for the Italians in satellite-building. The satellite ended up in a much lower orbit than planned, 40 km less than the planned 280 km, but this had the happy effect of leading to the unexpected discovery of concentrations of argon gas over the south pole and of higher temperatures among the particles over the south pole than the equator, because of solar heating near the poles.

Two ESRO projects never came to fruition. These were the Large Astronomical Satellite (LAS) and the comet fly-by. Both were considered during the mid-1960s, but ran up against the problem of limited budgets. The LAS eventually emerged many years later as the international ultraviolet *Explorer*, while the comet fly-by idea eventually became the *Giotto* probe to comet Halley.

2.25   ESRO IN DISARRAY

ESRO was not spared the controversies that ultimately undermined ELDO. Many of the problems of the latter spilled over to affect the former, and it is a tribute to ESRO scientists that they kept the scientific projects on track despite the political controversies swirling on around them.

From 1966, though, projects began to get behind and sought additional funding, which member states were slow to provide. The Causse review of European space activities was then under way, recommending a fusion of ELDO and ESRO. The European Space Conference at the end of 1968 set up a senior officials committee in March 1969 to progress Causse's idea. The committee was chaired by G. Puppi of Italy.

Even as Puppi's group struggled with the problems of ELDO and ESRO, ESRO came under much strain in 1970. That July, the French formally proposed that ESRO consider the building of applications satellites, especially communications and weather satellites. Belgium and Germany supported the proposal, but other countries felt that ESRO should be limited to a scientific brief and could and should not attempt to tackle the huge area of activity represented by applications satellites.

France then faced the prospect of either dropping this area of work or else proceeding on its own. In an attempt to provoke a favourable response, France threatened to pull out of ESRO, giving a date for its withdrawal as 1972. The French argued that they could not pay for both ESRO and ELDO and a national programme of applications satellites on its own.

By the end of that year, French and British objectives had so diverged that ESRO appeared to be in the process of disintegration [31]. Granted the state of ELDO at the same time, the picture of European cooperation in the early 1970s was not a pretty one.

Puppi's proposals were eventually presented in May 1971. He proposed:

- that ESRO now deal with applications, communications and meteorology;
- that the scientific programme continue at a 20% reduced level, not less than €35m a year, enough to support a medium-sized mission every 18 months;
- that the science programme, while smaller, remain mandatory;
- that ESRIN and ESRANGE be closed.

He proposed a budget of €74m for 1971, rising to €150m a year from 1974. In effect, ESRO's brief was extended to applications satellites, the areas of work agreed being communications, maritime communications and weather forecasting [32].

The following month, June 1971, the Puppi proposals were endorsed at a summit between French President Georges Pompidou and British Prime Minister Edward Heath. They had met to discuss British entry to the European Economic Community. Space cooperation was one of the flanking areas covered, and there was agreement between them for ESRO to be divided between mandatory and optional programmes. Those member states wishing to proceed with applications programmes could do so, but they were not binding and would not cost those states that did not wish to do so. The principle of applications programmes was agreed, which must have pleased the French; and ESRO would proceed with three initial applications programmes in the area of communications, air traffic control and meteorology: *OTS, Aerosat* and *Meteosat*. With minor changes, Puppi's proposals were accepted at the ESRO council meeting of December 1971.

The ESRO situation quickly became calmer, and the following year saw the successful launch of *HEOS 2, TD-1* and *ESRO 4*. The 1973 budget was set at €200m. Despite this progress, in London *The Times* was not impressed (7 March 1972):

The results [from ESRO] have been meagre, considering the time and money which have elapsed. In many ways the various countries are farther from agreeing on a coherent policy and programme of work than they were at the first debate. What little success there has been has come from ESRO with the brunt of failure borne by ELDO. Unfortunately the troubles from the launch development organization have spilled over into ESRO. The member countries have exerted all kinds of pressures at the negotiating table by threatening and eventually withdrawing support from one agency or another to try to force support for their particular proposals. Not surprisingly, most of the bitterness has been between the three biggest contributors, France, Britain and Germany. In spite of and certainly not because of national aspirations, ESRO has launched five satellites. All ESRO satellites have been launched by the United States and there is no reason to believe that a second source of launch vehicles could be created within the near future.

2.26   ASSESSMENT AND CONCLUSIONS

Although ESRO and ELDO are regarded, across the span of history, as having been failures, this would be a severe judgement. Seven ESRO satellites were put in orbit between 1968 and 1972 and these performed well, returning useful scientific data. They played an important part in European scientists, companies and governments learning how to build satellites together.

Of the two organizations, ELDO was the less successful. The withdrawal of Britain over 1965–8 had much more to do with British domestic politics than it had to do with the nature of the programme, and it was harsh of the withdrawing party to lay the responsibility on its former partners. Any project will come under enormous strain should its main driver pull out and reverse its launcher policy.

Ultimately, ELDO's failure was a management one rather than a technical or engineering one. This had dawned on ELDO when the Causse review was established in 1967 and was more starkly apparent when the F-11 commission was set up. ELDO had misjudged the importance of management in driving such a complex project as a three-stage rocket, never mind an international one. Quality management had been very much at the core of the NASA Moon landing programme. Conversely, the Russians had lost the Moon race through chaotic management, rivalries between designers and political interference. Even as *Europa II* was closed down, the Soviet Union was cancelling its own superrocket, the *N-1*, which had likewise failed to make a successful flight.

Although the Causse and subsequent committees pointed the way toward a new approach in which Europe could overcome these difficulties, the obstacles seemed formidable. ELDO had become embroiled in one dispute and difficulty after another and seemed unable to agree a coherent view on any major undertakings, be they launchers, laboratories or applications. So severe were the fault lines of disagreement that ministerial meetings could not be held for months on end. The prospects for a successful European launcher, never mind European cooperation across a broad range of activities, could not have appeared to be more bleak. Maybe European cooperation in space was an unrealizable ambition after all. What about the national space programmes? These are reviewed next (Chapter 3) before we return to the European Space Agency (Chapter 4).

3

The national space programmes

As we have seen, 1960 saw the agreement of Europe's leading industrial and engineering nations to pool their efforts in a joint launcher and satellite programme. By 1973 these efforts had unravelled. Britain had pulled out of launcher development and France was trying to save something from the untidy collapse of ELDO and the demoralization of ESRO. It was just as well in many ways that these countries had kept some national space programmes going. Even as Britain abandoned ELDO, it kept a modest national rocket and satellite programme going. France had been working on its own rockets, and Italy began to specialize in a number of key areas of space technology. Other, smaller countries, like the Netherlands and Sweden, developed their own satellite programmes. Others found a niche in the expanding European space industry.

Here we interrupt the narrative of the European space effort to look at the national programmes, starting with Britain and France, then Italy and Germany and then the smaller countries. Space activities do not lead themselves to easy categorization or putting things neatly in boxes. Not only are there European programmes (ELDO, ESRO, ESA), but also national programmes and bilateral or multilateral programmes between European countries (e.g., France, Sweden and Belgium joining for *SPOT*). In addition, there are bilateral programmes and projects between individual European countries and the United States or the Soviet Union.

3.1 FRANCE'S NATIONAL PROGRAMME

France was the third country in the world and the first European country to launch its own satellite. Preparations for France's first indigenous satellite took the full four years allocated from the initial decision of 1961. France took the decision to build and launch a domestic satellite indigenously, and at the same time have a team build a satellite for the Americans to put in orbit on the *Scout* launcher. Two teams thus worked in parallel. In the event, both satellites were launched only weeks apart, the indigenous French one just winning the race.

Domestically, the main problem was less about building a satellite, more about making the rocket sufficiently reliable to warrant committing a satellite. *Diamant* was developed at the Laboratoire de Recherches Ballistiques et Aérodynamiques (LRBA) at Vernon in Eure. The *Diamant* was 18.75 m tall, weighed 17.97 tonnes and had the ability to put 80 kg into a 300-km orbit. The first stage was called *Emeraude* with Vexin liquid-fuelled engines able to provide 28 tonnes of thrust for 88 seconds. The second stage *Topaze* was solid-fuelled, with a polyurethane motor set in steel casing, giving 14.5 tonnes of thrust through four nozzles. The third stage provided 5.3 tonnes of thrust for 44.5 seconds through a single nozzle. There were three failures of the *Emeraude* rocket, which held things up.

Under the agreement with the United States, France launched two *Aerobee* sounding rockets in the first phase. The second phase involved the construction by French engineers of the French satellite FR-1. The French team in America was told to develop the parts of FR-1 in France if possible, but in America if not. The emphasis was on acquiring know-how as expeditiously as possible.

Even as the American team worked on its project, the domestic French team tried to get France's first satellite and rocket ready in France and Algeria. France's original launch centres were in the Sahara desert. Two were set up there in the 1950s: one in Hamaguir and the other 110 km away on the plateau of Colomb-Béchar. Launches took place beside a lighthouse-shaped, cement blockhouse, with launch attachments scattering like shard on take-off. A tracking system was set up to follow French satellites. There were stations in:

French tracking stations

Brétigny	France (the main tracking site)
Pretoria	South Africa
Brazzaville	Congo
Ouagadougou	Upper Volta
Hamaguir	Algeria (later moved to Canary Islands)
Beirut	Lebanon (mobile site)

The first satellite was simply called A-1 for the armed forces, being built by Matra. The scientists had wanted to called it Zebulon or Zebby for short after a television show puppet. Officialdom felt that this was trivializing the satellite and chose a more Gallic hero *Asterix* (from the comic strip), which also matched the A of A-1. The Centre Nationale d'Etudes Spatiales (CNES) took the opportunity of announcing that satellites would have names from the date of the commencement of a project and would never again be put in the position of having to find names once a launch was due or had taken place.

On 26 November 1965, the three-stage *Diamant* put the *Asterix 1* satellite into orbit from the Hamaguir launch site, in the Algerian Sahara desert. Pictures showed

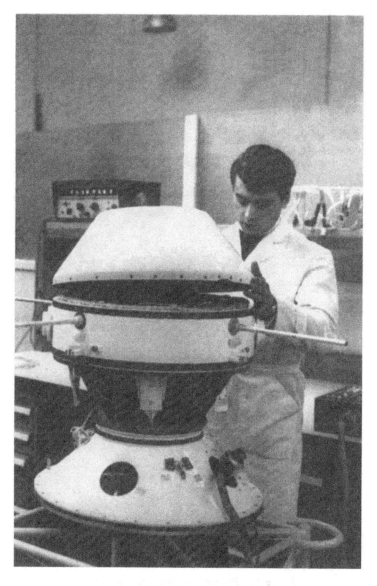

Figure 3.1. *Asterix.*

the needle-shaped rocket lifting off the desert site, sending launch débris flying as it headed up. Everything went perfectly. *Asterix* weighed 41.7 kg and entered an orbit of 528–1,768 km, 34°, 108 min. It was top-shaped, 50 cm in diameter, with four protruding antennae. The satellite transmitted for two days, but has an orbital life of 200 years. *Asterix* had a simple telemetry system to relay acceleration, velocity and temperature with a radar transponder for tracking. It was made of fibreglass and had

Figure 3.2. France, the third space power.

radio beacons to test both the French tracking network and a seaborne station on a naval destroyer in the Mediterranean, the *Guepratte*.

The launching of *Asterix* caused anguish in Britain. Writing in *Flight International*, Geoffrey Pardoe of Hawker Siddeley Dynamics commended the French success, drew attention to its economic benefits to France and lamented how Britain had now fallen behind. Should the European ELDO launcher fail, he said, Britain could still develop its own national programme with *Black Arrow*, which was after all larger than *Diamant* [33].

Diamant A

Dimensions	19 m tall × 1.4 m diameter
Weight	18.4 tonnes
Capability	80 kg into 300-km orbit

	First stage *Emeraude*	Second stage *Topaze*	Third stage *Rubis*
Dimensions	10 m × 1.4 m	4.7 m × 0.8 m	2.06 m × 0.66 m
Weight	14.72 tonnes	2.93 tonnes	0.71 tonnes
Fuels	Turpentine and nitric acid	Solid	Solid
Thrust	28 tonnes	14.5 tonnes	5.3 tonnes
Burn time	88 seconds	44 seconds	44.5 seconds

First satellites in orbit

October 1957	Soviet Union (*Sputnik 1*)	84 kg
January 1958	United States (*Explorer 1*)	14 kg
November 1965	France (*Asterix*)	40 kg
February 1970	Japan (*Osumi*)	38 kg
April 1970	China (*Dong Fang Hong 1*)	173 kg
October 1971	Britain (*Prospero*)	66 kg
July 1980	India (*Rohini 1B*)	35 kg
September 1988	Israel (*Offeq 1*)	156 kg

France's space launches

26 November 1965	A-1 (*Asterix*)	*Diamant A*	Hamaguir
17 February 1966	D-1A (*Diapson*)	*Diamant A*	Hamaguir
8 February 1967	D-1C (*Diadème*)	*Diamant A*	Hamaguir
15 February 1967	D-1D (*Diadème 2*)	*Diamant A*	Hamaguir
12 January 1970	*Péole*	*Diamant B*	Kourou
10 March 1970	*Wika/Mika*	*Diamant B*	Kourou (for Germany)
15 April 1971	D-2A (*Tournesol 1*)	*Diamant B*	Kourou
5 December 1971	D-2B (*Polaire*)	*Diamant B*	Kourou (fail)
21 May 1973	*Castor/Pollux*	*Diamant BP-4*	Kourou (fail)
6 February 1975	*Starlette*	*Diamant BP-4*	Kourou
17 May 1975	*Castor/Pollux*	*Diamant BP-4*	Kourou
27 September 1975	D-2B (*Aura*)	*Diamant BP-4*	Kourou

What of the American-based team? CNES was responsible for the design and testing of the satellite, which was built by 14 companies, 7 of which were French. The satellite was duly launched from Vandenberg Air Force Base on a *Scout* rocket on 6 December 1965. The CNES press conference the next morning was quite

Figure 3.3. Analysing *Asterix*'s signals.

subdued: although it was certain that the satellite had entered orbit, no-one could pick up any signals! Thankfully, before the loss of the mission was announced, the signals at last came through. Although designed to operate for three months, FR-1 worked for two and a half years.

FR-1 weighed 60 kg and had instruments for measuring the propagation of very long radio waves in the upper atmosphere, the lines of the Earth's magnetic field, natural radio waves and the distribution of ions in the atmosphere. The satellite itself was spheroid, but with many spider-like antennae emerging from the top, bottom and side. The experience that they gained was later put to good use in the building of the indigenous *Diapson* satellites.

FR-1 was the beginning of a long period of cooperation with the United States. Over the following years, French experiments were installed on a series of American spacecraft. *Eole*, the first American-flown mission after FR-1, was a data relay test satellite, designed to pick up information from up to 500 weather balloons over a five-month period and then relay their information back to ground stations. This was an unusual kind of satellite for its time and the brainchild of two American meteorologists. They figured that satellites would be a much more reliable means of collecting information from high-altitude balloons (such as their speed, direction and latitude) than any other method. *Eole* was duly launched on 17 August 1971 on a *Scout* from Wallops Island (Virginia), with the balloons released in the southern hemisphere at the same time. CNES set up 20 ground stations to receive data. In the event, *Eole* transmitted for 33 months. *Eole* was put to use in the 1972 transatlantic yacht race when Alain Colas' boat *Penduick 4* used *Eole* to give him his location to an accuracy of 250 m over 20 days of his journey. Ultimately, *Eole* paved the way for maritime navigation systems, a big outcome for a small project.

France's early satellites on American launchers

6 December 1965	FR-1	*Scout*	Vandenberg
17 August 1971	*Eole*	*Scout B*	Wallops Island

American spacecraft with French experiments or equipment

14 October 1965	Orbiting Geophysical Observatory 2
28 July 1967	Orbiting Geophysical Observatory 4
4 March 1968	Orbiting Geophysical Observatory 5
5 June 1969	Orbiting Geophysical Observatory 6
16 April 1972	*Apollo 16*
6 December 1972	*Apollo 17*
14 May 1973	*Skylab*
21 June 1975	Orbiting Solar Observatory 8
15 July 1975	*Apollo Soyuz* Test Project
20 August 1975	*Viking 1*
20 August 1977	*Voyager 2*
22 October 1977	*ISEE 1, 2*
8 August 1978	*Pioneer Venus 2*
12 August 1978	*ISEE 3*
20 September 1979	*HEAO 3*
14 February 1980	Solar Maximum Mission
5 October 1984	Earth Radiation Budget Satellite
5 May 1989	*Magellan*
18 October 1989	*Galileo*
14 September 1991	Upper Atmosphere Research Satellite
25 September 1992	*Mars Observer*

Returning to the indigenous programme, *Asterix* was followed by *Diapson* and *Diadème* that both had radios with a very stable frequency for geodetic research. *Diapson* was a 19-kg technology proving satellite, with a golden-coloured cylinder 51 cm in diameter and 20 cm tall. It had four little antennae at the top end, two transmitters and four rectangular solar panels with 2,304 cells at the base that were able to generate 4.5 watts of power supply.

Diadème 1 had laser reflectors and a magnetic stabilization system. However, its orbit was lower than planned, making the planned laser-tracking from Algeria, France and Greece difficult. *Diadème 2*, the last Hamaguir launch, was more satisfactory, providing laser and Doppler data for three months. *Diadème 1* and *2* were to carry out geodesy experiments using three triangulation points: St Michel in France, Stephanion in Greece and Colomb-Béchar in Algeria. They were equipped with clock and laser reflectors and a passive, magnetic, stabilizing system.

3.2 *DIAMANT B*

This concluded the *Diamant A* programme. CNES was already working on a more powerful *Diamant B*, a 24-tonne launcher able to reach a 500 km orbit, a performance comparable with the American *Scout*.

Diamant B was developed at the same time as launch operations were moved from Hamaguir to Kourou. It had a much larger first stage, making it a third taller and able to carry much more fuel and oxidizer, increasing the launch weight to 24.948 tonnes. Payload capacity was raised from 85 kg to 115 kg. *Diamant B* was 23.54 m long, its first stage had a 35-tonne thrust for 112 seconds and had the same upper stages as *Diamant A*.

Diamant B's first launch was 10 March 1970. Germany was anxious to test two capsules in orbit, *Mika* and *Wika*, and an intergovernmental agreement was signed in February 1969 for them to be put in orbit by *Diamant B*. They were duly launched on 10 March 1970 on *Diamant B*, the first launch from the Kourou space centre. Regrettably, vibration in the first stage meant that *Mika* failed to function properly.

The second *Diamant B* launch was *Péole* on 12 December 1970. *Péole* was equipped with laser reflectors and made tests of a worldwide network of reflectors. *Péole* was also a test for *Eole*, the balloon-tracker. The next D-2A *Tournesol* was sent up by *Diamant* on a six-month mission carrying five experiments to study solar radiation in the ultraviolet range and the distribution of water, methane and atomic hydrogen in the upper atmosphere – the event functioning well for over two years. *Tournesol* was so-called because of its yellow colour and because it would be permanently oriented toward the Sun. It had an attitude control system with cold gas microthrusters.

Unhappily, the last two launches in the national programme were failures. The second stage of the *Diamant B*, carrying D-2B for solar radiation studies and gamma ray bursts, failed after 147 seconds on 5 December 1971 while *Castor* and *Pollux* crashed in to the sea when the third stage failed to ignite and fell back into the Atlantic. They had been intended to test thrusters (*Pollux*) and examine orbital perturbations (*Castor*). Despite these setbacks, *Diamant* had succeeded on 10 of its 12 launchings, making it the most successful first launcher programme of any nation [34]. The D-2B backup was later launched on 17 June 1977 from the Soviet Union as *Signe 3*.

The third *Diamant* type was the *Diamant BP-4*, which went into development in 1972. The first and third stages came from the *Diamant B*, but there was a new and more powerful second stage called the P2.2 and a fairing based on the British *Black Arrow* design. The first launch of the *BP-4* was *Starlette*, a small 48-kg geodetic sphere 24 cm across carrying 60 laser reflectors still in operation many years later. *Starlette* studied terrestrial gravitation, the elasticity of polar movement and tried to find the precise location of observation systems. *Castor* and *Pollux* were eventually launched on 17 May 1975. The final national French satellite was *Aura* launched on 27 September 1975, which made solar and stellar ultraviolet observations.

French work on a civilian space programme was paralleled by a military rocket programme. Work on the *Diamant* programme was connected to development of the *Saphir* medium-range ballistic missile, which was fired successfully 13 times from Hamaguir. An important aspect of De Gaulle's programme for government as President of France (1959–69) was an independent French nuclear deterrent, which required both a French atomic bomb and a means of delivering it. Accordingly, France developed both a land- and sea-based missile system. The land-based

Figure 3.4. Preparing *Diamant* launch.

system involved the construction of rocket silos in Vaucluse and Drome, which first became operational in 1971, and a submarine-based system like the American *Polaris*. This first became operational with the submarine *Redoubtable* in 1971, the fleet being based in Brest in Brittany.

3.3 DEVELOPMENT OF SPACE FACILITIES

The *Diamant* had introduced the new French launching centre in Kourou, French Guiana and began to put it on the map. Even as it was doing so, there was a huge expansion of the ground facilities in France itself.

Although CNES headquarters are in Paris, this was a relatively small management function and the substantive work of the centre came to be done in Toulouse at Centre Spatiale Toulouse (CST), sometimes dubbed 'space city'. The reason for the Toulouse location lies in politics and administration. In 1955 the French government formally committed itself to a policy of decentralizing scientific departments from France to the regions. Little happened until 1960, when Toulouse was designated a centre for aeronautics and civil aviation. In 1963 the government decided that CNES should go there, even though it had only just been founded. This was news to the Paris staff, whose contracts of employment said nothing about a transfer to Toulouse. The government was relentless in pushing through the process, which was completed with the closure of Brétigny-sur-Orge in 1974. Between a third and half the CNES Paris staff refused to go and resigned. CST was officially opened in October 1973. What was a 72-ha greenfield site is now covered with administrative buildings, test centres, car parks, archiving and data centres and design offices. Although Toulouse is a CNES facility, a small number of ESA teams have also been located there, such as *Meteosat* at one stage. France now advertises Toulouse as the 'space capital of Europe' and has given it a futuristic image. Many of France's leading aerospace companies, large and small, have relocated there and the city and region are considered to be among the country's main, high-tech success stories. Matra was the first in 1968, followed by Thomson (now Alcatel). Companies locating there found it easier to win contracts and the rest soon followed.

Once Toulouse was operational, the worldwide French tracking network was restructured. Toulouse became the heart of the network, supplemented by new stations in Kourou (French Guiana), Hartebeestehoek (South Africa) and Kerguelen Island in the southern Indian Ocean. The old stations in Las Palmas, Pretoria and Ouagadougou were closed. In June 1998 the centre opened the Cité de l'Espace, an indoor and outdoor educational science park, complete with full-scale models of *Mir* and *Ariane 5*, in no time attracting over 300,000 visitors a year.

CST includes the full range of satellite-testing equipment necessary for a modern space programme. Several of the Brétigny-sur-Orge facilities were removed wholesale (Soviet-style) and relocated in Toulouse. The present facilities include vacuum, thermal, acoustic, anechoic and vibration test chambers and simulators.

Figure 3.5. Toulouse – greenfield site.

The exception to the move to Toulouse was the launcher section (the Launch Vehicle Directorate). CNES argued that the main companies dealing with launchers (e.g., SEP) were all located in or near Paris and that it made sense to keep the relevant part of CNES there. The French department for the regions kicked up a huge fuss and the matter went to Prime Minister Pierre Mesmer. He called in American consultants to arbitrate and they took CNES's side. So the Launcher Directorate duly moved to the new town of Evry where Rond Point des Poètes

Figure 3.6. Toulouse – modern space centre.

was renamed Rond Point de l'Espace, where the Launcher Directorate now shares two buildings with Arianespace.

3.4   COOPERATION WITH THE USSR

Cooperation with the United States was a feature of the early French space programme. Despite much French rhetoric about building up European autonomy in launchers and satellites independently of the United States, this has never got in the way of a high level of cooperation between the two countries. France was also one of the first Western countries to do business with the Soviet Union, with the USSR carrying French equipment on a variety of scientific missions. France also flew many of its early scientific satellites on American and Russian launchers before moving over to use Europe's own launcher *Ariane*. Here is a list of the principal missions on Soviet launchers:

France's satellites launched by the USSR

27 December 1971	*Aureole 1*	*Cosmos*	Plesetsk
4 April 1972	*SRET 1*	*Molniya*	Plesetsk
16 December 1973	*Aureole 2*	*Cosmos*	Plesetsk
5 June 1975	*SRET 2*	*Molniya*	Plesetsk
17 June 1977	*Signe 3*	*Cosmos*	Kapustin Yar
21 September 1981	*Aureole 3*	*Cosmos*	Plesetsk

Soviet–French cooperation can be traced back to 1964 when the French Minister for Scientific Research, Atomic Energy and Space visited the USSR. There were further contacts the following year, and on 30 June 1966 De Gaulle signed an agreement in the Kremlin. France was anxious to be even-handed in dealing with the space superpowers, prepared to demonstrate its independence and recognized that the Soviet space programme offered opportunities not available from the United States. Twice a year conferences on cooperation were organized, rotating between France and the USSR.

The first joint project was the *Roseau* satellite, a 150-kg probe to fly in an eccentric orbit halfway out to the Moon and explore the ionosphere. *Roseau* ran into problems when the Russians refused to give any details about their launcher, which was secret. Then the Russians refused to let the French integrate their satellite with their launcher at the pad. The embarrassed Russian scientists explained that they could not integrate their own satellites at the pad either and they appeared to know as little about their own launchers as the French. They always handed over their satellites to the military in Moscow, who took over from there. Then in May 1968 the French space agency suffered a number of budget cuts and the French had to pull out of the project. The Russians, who never had to struggle with these kinds of budgetary problems, refused to believe the reason and always felt that there was a political problem.

In the event these difficulties did not last. In 1970 the USSR and France signed

Figure 3.7. *Signe 3* team at cosmodrome.

an agreement called SRET (Satellites de Recherche et d'Etudes Technologiques), whereby small French satellites would fly piggyback on larger payloads in mainstream Soviet launches. *SRET 1* was duly launched 4 April 1972, the first Western probe to be launched by the USSR, *SRET 2* following in 1975. They weighed less than 20 kg and were flown piggyback with Russian *Molniya* communications satellites. *Molniya*s orbited in high, curving 12-hr orbits over the northern hemisphere, travelling in and out of the radiation fields twice a day. SRET followed in their paths, measuring how radiation, vacuum and temperatures degraded solar cells, lessons put to vital use in the design of the *Meteosat* satellites.

The SRET payloads were preceded by *Aureole*, a 660-kg satellite launched on a *Cosmos* rocket from Plesetsk on 27 December 1971. The satellite was a joint project called ARCAD (Arctic Auroral Density) between the Centre for the Study of Space Radiation in Toulouse, France and the Institute for Space Research in Moscow (IKI). The focus of the mission was the polar lights, or the aurora borealis. *Aureole 3* weighed 1,000 kg and carried instruments to measure protons, electrons and the magnetic field.

Signe 3 marked the first time a French spacecraft had been the sole payload on a Soviet launcher. Its aim was to study the effect of the solar wind of the upper atmosphere. *Signe* was the back-up of the hitherto unlaunched D-2B satellite.

Equally significant was French equipment carried as individual experiments on Soviet scientific and deep space missions. Despite the portrayal in the Western media of running an entirely secretive programme (some parts were very secret, it was true), the USSR invited a range of countries to be involved in Soviet space missions. Again, it was France that responded most to these offers. France flew equipment on the *Mars 2* and *Mars 3* missions to the red planet in 1971, *Mars 3* becoming the first probe to soft-land there. This was the first European equipment flown on a deep space probe. French lasers were carried to the Moon on the *Lunokhod* moon rovers in 1970 and 1973 to measure the precise distance between Earth and the Moon. Equipment to detect gamma rays was flown on the *Prognoz* series of solar observatories (*Prognoz 2, 6, 7, 9*). French equipment to detect gamma ray bursts was flown on the Venus probes *Venera 11, 12, 13* and *14*.

Perhaps the biggest collaborative project was *Vega*, the two probes to Venus and comet Halley in 1984–6. Here, the French supplied two TV cameras (narrow and wide angle), an infrared spectrometer and a plasma detector. A French spectrometer flew on the *Astron* ultraviolet observatory, 1983. For the Soviet *Granat* observatory, based on a *Venera* bus and launched on 1 December 1989, France made the main Sigma telescope, which weighed a tonne. The telescope was 1.2 m across the base and 3.5 m tall. A second French instrument, a gamma ray burst detector, was also installed. A *Proton* rocket put the satellite into a highly elliptical orbit of 2,000 by 200,000 km. France collaborated with the *Gamma* observatory, launched 11 July 1990 on a *Soyuz* rocket, providing the cameras and data-processing system. Four French instruments were carried on the *Interball* mission into the magnetosphere in 1995.

Following the success of *Vega*, the USSR invited cooperation on an ambitious project to send two probes to land on the little Martian moon Phobos. Twelve countries joined, making the foreign contribution worth R60m of the R272m project. France was the leading European contributor, providing instruments for the spectro-imaging of Mars and for the detection of solar oscillations. The value of the French contribution was about €5.3m. Even the smallest countries participated: for example, Ireland contributed SLED (the Solar Low Emission Detector), which detected a weak magnetic field around the planet. Sadly, *Phobos 1* was lost soon after launch (a ground controller accidentally sent up a command to close the probe down) and *Phobos 2* failed in the final stages of closing in on the moon, though not before sending back photographs and returning a modest amount of scientific data.

Despite the disappointing outcome to the Phobos mission, Russia (previously the Soviet Union) invited international cooperation on its next Mars probe, *Mars 8* (it was also called *Mars 94* and then *Mars 96* when it was delayed two years). *Mars 8* was an ambitious mission, a 6-tonne probe carrying two landers, two penetrators, an orbiter and equipment from 20 countries – in fact the largest, biggest, heaviest Mars probe ever built. Among the European equipment was a high resolution stereo camera (Germany) and on the lander there were a meteorology sensor (Finland), an alpha proton X-ray spectrometer (Germany) and a magnetometer (France). The French financial contribution to the project was in the order of €91m.

Many of the international collaborators were in Baykonur for the launch on 16 November 1996. The *Proton* rocket lit up the cold night sky, pushing its cargo into Earth-parking orbit for the trans-Mars injection. On the ground the vodka corks popped open to celebrate the successful launch. They should have waited. Over the south Atlantic the upper stage of *Proton* (called Block D) should have fired the probe out of Earth orbit on a curving trajectory to Mars. In what we can only presume to be a random failure, Block D failed to fire. *Mars 8* thought it had, separated from Block D and fired its engine to give it what would have been a final trans-Mars kick. In the event the manoeuvre pushed it into an elliptical orbit of Earth and the probe crashed into the Andes two days later. Block D ended up in the Pacific off Easter Island. This was the dispiriting end of the Russian Mars programme for the time being.

3.5 EARTH RESOURCES: *SPOT*

Returning to the domestic French space programme, France was the first country in Europe to develop Earth resources satellites. French interest in an Earth resources satellite was treated sceptically by other European countries, a view the latter may have since come to regret. The series was called *SPOT*, or Satellite Pour L'Observation de la Terre (Satellite for the observation of the Earth) and was adopted in 1978. The aim of the *SPOT* satellites was to take images of the Earth in different light bands that would be of value in measuring agriculture, water resources, renewable and non-renewable fuels and in contributing to an understanding of the oceans, climate and erosion.

   SPOT was originally proposed by France in February 1977 as a European project. To French disappointment, only two other countries backed the project, Belgium and Sweden. This put a real question mark over whether the project could continue, but a year later the French government adopted *SPOT* as a national project. Partly as a thank-you for their support earlier, Belgium and Sweden were invited to join on a bilateral basis (they contributed 4% each).

   In advance of the mission, CNES set up a subsidiary to market and sell *SPOT*'s images, the world's first satellite-based Earth resources company. CNES was a 39% shareholder in *SPOT* Image, with other investors from Belgium, Sweden and Italy. Images from *SPOT* were sent down to the Station de Reception des Images Spatiales in Toulouse and to a ground station in Kiruna, Sweden.

   SPOT 1 was duly launched by *Ariane 1* on 22 February 1986 and was soon returning high-quality data. It weighed 1,830 kg . *SPOT* was shaped like a box with a solar panel and operated in an orbit of 824–829 km, 98.7°, covering the same area of ground every 26 days, with altitude being adjusted every two months or so. *SPOT* had high resolution (10 m) visible imaging instruments, recorders and transmitters. Pictures were transmitted to the two ground stations, CNES in Toulouse and Kiruna in Sweden, sending down 1,500 pictures a day in both 10-m and 20-m resolution. *SPOT 1*'s pictures were intended primarily for map-making, forestry, town-planning, land development, prospecting, agriculture, land utilization and coast-line surveys.

   Within days, it had sent back excellent images of southern France and northern Italy, well justifying the hopes of its builders. *SPOT 1* was one of the first satellites to photograph the aftermath of the nuclear explosion at Chernobyl, including thermal images of overheated areas. The stereoscope quality of the images, combined with 10 m resolution, was stunning. Sceptics of the *SPOT* project quickly melted away.

Spacewatchers called up pictures it had taken of the Baykonur cosmodrome in Kazakhstan, where they found that preparations for a Soviet space shuttle were at an advanced stage and a landing runway now completed. By this stage *SPOT* Image already had contracts to sell the pictures in 40 countries. By July *SPOT* had sent back 5,000 images, the baseline set of scenes to be sold on to users. Customers could ask for either multispectral or visible images, overhead or oblique, the costs being between €500 and €1,500 per image. Within the first year 200,000 images had been sent back and of these 48,000 had been archived for sale. France accounted for 54% of sales. By 1989 *SPOT* Image's business had reached €3m and was growing at 25% a year. The main areas of user interest were mapping (30%), vegetation studies, farming and forestry (20%) and geology (18%).

SPOT 1 was taken out of service in January 1991, but was pressed back into operation in March 1992 when *SPOT 2* became overloaded. *SPOT 2* was operational within a week of reaching orbit on 22 January 1990 on an *Ariane 4* (Mission V35), sending back commercial images by the end of March. *SPOT 2*'s first image was a colour shot of Lake Garda in northern Italy, showing clearly the town of Brescia to the south and the jagged shapes of the Alps to the north.

Eventually, *SPOT 2*'s tape recorder broke down, probably from overuse. *SPOT 3* was put into orbit 25 September 1993 on *Ariane V59* (accompanied by five minisatellites). This time much attention had been given to the tape recorders, with improvements to the tapes, heads and guides. The launch was broadcast live to over 2,000 guests to the French space centre in Toulouse. After a 24-hr weather delay, the second countdown had to meet an 18-min launch window. There were three holds during the countdown, causing a growing level of tension as the time began to run out. *SPOT 3* just made the last minute of the window and the viewers cheered as V59 climbed into the night sky, before celebrating in the traditional French manner (champagne). *SPOT 3* operated until November 1996 when it abruptly ceased operations. A software fault appears to have left the spacecraft tumbling less than halfway through its design lifetime. Attempts were made to bring the old *SPOT 1* and *2* satellites out of retirement, *SPOT 1* resuming service again from January 1997.

By 1994 *SPOT* pictures had sales of €33.8m. Half a million pictures were sent down for processing each year. Real-time images were made available to ground stations as far apart as Prince Albert and Gatineau (Canada), Maspalomas (Spain), Cuiaba (Brazil), Lad Krabang (Thailand), Hatoyama (Japan), Islamabad (Pakistan), Hartebeestehoek (South Africa), Riyadh (Saudi Arabia), Alice Springs (Australia), Tel Aviv (Israel), Taipei (Republic of China), Pare Pare (Indonesia), Fucino (Italy) and Singapore.

SPOT 4 was approved in 1989. When *SPOT 3* went out of action, the launching was brought forward. *SPOT 4* was put up by *Ariane 4* on 24 March 1998. *SPOT 4* was a jump ahead, with the addition of another waveband (infrared) and a longer lifetime (five years). It carried a new vegetation-monitoring instrument with a low resolution (1 km) and a high-speed laser communications system. *SPOT 4* weighed 2,750 kg. The solid state tape recorder had double the capacity and was made more robust to guard against failure. *SPOT 4* carried additional experiments for the monitoring of the ozone layer and radio-positioning. By this stage the number of direct-receiving stations for *SPOT* data had risen to 21, from China to Canada. *SPOT 4*'s first pictures were spectacular, one showing the whole subcontinent of India ringed by the snowy Himalayas to the north; the other showing the new Vasco da Gama bridge across the Tagus in Lisbon. The vegetation instrument, part-funded by the European Union and designed to react to the colour signature of plant cover, proved to be exceptionally popular and was posted on the Internet for free availability. Its images showed crop production, flooding, forest fires and seasonal stages, the data being so rich that a conference was called in 2000 to see how the instrument could best be developed in the future. This became *Vegetation 2* on *SPOT 5*.

SPOT 5 was approved by the French government in 1994 and was launched by *Ariane 4* in May 2002. This was another big spacecraft and it only just fitted into the Airbus Beluga transporter that carried it to French Guiana. The Beluga super-transporter was an adapted Airbus with a huge hangar-like cargo hold on the top, with the nose of the plane peeping out below the front. Ugly and functional. *SPOT 5* offered significant advances in quality and data-processing – and it was the last of the big French Earth resources satellites. Its first pictures showed Greece in stunning detail, delighting the operators.

Its quality of imagery was a fourfold improvement over *SPOT 4*. The new 3-tonne *SPOT* carried a 2.5-m resolution high-resolution geometric instrument with three spectral bands and a 10-m resolution high-resolution stereoscopic imager capable of taking oblique as well as overhead images. *SPOT 5* could supply 550 images a day (compared with 400 on *SPOT 4*), and *SPOT* Image aimed to have photographs catalogued and available for commercial users within an hour of receiving them. *SPOT 5* also carried a vegetation instrument, orbital tracker, star tracker and 90 GB of solid state memory. The satellite was built by Astrium. In order to ensure the commercial success of the series, SPOT Image teamed up with a series of new distribution partners in Japan, Australia, Canada and China and also began a mapping project with the French National Geographic Institute to provide a detailed space-based map of a third of the world's surface.

SPOT 5 is intended to be the last of this pioneering series. Its place will be taken by a Franco-Italian project, Cosmo-Skymed-Pleiades, of six small 1-m resolution satellites to start flying from 2006.

SPOT series

22 February 1985	*SPOT 1*	*Ariane 1*
22 January 1990	*SPOT 2*	*Ariane 40*
26 September 1993	*SPOT 3*	*Ariane 40*
24 March 1998	*SPOT 4*	*Ariane 40*
3 May 2002	*SPOT 5*	*Ariane 42P*

3.6   OTHER FRENCH EARTH OBSERVATION SATELLITES

France has followed the *SPOT* series with two other missions concerned with more specialized aspects of Earth observations. These were *Topex-Poseidon* and *Jason*, joint ventures with the United States.

Topex-Poseidon was a project with the United States to develop an oceanographic satellite. *Topex* and *Poseidon* began as separate missions in the two different countries in 1980. *Topex* was originally the successor to *Seasat*, a successful satellite that had mapped the oceans in the early 1980s, *Poseidon* being a proposal to develop an oceanographic satellite building on the *SPOT* experience. They had so many similarities that they were merged in 1984, with a formal intergovernmental

Figure 3.8. *Topex-Poseidon.*

agreement in 1987 [35]. The projects were merged to save money on both sides, with the French share being 30%. The aim of the joint mission was to use precise radar altimetry of the ocean's surface to compile a map of the circulation of the Earth's oceans. Its aim was to measure wave heights, ocean currents and climatic change. Eventually, *Topex-Poseidon* involved 38 principal investigators from nine countries and 200 researchers. *Topex-Poseidon* looked like a box with a dish aerial and one solar panel. It was 1.8 m tall, 2.8 m across, 5.5 m wide and weighed 2.38 tonnes.

Launched on *Ariane V52* on 11 August 1992, *Topex* orbited in a rigidly circular 1,331-km orbit at 66°. Thrusters adjusted the orbit within a kilometre. The satellite covered 90% of the Earth's oceans every 10 days. *Topex-Poseidon* was still working eight years later, its coloured maps tracing the movement of cold and warm waters across the Pacific Ocean. It was possible to calculate that the sea level globally was rising by between 1 mm and 2 mm a year, or 10 cm to 20 cm per century. *Topex-Poseidon* attracted 500 regular users all over the world, some claiming that it caused a revolution in modern oceanography.

Jason was the successor to *Topex-Poseidon*. France invested €76m in the project, providing the Alcatel-built Proteus bus while the Americans provided the launcher and part of the payload (GPS [global positioning system] receivers, microwave radiometer and laser reflector). *Jason* was an interesting example of how smaller, cheaper, faster technologies have benefited Earth resources satellites. *Jason* was much smaller than *Topex-Poseidon* (500 kg compared with *Topex-Poseidon*'s 2.38 tonnes), longer lasting (12 years compared with 2), a third the cost and radiation-hardened.

After a number of delays, *Jason* was launched into orbit on 7 December 2001 on the 100th *Delta II* launcher from Vandenberg Air Force Base in California, with a subsidiary American payload. *Jason* flew in a phased formation with *Topex-Poseidon*, about one minute ahead and on the same ground track. First checkout found *Jason* to be in good condition, and by April 2002 CNES had published a map of spring wave heights throughout the world. The north Atlantic glowed red, showing its agitated state, with a deep blue for the calm tropics of the central Pacific, eastern Atlantic and western Indian Ocean. *Jason 2* will follow in 2005.

French oceanographic satellites (with the United States)

11 August 1992	*Topex-Poseidon*	*Ariane 42P*	Kourou
7 December 2001	*Jason*	*Delta II*	Vandenberg Air Force Base, California

3.7 FRENCH COMMUNICATIONS SATELLITES

Communications satellites were first pioneered by the United States in 1958. The *Syncom* satellite in 1963 was the first to use the 24-hr orbit, circling the Earth at the same speed as the Earth's rotation and thus appearing to hover in the same position all the time. The first operational 24-hr communications satellite *Early Bird* was

Figure 3.9. *Telecom* apogee motor.

orbited in 1965, and within a few years news, telephone and voice were being relayed on a regular basis through satellites in 24-hr orbits. The Soviet Union built its own system *Molniya* using a different type of orbit in the same year. After that, it was only a question of time before Europe developed its own 24-hr communications satellites.

The idea of a domestic French communications satellite programme was approved in 1979 by the French state telephone company Telecom and soon thereafter by the government. The satellite itself was accurately, if unimaginatively called *Telecom*. The series aimed to provide television, telephone, data and military links for France and its overseas territories, with an encrypted system for military links. The *Telecom 1* series of three satellites was based on the European Communications Satellite (ECS) series and was launched over 1984–8. The objective was to have two operating at a time.

Telecom 2 was a second generation improvement, with 10 transponders instead of 4, resulting in an extension of the transmission area to other European countries. Overall, the second generation doubled the capacity of the first. Each *Telecom 2* weighed 2,274 kg (half of that on arrival in position), was 3 m tall, with a 22 m solar wingspan and 3.75 kW of power. They were built by Matra and Alcatel.

Telecom series

4 August 1984	*Telecom 1A*	*Ariane 3*
8 May 1985	*Telecom 1B*	*Ariane 3*
11 March 1988	*Telecom 1C*	*Ariane 3*
16 December 1991	*Telecom 2A*	*Ariane 44L*
15 April 1992	*Telecom 2B*	*Ariane 44L*
6 December 1995	*Telecom 2C*	*Ariane 44L*
8 August 1996	*Telecom 2D*	*Ariane 44L*

Telecom satellites relayed telephone and broadcast channels to telephone and broadcasting companies across long distances for redistribution through national networks by traditional methods. Direct broadcasting was a step further. A television channel would be sent up to the satellite, but it would beam the signal direct into people's homes through rooftop receiver dishes. Direct broadcasting clearly required a powerful satellite strong enough to transmit to small dishes. As time passed, direct broadcast satellite makers tried to make the dishes ever smaller (for reasons of cost, convenience and planning laws) and this required the broadcasting satellites to be correspondingly ever more powerful.

France approved the concept of a direct broadcasting system with the TDF (Télédiffusion de France) series, approved as early as 1975. Originally, the satellites were to broadcast radio, but television grew in importance. The idea was merged with German developments for direct broadcasting satellites. It took until 1980 to define the project with the Germans, and the outcome was that each country would launch one satellite – *TDF* in France and *TV-Sat* in Germany. Each satellite was designed to carry five TV channels and transmit to 45-cm dishes.

Unfortunately, when the 2-tonne *TV-Sat 1* reached orbit on *Ariane* V20 in November 1987, one of its two solar panels failed to open, thereby reducing its broadcasting capacity. Then *TDF 1*'s thrusters leaked, causing a transponder to fail. This was followed by *TDF 2* in 1990: although the orbit was all right, two channels failed. The project was not a very successful one, for only 35,000 people bought TDF dishes and a third satellite was then cancelled. Direct broadcasting exploded in the 1990s with Luxembourg's *Astra* satellite, so it is possible that the TDF project was ahead of its time.

TDF series

28 October 1988	*TDF 1*	*Ariane 2*
24 July 1990	*TDF 2*	*Ariane 44L*

In 1995 France approved the construction of an experimental communications satellite called *STENTOR*, designed to develop cutting-edge technologies so that France could compete more effectively with Japan and the United States in the development of new comsat technologies. *STENTOR* stands for Satellite de Télécommunications pour Expérimenter les Nouvelles Technologies en Orbite. The programme, costing €400m, involved placing the satellite at 19°W for a series of tests over two years. Station-keeping was to be carried out by plasma thrusters, and power would be supplied by powerful new gallium arsenide solar cells and lithium batteries. Eighty-five percent of *STENTOR* develop new technologies. Sadly, it was lost on launch in December 2002.

France is in the process of developing a military satellite communications system, one equivalent to the British *Skynet*. This is called the *Syracuse* programme. From the 1980s Syracuse used leased lines on the French *Telecom* communications satellite, but was not a satellite in itself. This changed in 2000 when the French military commissioned a dedicated satellite, to be called *Syracuse 3*, to be built from the Alcatel Spacebus design. Two more satellites would follow between 2003 and 2018, at an overall cost of €1.4bn. France has also put forward the idea of a radar military reconnaissance satellite called *Osiris*.

3.8   FUTURE FRENCH SPACE PROJECTS

When Daniel Goldin became administrator of the American space agency NASA (National Aeronautics and Space Administration) in the 1990s, he committed the agency to building fewer larger spacecraft, instead constructing many smaller space-craft that could be launched more frequently for the same price. This was the 'faster, cheaper, better' philosophy. France, like many other countries, has been influenced by this trend. It was Britain that gave a lead in the development of very small spacecraft, able to take advantage of the computer revolution that miniaturized spacecraft and their components.

France had already launched, or had put in orbit, a number of quite small spacecraft, as the cooperative history with the USSR demonstrated. Since then, *Sara* was put into orbit piggyback on *Ariane* in July 1991, *S80T* in August 1992 and *Stella* in September 1993. *Sara* was a 26-kg amateur radio satellite costing €460,000 aimed at observing radio emissions from the planet Jupiter. *S80T* was a 50-kg minisatellite designed to test technologies for messaging systems. Stella was a small 36-kg geodetic satellite built by CNES, the French Atomic Energy Authority and Arianespace.

Small French scientific satellites on European launchers		
17 July 1991	*Sara*	*Ariane 40*
11 August 1992	*S80T*	*Ariane 42P*
26 September 1993	*Stella*	*Ariane 40*

Several more small satellite missions are planned. These include *Demeter*, a 100-kg microsatellite to study the ionosphere, and *Picard*, a satellite of similar dimensions to study the Sun.

Some other, somewhat larger probes are in the pipeline, like *COROT*. *COROT* stands for COnvection, ROtation & planetary Transients and is a 420-kg probe based on the *Jason* spacecraft, but with a quite different objective: to study stellar

Figure 3.10. *COROT.*

seismology (the inner structure of stars) and to detect planets. For the former, it will examine 50 stars in detail, but in the search for planets will cover at least 6,000 stars. *COROT* will operate from a polar orbit. It will also pave the way for Europe's Eddington mission (Chapter 7).

Megha-Tropiques is a cooperative project with the Indian Space Research Organization to study the water cycle in tropical regions. It hopes to model the origin, development and functioning of tropical storms and the weather's energy balance in the tropical regions by examining clouds, water vapour, precipitation and evaporation.

3.9 FRANCE TO MARS

Although France had a high profile for collaboration with the Soviet Union and Russia, cooperation with the United States has been important as well. Following the failure of the Mars 1996 mission, France seems to have made a decision to cooperate with the United States, instead, in the fulfilment of its interplanetary ambitions. According to CNES, Russia no longer had the ability to put forward a credible short- or medium-term Mars programme [36]. The French minister responsible – he was Minister for Education, Research and Technology – Claude Allègre was not an enthusiast for manned spaceflight. He preferred to see what robotics could do for less cost. Accordingly, he redirected €381m from the CNES budget into a collaborative programme of Mars exploration with the United States. Allègre convened a French seminar on solar system exploration in Arachon in March 1998 and an international conference in Paris in February 1999. These explored the possibilities and priorities.

Following them, France signed an agreement with the United States on 26 October 2000 for cooperation on Mars missions, specifically a sample return mission. French collaboration was attractive to the Americans because of France's ability to offer the *Ariane 5* launcher, which is much more powerful than their own

Delta II (the Americans do have the Titan, which would be powerful enough, but is too expensive).

In advance of the sample return mission, the French space agency CNES designed a €500m precursor mission called Mars First (*Mars Premier*) to fly with the proposed American Mars orbiter for 2007. Just as American Mars plans have been through numerous evolutions over the years, the role of *Netlander* has also been modified. The idea of *Netlander* is to land small probes on the surface of the planet Mars to form an interconnected set of stations, especially valuable if marsquakes are to be detected. The *Netlander* mission is managed by CNES, but also involves the Italian Space Agency ASI (Agenzia Spaziale Italiana), the American space agency NASA, the Finnish Meteorological Institute and the German space agency DLR (Deutsches Zentrum für Luft & Raumfahrt). Although France made it clear that the United States would now be its main partner, Alcatel signed a deal with Lavotchkin in 2001 for Russia to provide descent modules for the Netlander mission (*Netlander* is about the size of Lavotchkin's *Luna 9* that soft-landed on the moon in 1966). Other countries involved in *Netlander* are Belgium, Finland, Germany and Switzerland.

In its current form, the 2007 Mars expedition will comprise an American lander, an Italian orbiter and four *Netlander* landers. Four 240-kg *Netlanders* will separate from an orbiter as it approaches Mars. They will shoot through re-entry protected by a 90-cm heat shield. A parachute will open to slow the descent craft through Mars's thin atmosphere. A bouncing airbag will cushion the landing on the Martian surface. Once the probe rolls to a halt, the airbag will deflate and solar petals will open to charge the *Netlander* batteries. Once they land they will begin to sample the atmosphere, subsurface structure and climate. The instruments will pop out: camera, weather sensor, water detector. The crucial instrument is the seismometer, which must implant itself in the surface far enough away from the lander not to record the lander operations. Once each of the four *Netlanders* is operating, there will be a network of linked stations across the surface, able to triangulate weather and seismology. The instruments should also indicate the presence of subsurface water and ice, the geology and mineralogy of the landing site, the tilt of the planet's rotational axis and changes in the atmosphere. The weather station will measure pressure, wind, temperature and humidity. The *Netlanders* will come down between 2,000 km and 3,000 km apart. One will land in the Hellas basin, another in the Tharsis volcanic plain. Information from the *Netlanders* will be relayed back to Earth via the Italian orbiter.

The sample return mission will be some time between 2009 at the earliest and 2016 at the latest. 2007 is now being discussed as a 'validation' mission for the later sample return flight and in addition to the *Netlanders* may be used to demonstrate aerocapture techniques, precision landings, orbital relays and Mars orbit rendezvous. To carry out a Mars orbit rendezvous, the searching spacecraft must be able to find the canister in a distant orbit, using only its homing signal. Laser rangefinders may be used from a closer distance. For predecessor designs, reference may be made to the Russian *Luna* capsules that took samples from the Moon over 1970–6 and a Soviet design for a Mars sample return mission called Project 5M in 1978.

The design of the sample return landers is still some time away. The first step of a sample return mission will be to launch a Mars lander, probably American on a *Delta IV*, equipped with a 55-kg rover to drill for samples at appropriate and interesting points on the surface and bring them back to the lander. A small 120-kg rocket at the top of the lander, filled with 300 g to 500 g of samples, will then be fired in a canister into a 600-km-high Mars orbit with a tracking beacon so that it can be found. Another possible approach – the plan has not been finalized – is to place the return canister on the rover. The second stage will be for an *Ariane 5 ESC-A* to launch a Mars orbiter from French Guiana. Once it enters Mars orbit eight months later, it must locate the canister, a process that could take several months. Having rendezvoused and docked with the canister, the time will come for the crucial Mars–Earth injection, sending the 30-kg canister back toward Earth. A further eight months later, it will enter the Earth's atmosphere and come down in the Utah desert [37].

When Claude Allègre was replaced by Roger-Gérard Schwarzenberg in 2000, there was some concern that the Mars programme might be downgraded. This was not the case and Schwarzenberg since pressed ahead. The Mars programme became a firm part of the CNES long-term plan, restating the commitment of up to €500m. Several problems now appeared on the American side. The United States lost both its 1999 Mars probes: the orbiter crashed because one team of controllers was using metric units and the other imperial units; the lander crashed when the opening of the landing legs triggered a touchdown vibration detector and cut off the engine early. As a result, future missions were set back and rescheduled.

3.10 FRANCE'S NATIONAL SPACE PROGRAMME TODAY

As this description has shown, France has a broad-ranging space programme, from small satellites to scientific and interplanetary missions and has pioneered space-based Earth resources observations. Cooperative programmes have been carried out with the United States and Russia. This chapter understates the French contribution, for the story of *Ariane* and manned exploration has yet to come.

France's space industry is by far the largest in Europe. France has an annual space budget of €2,353m, of which €670m goes to ESA and €1,683m to national programmes (2001 figures). The military budget may be in the order of a further €454m. Most comes as a government grant, but significant revenues come from charges for the use of the Kourou base, income from Earth resources images and profits from *Ariane*.

The French space agency CNES has 2,488 staff, over two-thirds of whom are located in Toulouse, the balance in Paris, Evry and French Guiana. With two exceptions, CNES does not maintain in-house research and development facilities, preferring to support existing laboratories in the universities. The exceptions are Evry, or the Centre Spatial d'Evry, responsible for the development of *Ariane* for ESA (European Space Agency), and the CST. CNES also cooperates with two state companies for space development, such as the Société Nationale des Poudres et

Explosives (SNPE) (Company for Solid Fuels and Explosives). SNPE is a large company with 5,000 staff making solid rockets and the storable fuels for liquid-fuelled rockets – a commitment that involves it producing 6,000 tonnes a year of ammonium perchlorate sludge to keep the *Ariane 5* flying.

France has the most extensive infrastructure of all the European countries. France's main space facilities are located in Toulouse, which has been termed 'space city'. These include thermal, electrical and acoustic vacuum chambers. Here, there are 20 laboratories dealing with such diverse aspects of space research as controlling missions, launching balloons, Earth observations and interpretation and mission control centres for *SPOT* and *TDF*. CNES has spun off an affiliate, SCOT Conseil (Service de Consultation en Observation de la Terre, or 'Consulting Services for Earth Observation'), which aims to promote, market and sell images of the Earth taken by *SPOT*.

Over 10,000 people work directly in the space industry in France. Between 100 and 150 major companies are involved, the main ones being Aérospatiale, Alcatel, Matra Marconi and SEP. Some of the French space industrial companies are large. For example, EADS Launch Vehicles employs over 3,000 people building rockets in its production plants in Les Mureaux (Paris), St Médard en Jalles (Bordeaux) and in Kourou. Sales in 2000 were worth €1.043bn. EADS is prime contractor for *Ariane 5* (cryogenic and solid fuel rockets) with responsibilities for the Automated Transfer Vehicle for the international space station (system engineering, guidance, control, testing), the ARD capsule and the heat shield for the *Huygens Titan* probe.

Examples of other leading French companies are Alcatel (communications) and Snecma (engines). Snecma built the Viking and Vulcain engines for *Ariane* and subsequently became involved in the new Vinci engine for the *Ariane 5* successor upper stages and the P-80 for the smaller *Vega* rocket. Alcatel is one of the world's leading communications satellite makers, with 5,600 employees and a turnover of €1.4bn a year. Its best known area of expertise is in communications satellites where it developed a powerful common platform called the Spacebus series (3000, 4000, etc.) that could be adapted for specific payloads. Alcatel has also been involved in navigation (*Galileo*), observation (*Helios* and *SPOT*), meteorology (*Meteosat*) and space science (*Herschel Planck*), becoming prime contractor for such prominent missions as the Infrared Space Observatory and *Huygens*. Although its French bases are in Toulouse, Cannes and Nanterre, Alcatel has a presence in seven other European countries. TDF is the main company operating communications satellite channels for television, radio and direct broadcasting. Apart from all this, there is a significant space defence industry, led by ONERA (Office National d'Etudes et de Recherches Aéronautiques), which operates under the Ministry for Armaments. It has facilities in different parts of the country, especially Toulouse and Modane.

France's space programme extends far beyond France itself. When the Soviet Union broke up in 1991 and private enterprise was introduced, France was quick to set up strategic alliances with leading Russian space enterprises, state or privately owned. The Americans had moved rapidly into the Soviet space industry, Lockheed Martin setting up a joint company with Khrunichev to market the *Proton* rocket while Boeing set up a company for the *Zenit* launcher with

Energiya space corporation. France's alliance was with the Progress plant in Samara, formerly Kyubyshev. This was the producer of the original Russian orbital rocket, the R-7, now called the *Soyuz*. The resultant company was called Starsem. *Soyuz* eventually became part of the *Ariane* launcher portfolio. Starsem was owned by EADS (35%), the Russian Space Agency (25%), the Samara plant (25%) and Arianespace (15%). Starsem provided capital to improve facilities at Samara, but especially at the Baykonur cosmodrome. The old and increasingly run-down assembly hall for the R-7, which dated to *Sputnik*, was replaced by new facilities in the Energiya assembly building. Here, Starsem invested €35m in three clean rooms of world standard – a payload preparation facility, hazardous processing facility and integration facility – and a hotel for staff on-site. Later, additional resources were invested in improving the two *Soyuz* launch pads, Pads 1 and 31. By 2000 Progress could look forward to ramping up its production line. In 2001 the French government approved €300m plans for the construction of a *Soyuz* launchpad at French Guiana, to be built on the site of the dismantled ELA-1.

On an educational note, France also became home to Europe's first space university, in Strasbourg. The International Space University (ISU) was an itinerant summer school formed in 1987 to provide 10-week courses for current or aspiring space professionals. In 1993 the ISU found a permanent home in Strasbourg, France and by its 10th anniversary had clocked up a thousand students from 60 countries. In 1996, the university introduced its Master in Space Studies, under which students spend a year learning the theory and practice of spaceflight, including a three-month placement in leading European space institutes or space industrial companies. The cost of the Masters is in the order of €30,000 and about 30 join the course each year. Many go on to companies like *Ariane*space and SEP.

On 18 December 2001, France celebrated the 40th birthday of CNES. Sadly, the first director, Robert Aubinière had died only days before the party. All previous presidents, directors general and overseeing ministers were invited. French President Jacques Chirac reminded them of CNES's achievements and how they had fulfilled General de Gaulle's mandate to build national independence through autonomous access to space. Chirac urged his colleagues on to ever greater things: 'We will never become vassals of the United States,' he swore.

Now we turn to France's neighbour and co-founder of European space co-operation in ELDO (European Launch Development Organization) and ESRO (European Space Research Organization). What happened to the British rocket programme?

3.11 BRITAIN'S NATIONAL SPACE PROGRAMME

As we saw earlier, Britain put most of its space effort in the early 1960s into ESRO and into the *Blue Streak*-based *Europa I* launcher. Originally, Britain hoped that its own *Black Knight* launcher could serve as the second stage of *Europa I*, but this was unacceptable to the French who felt that they must provide the second stage if they were to have a meaningful role at all. *Black Knight* was therefore sidelined and out of the equation.

Despite this, the Royal Aircraft Establishment (RAE) was not prepared to give up. The mission of the RAE was to test cutting-edge technologies and the Establishment saw no good reason why *Europa* should put small British rockets out of business. The RAE designed an upgraded version of *Black Knight* able to put a small, 144-kg satellite into orbit. The RAE's idea was that Britain would thereby develop its own small launcher and have the capability of testing out new satellite technologies in orbit. In the autumn the RAE convinced the Aviation Minister Julian Amery to support the proposal.

In September 1964 the government decided to develop a modest rocket as a national launcher vehicle, one which would enable Britain to retain rocket-launching technologies in a number of key areas. A three-stage rocket was developed on the basis of the *Black Knight* sounding rocket, which had already flown successfully 22 times. Britain would thus run a small national launcher programme alongside *Europa* – just like the French, in fact.

No sooner had the Amery proposal come through than his government lost the October 1964 general election. The incoming Wilson government was much less enthusiastic about rockets and was confronted by a growing economic crisis. The new launcher, still without a name, was kept going on a series of shoestring budgets and three-month contracts. This made it unlike the French programme, which in 1965 reached orbital flight.

3.12   BLACK ARROW

Not until the Labour government was re-elected in March 1966 did the programme get the final go-ahead and a name *Black Arrow* was acquired in March 1967. The programme was given a remarkably low budget of £3m a year, enough to produce one rocket a year. The RAE was told that there could be three test launches. The project had been sold to the government and subsequently approved on the basis that it would make use of existing technologies, thereby keeping costs down, so the government was holding the Establishment to these promises.

Black Arrow used an unusual fuel, hydrogen peroxide, or high-test peroxide (HTP). This was much easier to handle than the liquid oxygen of the *Blue Streak*, which had to be kept at very cold temperatures. HTP could be kept at room temperature and stored for a long period. At that time many other countries were developing room temperature storable fuels, especially those based on UDMH (Unsymmetrical Dimethyl Methyl Hydrazine) mixed with nitric acid. These storable fuels could be kept ready in their rockets at room temperature for days or weeks on end – ideal for missiles or conventional rockets facing the type of countdown delays that were frequent in the early days of space travel. While convenient to handle, these UDMH and nitric fuels were extremely toxic. Nitric acid leaks could – and subsequently did – cause fatal poisonings around launch pads. Worse, a handling error with nitric acid at the Baykonur cosmodrome in October

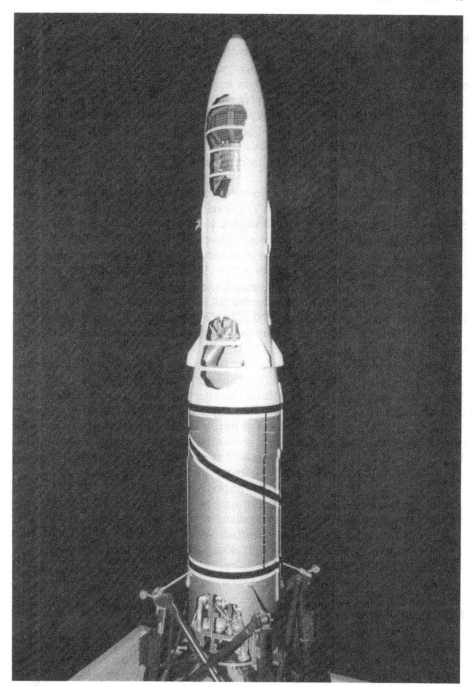

Figure 3.11. *Black Arrow.*

1960 caused the worst accident in rocket history, with over a hundred people incinerated.

HTP had one advantage over nitrogen-based fuels: it did not require the mixing of a fuel with an oxidizer. It was a monopropellant, requiring one tank and a means of igniting the rocket (metallic filings were inserted). So, there was no need to mix the product of two tanks in a very precise ratio to get the desired thrust. Nor was HTP toxic, but it could be equally dangerous in another way. HTP must be kept in absolutely pure tanks and fuel lines, otherwise it will decompose or, if mixed with particular impurities, would explode. HTP was later used to fuel the torpedoes on the Russian submarine *Kursk*, with disastrous results when they exploded in August 2000.

HTP engines had been pioneered by the Germans during the war, but by a team different from the Peenemünde group. The developer was Dr Hellmuth Walter, who developed HTP engines for the Messerschmitt 163 *Komet* rocket interceptor and for high-speed U-boats. After the war he went to work for the Royal Navy in Barrow-in-Furness while his other colleagues went to the RAE in Farnborough and to the Guided Projectile Establishment in Westcott. Walter developed a series of HTP engines called the Alpha, Beta and Gamma, the Gamma becoming the engine for the *Black Knight* and the rocket-powered SR53 interceptor. Several versions of Gamma had been developed by 1960. The Gamma became the engine for the *Black Arrow*. The guidance system was taken from Britain's most advanced aircraft, the later cancelled TSR-2.

Black Arrow was probably the shortest rocket ever to put a satellite into orbit, though by no means the thinnest (that honour goes to the Japanese *Lambda*). *Black Arrow* was 12.9 m high, with a diameter of 2 m and weighed less than 20 tonnes. The first stage had a Rolls-Royce Gamma 8 engine able to swivel in all directions, generating 22.68 tonnes of thrust, using kerosene and HTP fuel. The second stage, also liquid-fuelled, had a Gamma 2 engine of 6.94 tonnes thrust. The third stage was a small solid propellant motor (Waxwing) developed from the *Black Knight*, with a capacity of 110 kg into polar orbit.

Three launch sites were considered. The preferred choice was in Norfolk, but the burgeoning North Sea oil industry with its rigs presented a problem. North and South Uist in the Outer Hebrides of Scotland were considered, but adjudged to be 'too remote', so Australia was chosen instead. Woomera was already available, with a fully developed infrastructure, the only disadvantage being that it was 16,000 km away and it took ships a month to get there.

3.13   LAUNCHING *BLACK ARROW*

The first *Black Arrow* was eventually ready for launch on 23 June 1969, using a modified Woomera pad from the *Black Knight* period. The countdown clock failed and by the time they got it going again, low cloud had moved in and they had to reschedule. Eventually, the first *Black Arrow* was launched on a suborbital mission on 27 June 1969. No sooner was it airborne than it began to twist and

Figure 3.12. Gamma engine.

corkscrew visibly and violently. The top stage was thrown clear. At an altitude of 8,000 m it keeled over and the range safety officer had no option but to inject manganese dioxide into the fuel tanks and blow what was left of the launcher apart. Given the good record of the preceding *Black Knight*, this was a disappointing start. The inquiry found that a broken wire was probably to blame. The attitude reference unit had been signalling the engines to move. The broken wire meant that the receipt of the signal was not acknowledged, with the result that the unit kept commanding the engines to swivel from one extreme to another, hence its erratic behaviour.

The failure of the first rocket was a real problem. At a general level, the programme had no financial reserves and could ill afford failure. The second mission was to have been an orbital attempt, but it would now have to carry out the intended suborbital mission of the June flight. At a practical level, the second *Black Arrow* was back in the Isle of Wight so the team had to return to Britain to make the necessary fixes. It was eventually got ready on 4 March 1970 and launched early that morning. This was entirely successful, impacting 3,050 km away in the Indian Ocean 15 minutes later. However, the engineers can't have failed to notice that the delays had cost them dearly. Japan became the fourth country in space that February and China the fifth that April. Britain was now aiming for sixth place rather than fourth.

The third flight was the first attempt at orbit, carrying a small 82-kg X-2 satellite called *Orba* with a small transmitting package to measure air density. It was to be accompanied in orbit by an instrument package. The engineers had hoped to go sooner, but they had to wait a month for a transport ship to be available. The first attempt was made on 1 September 1970, but one of the tracking stations went down at 35 seconds to go. By the time the problem was sorted out, they had to delay for another day. The second attempt was on 2 September 1970. All seemed to go well at first, but the second stage lost thrust and then shut down 30 seconds early. The third, the Waxwing stage, worked perfectly, but the engineers knew it would have insufficient velocity to reach orbit. It ended up in the Gulf of Carpentaria. The postmortem revealed that the pressurization system had vented nitrogen accidentally and thrust had been lost. The government put a brave face on it, declaring that the mission had been only partially unsuccessful. An extensive programme review was carried out and this delayed the next mission. The review came to the conclusion that the project was fundamentally sound and only minor modifications were necessary to bring it to fruition.

Even as its first launch was nearing in summer 1971, the government cancelled the programme that July, making it clear the cancellation would stand regardless of the outcome of the last attempted launch. Aerospace Minister Frederick Corfield was quoted as saying that the rocketry business was too expensive. The House of Commons Select Committee on Science and Technology criticized the *Black Arrow* programme for using money that could be better used on building satellites that could be launched on cheaper American rockets. The American *Scout* rocket, for example, was less expensive and had won a reputation for reliability. As a precaution, the next satellite X-4 had been constructed so that it could fit either *Black Arrow* or *Scout*. The X-4 would not be ready for three years and it did not make sense to keep the *Black Arrow* production line open until then just to fly one more satellite.

According to the government, Britain would now concentrate on space applications (principally communications satellites to fly on American launchers), science missions (again with American launchers) and microgravity (through the *Skylark* sounding rocket programme). Britain signed an agreement with the United States in January 1973 for the US to launch future British satellites at agreed rates of reimbursement. The government view was that neither Europe nor Britain could afford to build their own launchers and that doing so was wasteful when relatively economic launch prices were available from the United States.

3.14   *PROSPERO* PROSPERS

The heavy-hearted engines saw the programme out to what must have been a bitter end. *Black Arrow* was duly launched from Woomera, Australia, on the hot afternoon of 28 October 1971. The hydrogen peroxide engines left almost no mark as the rocket accelerated into the clear blue sky. The second stage brought the rocket up to the edge of space, soaring to 360 km. The second stage fell back, probably coming down

Figure 3.13. *Prospero.*

in the Pacific off New Guinea. For nine minutes the third stage coasted, reaching 936 km. Then the apogee motor blazed for 55 seconds to achieve orbital insertion. X-3 was cast free and 39 minutes later the ground station in Fairbanks, Alaska picked up its signal. Britain was in orbit, at long last!

The satellite itself, called *Prospero*, sometimes called R-3 and X-3, was a pumpkin-shaped 66-kg box, 71 cm tall, 114 cm in diameter. The purpose of the satellite was to test the technologies involved in communications satellites, such as telemetry, power systems, solar cells, thermal systems and electrical systems. There was a micrometeoroid detector to measure dust particles. It had a modular design, making it easy to inspect from the outside. *Prospero* reached an orbit of 547–1,582 km, 82.06°, period 106.5 min, enough to keep it circling the Earth until around 2070. The on-board tape recorder worked for almost two years (one more than anticipated) and real time data were still being received two years later.

No film of the launching was released, and a still photograph of the *Prospero* launch was not exfiltrated until many years later. Coverage in the British press was, to say the least, muted. The engineers returned to Britain for the last time and the pad area was cleared away. Britain's distinctive hydrogen peroxide tradition disappeared with the rocket itself. The next *Black Arrow* was fated for scrap, but its makers gave it instead to the Science Museum in London where it was eventually displayed many years later. Its many admirers may puzzle at why, in the words of the history of the project, Britain became the sixth nation to launch its own satellite, but the first to abandon that very capability [38].

Britain's *Black Arrow* space launches from Woomera

27 June 1969	X-1	Suborbit (fail)	
4 March 1970	R-1	Suborbit	
2 September 1970	R-2	X-2/*Orba*	Orbit (fail)
28 October 1971	R-3	X-3/*Prospero*	Orbit

In the event, a successor satellite to *Prospero* did make it into orbit. The 93-kg *Miranda* or X-4 flew on an American *Scout* on 28 February 1974 to continue further tests of the equipment originally flown on *Prospero*. *Miranda* carried five experiments, one of which was of improved solar arrays (rather than cells), the other four experiments testing out new and more advanced systems of attitude control, gyroscopes, sun sensors and nitrogen thrusters. *Miranda* cost £5m to build and another £2m to launch. Soon after getting into orbit it was transmitting signals to the NASA network, and the RAE in Farnborough and Lasham.

Woomera was abandoned in 1976 and the desert began to reclaim what was left of its rusting gantries . At the peak of the *Blue Streak* programme, 4,500 people had lived in the Woomera township, with more than 500 houses supplied with power and water across 160 km of desert. Only two satellites were ever launched into orbit from there: *Prospero* and an Australian satellite called *Wresat* on an American *Redstone* rocket in 1967.

The range temporarily reopened in August 1987 for sounding rockets. In 1995 Woomera was designated the recovery zone for a Japanese–German–Russian mission called *Express*, but the cabin came down in the jungle of west Africa instead. Some small sounding rockets are still flown from there. Many feasibility studies have been done for recommissioning Woomera. With its fine weather, size and infrastructure, it could still be a launch centre once again [39]. Woomera returned to the headlines in 2002 as a detention centre for refugees.

3.15   SOUNDING ROCKETS CONTINUE TRADITION OF BRITISH ROCKETRY

Despite the later achievements of British space science, the cancellation of *Blue Streak* and *Black Arrow* is, even today, still a sore subject for the older generation of British rocketeers. Although *Blue Streak* was cancelled as obsolete, *Blue Streak*'s rivals – *Atlas*, *Soyuz* – are still in business 40 years later despite being, presumably, equally obsolete. The 1990s saw the revival of the small launcher market and virtually every one of the payloads concerned could have been a candidate for *Black Arrow*. Leaving aside the industrial and technological issues, the commercial wisdom of the cancellations was ultimately proved to be wrong. Those among Britain's rocketeers who did not leave for retirement went to work for the

Americans (many played a distinguished role there) and a generation of rocket and engine-building knowledge was lost for ever. From being the driving force of European rocketry in the early 1960s, Britain had slipped by the 1980s into a poor fourth place in the European space industry, far behind France, Germany and Italy.

British rocket-making experience did not entirely die out, being maintained through the programme of *Skylark* sounding rockets and the manufacture of engines for American space probes. For example, British Aerospace made the Leros rocket engine used on the spectacular mission of the *NEAR* spacecraft to land on the asteroid Eros; and on the successful *Mars Global Surveyor* [40].

The *Skylark* rocket, which first flew in 1957, continued to fly despite the ups and downs of the rest of the national rocket programme. The *Skylark* engines were, appropriately, given bird names: Goldfinch for the first stage, Raven for the second, Cuckoo for the third and Stonechat for the fourth. The *Skylark* stood 11 m tall. New and improved versions were introduced. *Skylark 12* was a three-stage version, introduced in Andoya, Norway in November 1976. *Skylark 12* was able to carry a payload of up to 180 kg as far as 1,000 km high. By this stage, 354 *Skylark*s had been fired.

In December 1978 the German space agency DLR hired the *Skylark* for a launching campaign called Texus. The purpose of Texus, which was flown from the Kiruna range in northern Sweden, was to study the behaviour of materials, alloys, glass, ceramics and liquids in microgravity, in advance of the *Spacelab* missions. These missions were considered most successful in terms of scientific outcomes and cost-efficiency. In a scientific mission to study helium gas in space, the Germans fired a *Skylark 12* a record 834 km high from Barriero do Inferno in Natal, Brazil in 1979. Most of the missions have been microgravity research, with the payload parachuting back to Earth for recovery. *Skylark* was able to provide 6–7 minutes of microgravity at altitudes of up to 450 km [41].

Skylark launches resumed 25 August 1987 from Woomera, Australia, the rocket in question carrying a German X-ray telescope to study the Larger Magellanic Cloud. Most 1990s' firings were from the Swedish launch range of Kiruna. By 1997 *Skylark* had made 432 launches. Of these 266 had been for the UK, 82 for ESRO and 72 for Germany, and 259 had been fired from Woomera, 76 from Kiruna, 56 from Sardinia and 23 from Andoya. The final version of *Skylark*, *Skylark 17*, could reach 1,600 km and provide microgravity for 19 minutes [42].

3.16   THE CAMPAIGN FOR A BRITISH NATIONAL SPACE CENTRE

With its withdrawal from ELDO and the cancellation of *Black Arrow*, Britain lost the two most important pillars of its space programme. We now look at the evolution of British space policy to the present day in the absence of a rocket programme before returning to individual themes and projects in British space activities. The government argued that there was a meaningful role for Britain in building scientific satellites (which could be flown on both available and inexpensive

American launchers), subcontracting parts of European programmes, and in other forms of international collaborative projects. Indeed, British companies were both skilful and successful in winning such contracts.

Critics contested this: even successful subcontracting fell far short of the potential British contribution to space development. From the early 1980s they argued that the first step toward a more purposeful British space policy was a National Space Authority, like NASA or CNES. The British Interplanetary Society (BIS) had been pressing for such a body since 1960, pointing out that an assemblage of user projects was not the same as a discernible line of policy. The proposal for a national agency was repeated in 1973. Several years later, in 1980, the Royal Society proposed a national council for space. Despite the lack of progress, the BIS returned to the fray in 1985 [43].

When the BIS marked its 50th anniversary, the society lamented the low level of British participation in space activities. Twenty years earlier Britain had led European aerospace and was in third place in space research after the two super-powers. Cuts in public spending, especially in long-term investment programmes, had led to a long-time decline. Were it not for private industrial initiatives, Britain would have disappeared from space activity altogether [44].

The government eventually heeded these pleas. In January 1985, the Minister of State for Industry and Information Technology, Geoffrey Pattie, announced that the British National Space Centre (BNSC) would now be set up to bring a sharper focus and more long-term perspective to British participation. At the subsequent ESA council meeting held a few days later, he urged his colleagues to respond positively to President Reagan's invitation to join the American space station. Britain agreed to pay 15% of the project definition stage of the *Columbus* programme, well above the then-standard level of British funding support of 4%. Was this the long-hoped-for resurrection of British space prospects?

Roy Gibson, the first Director General of ESA, was selected to lead the new body. No person could have been better qualified or held in higher esteem. There was much optimism that his appointment would lead to a revival of British space fortunes, coming as it did with Britain playing a much more active part within ESA, raising the level of its financial contribution compared with previous years. He was assigned 30 civil servants to start with, a further 270 following shortly thereafter. During 1986 the centre recruited a series of senior director appointments. The new BNSC was seen as playing a role equivalent to CNES in France.

Roy Gibson announced that his first task was to draw up a coordinated space plan for the UK [45]. One of its first decisions was to provide £1.5m in a cash grant for proof-of-concept studies of the *HOTOL* space launcher. The next year saw a flurry of activity and upbeat attitudes by Britain to its participation in European projects. A collaborative agreement was signed with the Soviet Union in September 1986, paving the way for British participation in the *Kvant* module on the *Mir* space station [46]. A memorandum of cooperation was even signed with China. In November 1986 the BNSC concluded an 11-volume study on how Britain might best use the European space station module *Columbus*. Producing the report had been a positive experience for BNSC, for it confirmed that there was an extensive

Figure 3.14. Roy Gibson.

scientific and industrial community interested in such an involvement. Britain joined the *Hermes* spaceplane preparatory programme. Things were moving at last.

3.17 GIBSON'S RESIGNATION

Roy Gibson pledged that his coordinated plan for British space activities would be ready within six months, in June 1986. The plan covered 15 years. The main elements were believed to have been:

- Britain should raise its level of contribution to ESA to one commensurate with the proportion of gross national product spent on space by other European countries.
- Overall, Britain's space budget should rise from about £112m a year to £300m a year.

- The key investment areas for Britain were launch vehicles, the American space station, remote-sensing, space science and communications.
- Projects that Britain should pursue were *Columbus* and *HOTOL*.

Although Roy Gibson completed the draft plan on schedule, the trouble started thereafter. When it went to the Cabinet for approval, there was much difficulty. The government found it more and more difficult to meet existing commitments to ESA and some companies found themselves compromised. Following the June 1987 general election, the minister who had done most to promote the BNSC, Geoffrey Pattie, lost his Cabinet position, even though the Conservatives were returned to power. Kenneth Clark was the new minister responsible. Eventually the situation reached a démarche on 24 July 1987. Prime Minister Margaret Thatcher said that the government could not find the additional resources requested in the plan (which was not released). Any additional space investment should come from the private sector. Britain already spent £4.5bn on research and development a year: she was not prepared to take away from other areas to move it into spaceflight. There was a howl of protest from some parts of the British industrial, academic and scientific establishment – not so much for the decision itself, but for the boorish attitude that seemed to underlie it. The space plan was not published, and it was not until August 1999 that a formal policy statement eventually emerged (*The United Kingdom Space Strategy, 1999–2000 – New Frontiers*).

The Director of BNSC, Roy Gibson, resigned 4 August 1987, the clear reason being the lack of sufficient government funding for either national or ESA missions. The BIS described the government's approach as unresponsive, discouraging, non-committal and dismissive and the loss of Gibson as a glaring example of wasted opportunity and personal talents [47].

The government may have been taken aback by the reaction, for later in the month (28 August) it added a further £4m to the BNSC for Britain's participation in *Hermes* and *Columbus*. However, the respite was short-lived, for on 10 October Trade and Industry Minister Kenneth Clark announced that no further money would be available for new space projects. Kenneth Clark publicly attacked ESA as 'a hugely expensive club' to the considerable embarrassment of his officials who had been attending ESA meetings for years. At the autumn ESA council meeting in the Hague held the following 9–10 November, Britain formally opposed decisions to move ahead on key European programmes and refused to increase its annual contribution to ESA. Britain tried to trim European space budgets and halt the *Ariane 5* rocket, the *Columbus* module and the *Hermes* spaceplane, but, having failed to do that, became the only European country not to participate (even small countries like neighbouring Ireland joined the *Ariane 5* programme). Clark described these three projects as 'plans that led nowhere' and said that Europe was embarking on an 'expensive frolic at taxpayer's expense'. He refused to support even a 5% increase in the budget for scientific programmes. Because of the important issues at stake, this was a ministerial meeting, with ministers rather than their senior civil servants attending. The meeting was later politely described by diplomats as 'stormy'. Geoffrey Pattie expressed disappointment with his   government's stand while

Roy Gibson commented that Britain now had very little influence left in Europe. ESA Director General Reimar Lüst was overheard to remark that ESA could do better without Britain and 'didn't need membership like that' [48]. Eventually, in May 1988 Britain decided to join the *Columbus* project after all, contributing 5.5% of the project or £5m immediately or £250m over 10 years by 1998. However, Britain made clear that its interest was in the polar orbiting version of *Columbus* and was dependent on its lead role being guaranteed.

3.18 AFTER GIBSON

A new director was appointed for the BNSC, Arthur Pryor. Unlike most of its opposite numbers in other European countries, the BNSC did not have the authority to develop a British national space programme. Instead, it acted as a coordinating body for British companies interested in contracts for space projects. Space policy was determined by the Interdepartmental Official Committee on Space Policy. Roy Gibson later commented that the original centre had been established in too makeshift a way. Too much had been left to gentlemanly agreements while departmental sticking points had been unresolved. When the system came under strain, it fell apart [49].

Britain's space policy was defined by Arthur Pryor as being based on value for money, with a toughly selective approach to the applications areas to be approved. Space activity did not have a privileged place in the government's public expenditure plans and would be examined as rigorously as any other research, development and industrial spending. Funding would be pegged at around £130m a year [50].

Under the guidance of Pryor and his successors, the BNSC went on to play an enabling, coordinating, export promotion role rather than as a directing centre, encouraging large and small British companies to participate in European and other space programmes. It tries to ensure the highest possible level of coordination between government agencies involved in space research (viz. the various government departments, QinetiQ, the Natural Environment Research Council, the Meteorological Office, the Rutherford Appleton Laboratory and the Office of Science and Technology). The BNSC has 235 staff and it clearly does play an important role in publicizing space activities and encouraging British industrial participation therein. Britain's contribution to European space science is organized through the Central Laboratory of Research Councils, established 1995, based on the old Rutherford Appleton Laboratories. Here, the funding stream for British space science comes through the Particle Physics and Astronomy Research Centre, which pays for Britain's contribution to the ESA science budget.

BNSC represents Britain on ESA. Britain's space budget was €295m in 2001, of which €179m went to ESA and €115m to national programmes. In addition, Britain has a substantial military space budget, which at an estimated €80m is probably the biggest in Europe. This goes largely toward the *Skynet* military communications system, more of which shortly.

3.19   A FRESH PATH FORWARD?

Despite the debacle in the 1980s, there were signs by the late 1990s of a modest improvement in British space fortunes. In 1996 the British government's minister responsible Ian Taylor presented *Forward Plan for Space*. This articulated a broad set of aims, including participation in the Earth resources industry, ESA space science and small satellites. According to the government, the UK space industry was worth £750m, providing jobs for 6,500 people in 400 companies. A future programme should build on those areas where Britain had developed recognized expertise. For example, Britain is considered a leader in Earth observations and built the National Remote Sensing Centre at Farnborough, to which has been added the Earth Observation Centre in Farnborough, one of four ESA processing centres.

   The United Kingdom Space Strategy, 1999–2000 – New Frontiers is the current articulation of British space policy. This emphasizes the importance of innovation, the maximization of business opportunities, space science and applications especially in the environmental area. The policy saw a positive role in encouraging the further development of those areas where British expertise was strongest, such as space science, communications, instrumentation and small satellites. It did not envisage any changes in the institutional architecture for British participation in space activities, increases in funding, nor any change in the UK attitude toward the participation in manned flight or launcher programmes (the BNSC would advise the government on participation in future programmes as funds allowed).

   Until then the BNSC did not run space projects in its own right. In a break with precedent in 2000, the government provided £15m to support demonstration projects for small satellites. This was awarded to three projects: Topsat (for QinetiQ), the Disaster Monitoring Constellation and Gemini (both by Surrey Satellite Technologies Ltd). £2m was provided under the S@TCOM programme to help UK companies to compete in the international satellite communications market.

3.20   BRITISH SATELLITES: *ARIEL*

So much for the evolution of British space policy and the demise of the British rocket programme. What about satellites, projects and missions? Here, the story is a happier one.

   Long before *Prospero* reached orbit, Britain had developed a national satellite programme, by all accounts a most successful one. This programme had its origins in American–British cooperation inaugurated during the early MacMillan period. The *Ariel* programme was a joint American–British programme of small research satellites to study the density of electrons in the atmosphere. The satellites received a numeric UK designation, *UK 1*, *UK 2*, etc. until they received orbit, after which they were called *Ariel 1, 2* and so on. Six were flown altogether.

   Except for the first mission, *Ariel* used America's smallest rocket – the solid-fuel, trusty *Scout*, which flew 118 missions until eventually retired in May 1994. NASA

Figure 3.15. Surrey minisatellite.

had the wisdom, during its earliest days, of foreseeing the need for a small, inexpensive, reliable, solid-fuel launcher able to put 50-kg payloads into low Earth orbit [51]. Although the *Scout* had been intended for *UK 1*, its development fell behind and the Americans provided at their own considerable expense a *Thor Delta* instead.

Britain's satellite programme

26 April 1962	*Ariel 1*	*Thor Delta*	Cape Canaveral
27 March 1964	*Ariel 2*	*Scout*	Wallops Island
5 May 1967	*Ariel 3*	*Scout*	Vandenberg
11 December 1971	*Ariel 4*	*Scout*	Vandenberg
9 March 1974	*Miranda X-4*	*Scout*	Vandenberg
15 October 1974	*Ariel 5*	*Scout*	Formosa Bay, Kenya
2 June 1979	*Ariel 6*	*Scout*	Wallops Island

The first two *Ariel*s were built by British engineers with American assistance. *Ariel 1* was launched by the Americans in April 1962 and Britain had the satisfaction of being the third nation with a satellite to its name circling the Earth. *Ariel 1* was

Figure 3.16. *Ariel 1.*

devoted to examining solar radiation in the ultraviolet and X-ray parts of the
spectrum. *Ariel 1* was at one stage knocked out when, that summer, the United
States exploded the Starfish atomic bomb high in the atmosphere, many spacecraft
succumbing to the subsequent high levels of radiation. The solar cells were badly
damaged and the satellite operated somewhat erratically as a result. *Ariel*'s scientific
instruments were saturated with radiation, but most recovered and worked properly.
The radiation disabled the spacecraft timer that was supposed to close *Ariel* down
after one year. As a happy result, *Ariel* continued to send useful data long after the
period intended and the satellite eventually decayed in May 1976.

   Ariel 2 was the first to use the *Scout*. Observers gathered in Virginia saw an
orange flash as the *Scout* ignited and a white arc against the blue sky as it ascended.
They saw the rocket pause momentarily before the second stage took over. By the
time it was over Antigua, it was close to orbit. Later, it was transmitting successfully
to 14 ground stations in Britain and Australia. *Ariel 2* was modelled on American
Explorer small satellites and its mission was to study meteoroids, the distribution of
ozone in the atmosphere and galactic noise. *Ariel 2* was a 70-kg cylinder with four
paddle-type solar panels at the base and a series of extruding masts and antennae
(one 40 m long), with detectors for galactic noise, ozone and meteorites.

   Ariel 3 to *5* were radio astronomy satellites, *Ariel 3* also specializing in galactic
noise, *Ariel 4* the ionosphere and *Ariel 5* X-rays. *Ariel 3* was the first entirely British-

built satellite, took three years to construct and cost £1.5m. Its five instruments were designed to study low-frequency radiation, the density and temperature of electrons, galactic radio noise, natural radio noise and the distribution of oxygen at high altitudes. Information was recorded during the 95-min orbits and dumped in 3-min downloads to the tracking station at Winkfield, whence it was sent on to the Radio and Space Research Station in Slough for processing. *Ariel 4* had five instruments to probe radio noise, the ionosphere and charged particles. *Ariel 5* was regarded as one of the most successful of the series [52]. It was built by Marconi in Portsmouth and was only the second X-ray satellite to go into orbit when it was launched by *Scout* from Kenya in October 1974. *Ariel 5* had six experiments: high-energy scintillation telescope, low-energy proportional counter, three other proportional counters, spectrometer and all-sky monitor. It weighed 136 kg and was put into an orbit of 500–545 km, 2.9°, 95 min. Satellite operations were controlled from the Appleton Laboratory in Slough. The scientific haul from *Ariel 5* was enormous, leading to the publication of a hundred scientific papers within four years. The survey found 80 new X-ray sources and determined that others located by another X-ray satellite *Uhuru* had disappeared over the previous five years.

In April 1975 *Ariel 5* detected an X-ray nova before it flared into being the brightest X-ray source in the sky, observing it rise and then fall. *Ariel 5* also discovered what became called X-ray bursts, objects associated with black holes that emitted intense X-rays for short periods of 10 to 20 seconds. *Ariel 5* put Britain in the vanguard of X-ray astronomy in the world and set the scene for the following mission.

After a gap of four years, the 154-kg dome-shaped *Ariel 6* flew in June 1979. It cost £10m to build (Marconi in Portsmouth) and £2.7m for the *Scout* launch. *Ariel 6* had instruments to examine ultra-heavy cosmic rays, fluctuating X-ray sources, supernovae, black holes and discrete X-ray sources. The instruments were made by Bristol University, Mullard Space Science Laboratory, Leicester University and Birmingham University. *Ariel 6* was eventually switched off in February 1982 because its pointing accuracy had deteriorated below the acceptable due to the loss of control gas. Scientists pronounced themselves well satisfied with the outcome, with studies of heavy cosmic particles, 30 X-ray objects studied in detail, data on black hole candidates and neutron stars, observations of galaxies and the finding of a white dwarf.

Since *Ariel 6*, Britain maintained its expertise in X-ray astronomy through collaborative programmes; for example, in the Japanese *Astro-C* mission launched 5 February 1987. The X-ray detector, a 100-kg instrument, was built by the Institute for Space and Astronautical Science at the University of Tokyo, the University of Nagoya, the Rutherford Appleton Laboratory and the University of Leicester with the support of the Science Research Council.

In addition to its national satellite programme, Britain made substantial contributions to European satellites, such as *ESRO 2* and *4* and *GEOS*; and contributed components for the *Intelsat II, III, IV* and *V* series as well as the American *Nimbus* weather satellites. Individual components and experiments flew on a wide range of American satellites, almost always with distinction and success. For example, Oxford

University contributed the pressurized module radiometer for the *Pioneer Venus* orbiter mission of 1978, which scanned its clouds and did much to improve knowledge of the planet's atmosphere [53].

Britain contributed an important set of instruments to the *Kvant* astrophysical module attached to the *Mir* orbital space station. In 1986, following an agreement between Britain and the USSR, British and Dutch telescopic equipment (the UK part was made in the University of Birmingham) was carried on board the Soviet astrophysical module *Kvant*, which rendezvoused with the *Mir* orbital station in March 1987. The detectors, made at the University of Birmingham, picked up valuable information on the chemical elements of an exploding supernova in the Large Magellanic Cloud. In the first six months of operation, *Mir* cosmonauts made over 300 observations with the telescope. In October 1988 cosmonauts Musa Manarov and Vladimir Titov spacewalked outside *Kvant* to attach a new X-ray detector part ferried up by the *Progress 38* cargo ship. *Kvant* was still attached to *Mir* when the great orbital station was finally brought down in March 2001.

3.21 *SKYNET*

In the early 1960s the Ministry of Defence began to appreciate the value of satellite-based communications in linking Britain's defence forces, which were then located in many distant places and bases over the world. Originally, the Ministry of Defence had planned to develop a global military communications system by bouncing signals off the Moon and then off space junk, but eventually came around to the idea of communications satellites in 24-hr orbits [54]. As was the case with the *UK* or *Ariel* programme, the Americans were asked to provide launchers. At the time, no UK company had the ability to build communications satellites, so the first series was contracted to Philco Ford in the United States, but with Marconi involved in the construction for training and experience.

With *Skynet*, Britain became the first country apart from the two superpowers to develop a military communications capability, the aim being to provide teleprinter, voice, telephone, fax and data links between Britain and its forces on land, air and sea all over the world, some using small dish terminals only 1 m across. Three Earth stations were built or adapted to support the *Skynet* system. RAF Oakhangar with its 13-m dish was designated the prime domestic base for the system. It was staffed by 150 RAF officers and men of the 1001 Signals Unit (before that, it was a US Navy signals station). There were two other Earth stations – in Hong Kong and Cyprus. Receivers were also installed on the assault ships HMS *Fearless* and HMS *Intrepid* and at one stage the army also had two mobile receivers [55].

The *Skynet* programme had a difficult start. *Skynet 1* was a 249-kg satellite with a single transponder positioned over the Indian Ocean in 1969, but its wave tube amplifiers failed in 18 months. Worse was to come, for *Skynet 1B* failed to fire to synchronous orbit (the apogee kick motor exploded) and was stranded.

The *Skynet 2* series was built in Britain by Marconi with American help. *Skynet 2* had more power and two transponders. The run of bad luck continued when in

January 1974 the steering mechanism of the *Delta* rocket for *Skynet 2A* went hard over during the final stages of its ascent through the atmosphere. The US Air Force launching team told the British team that the satellite was destroyed. In fact, the American ballistic early warning system then found the satellite in a very low orbit. The British team tried to fire the solid-fuel engine to raise the orbit as the first step to retrieving the mission. However, the satellite could not get a proper orientation and did not know which way was up and which way was down. It was commanded to fire anyway, on the 50/50 chance that it might go up, but the toss went the wrong way. *Skynet 2A* ended up a spectacular fireball over the Himalayas.

Britain did not know success until *Skynet 2B* entered the correct orbit over the Indian Ocean on 22 November 1974. It made up for the bad behaviour of its predecessors and was the mainstay of British military communications for many years. The Navy used it increasingly on its ships as did the Army on its Landrover-based terminals. The *Skynet 2B* propellant supply, designed to last three years, in fact lasted for 18. Its transmitters were still operating in 1995.

By the time *Skynet 2B* was operational, Britain had withdrawn all its forces east of Aden and, apart from occasional military exercises, had little use of a good, but expensive global communications system. *Skynet 3* was cancelled in the financially strapped mid-1970s and the decision was taken to lease lines from American and NATO military communications satellites thereafter [56].

3.22   *SKYNET* – BRITISH PASSPORT TO MANNED SPACEFLIGHT?

That looked like the end of the road for *Skynet*, but it was not. A reconsideration of the value of military satellites took place in the late 1970s when it became apparent that they could, through small terminals, link not merely command stations with bases but also with individual ships, planes and very small military units. As it was, Royal Navy and Army demand for long-distance satellite-based communications exceeded the supply available on American and NATO satellites.

The issue was settled in spring 1982, with the war between Britain and Argentina. The war in the South Atlantic re-emphasized the importance of the Royal Navy having speedy access to a fast military communications system. American hesitation in supporting the British expedition to the South Atlantic meant that it should be an indigenous one as well. Britain had planned to use a projected *NATO IV* military communications satellite system, but when this was cancelled, the choice of a new national system became unavoidable. Parliament quickly approved the *Skynet 4* system.

Skynet 4 was a 1,433-kg box-shaped spacecraft with long solar panels giving 1,200 W of power, providing a range of channels for submarines, aircraft, in-flight refuelling tankers, maritime patrols, soldiers with 1-m terminals, all proofed against jamming, radiation damage and other interference. When the satellite was being prepared, tests were carried out with the terminals on Royal Navy ships, large RAF aircraft, Landrover-based collapsible terminals and even individual infantry backbacks. *Skynet* was designed with spot beams that could be swivelled to many different locations.

Skynet 4 launches were booked on the US space shuttle, rather than the European *Ariane*. The shuttle had become the main (it was intended as the sole) satellite cargo carrier in the American launch fleet. Satellites were ejected from the cargo bay of the shuttle and a specialist was normally assigned to each mission to ensure that this ejection took place smoothly. This provided a golden opportunity for a British military astronaut to accompany each *Skynet* into orbit. Accordingly, four British astronauts were selected and went to Houston for training, NASA charging only expenses.

The candidate astronauts were Nigel Wood (RAF), Richard Farrimond (Army), Peter Longhurst (RN) and Chris Holmes (Ministry of Defence). In April 1985 Nigel Wood was selected for the *Skynet 4A* deployment, set for 24 June 1986. Peter Longhurst was chosen to accompany the following *Skynet* (*4B*) into orbit. The mission would fly six additional experiments in such areas as radiation, space medicine and materials.

Squadron Leader Nigel Wood was almost ready to fly when *Challenger* exploded. It was quickly apparent that the shuttle would take some time to fix (over two years in the event). NASA decided that the shuttle would not be used again as a cargo carrier for ordinary communications satellites and the production line of expendable launchers was restarted. Nigel Woods lost his chance to fly and the team was disbanded.

Despite pressure from the French to use *Ariane*, *Skynet 4A* was rebooked on another American launcher. Because *Skynet 4A* had been rigged for a shuttle launch, it took some time to reconfigure. *Skynet 4A* was eventually put into orbit on a *Titan III* in January 1990, almost four years later than planned. *Skynet 4B* was made ready soonest and actually flew first, using the *Ariane* rocket. *Skynet 4C* flew in August 1990. *Skynet 4A* was located at 6°E, then 34°E; *Skynet 4B* at 1°W, then 53°E. The early points are probably for checking out the system, whereas 53°E, the operational position, is perfect for the military forces in the Gulf.

The last of the series *Skynet 4F* flew in February 2001 with the Italian military communications satellite *Sicral* and was positioned at 1°W. The idea was that *Skynets D*, *E* and *F* would over time replace *A*, *B* and *C*.

What of the British astronaut project? Britain tried to persuade the Americans to fly a British payload specialist on the *Spacelab 2* laboratory on the shuttle to operate its telescopes. The Americans did not respond. In the end Britain's first astronaut flew courtesy of the Soviet Union, when Helen Sharman spent a week on the *Mir* space station in 1991, described later. Perhaps the best known British-born astronaut was Michael Foale, who went to live and work in America. He made several shuttle missions, his most famous being a six-month tour on the *Mir* space station during the great crisis of 1997.

3.23   FUTURE *SKYNET*

Whatever doubts that might have existed about the value of satellite military communications were swept aside when the system proved its value during the

Gulf War of 1990–1. *Skynet* was considered such a successful series that NATO commissioned two *Skynet*-type satellites from Britain.

With the *Skynet 4* series nearing completion, consideration was given to a successor programme. Originally Britain decided to participate in a European military communications satellite programme with Germany and France (1997), but pulled out the following year. Britain then decided to go for a *Skynet 5* programme for two satellites, to be provided privately. The contract was worth £1.5bn (about €2.5bn), with an in-service date of 2005. Under the private arrangements, the UK military would lease the lines from the provider as required, but the company would be free to lease spare capacity to NATO or even for civilian commercial satellite links.

The *Skynet* programme had a peculiar footnote. In January 1987 the *New Statesman* weekly magazine published an article alleging that the next *Skynet* (then *Skynet 4C*), to be positioned over Asia, would in reality be a large, American-built, electronic eavesdropping satellite based on its Chalet and Magnum programmes, paid for and working for British intelligence. This was called Project *Zircon*. The large 15-tonne electronic surveillance satellite would sweep in electronic information collected over the Soviet block for General Communications Head Quarters (GCHQ) in Cheltenham. There data would be analysed by British military intelligence, some being passed on to the Americans in an intelligence-sharing arrangement. Analysis by the *New Statesman* of some inadvertent press releases concerning *Skynet 4C* suggested that both the spacecraft and its mission were quite different from the *Skynet* programme, though using it as cover.

Later, the Ministry of Defence was forced to admit to the existence of *Zircon* and that £70m had already been spent. By 1990 the total bill had come to over £1bn. It is reported that *Zircon* was eventually launched as a classified American payload on 4 September 1990 on a *Titan 34D* from Cape Canaveral. From the 1980s the Americans began to launch a number of these large, classified military payloads, releasing only launch dates and initial orbits, not even a payload name, however fictitious. The real *Skynet 4C*, its cover blown, flew a normal *Skynet* mission, as did its successors. Until the details of American military launches in the 1980s and 1990s are fully declassified, we will never know if one of them was indeed a *Zircon* used for British intelligence.

Skynet missions

1	22 November 1969	*Delta*	Cape Canaveral
1B	19 August 1970	*Delta*	Cape Canaveral
2A	19 January 1974	*Delta*	Cape Canaveral
2B	23 November 1974	*Delta*	Cape Canaveral
4A	1 January 1990	*Titan 3*	Cape Canaveral
4B	11 December 1988	*Ariane 44LP*	Kourou
4C	30 August 1990	*Ariane 44LP*	Kourou
4D	9 January 1998	*Delta*	Cape Canaveral
4E	26 February 1999	*Ariane 44OL*	Kourou
4F	7 February 2001	*Ariane 44L*	Kourou

3.24   BRITAIN: PIONEER OF MICROSATELLITES

Britain found an unexpected niche in the world space industry in the 1990s, courtesy of the University of Surrey in Guildford. These were small satellites. Even as communications and other satellites became larger and larger in the 1980s and 1990s, some engineers paradoxically spotted a niche for the construction of very small satellites able to take advantage of recent advances in computerization and miniaturization. These microsatellites soon demonstrated their versatility and the quality of their imaging was excellent. The payloads were so small that their promoters were able to go to the big launcher companies and persuade them to carry these microsatellites either for free or at very low cost alongside the main payloads. In 1988 Arianespace took the enlightened step of adding a structure to the upper stage of the *Ariane*, called the *Ariane* Structure for Auxiliary Payloads (ASAP), to facilitate the installing of microsatellites. Later, in the 1990s the availability of small rockets left over from the Cold War (e.g., the *Dnepr*) also facilitated the launch of small satellites.

Small satellites also held out the possibility – though this was probably apparent only later – for affordable satellites for emerging spacefaring nations. Small satellites should, to be technically precise, be broken down into a number of subcategories:

- small satellites, less than 1,000 kg;
- minisatellites, less than 500 kg;
- microsatellites, less than 100 kg;
- nanosatellites, less than 10 kg;
- picosatellites, less than 1 kg.

There had been small satellites before the arrival of Surrey Satellite Technology Limited (SSTL), but these were limited to amateur radio satellites – the OSCAR (Orbiting Satellites Carrying Amateur Radio) series of small radio amateur satellites that had been put piggyback into orbit over the 1970s on the top of larger American missions.

SSTL used miniaturization to cram into very small, typically 50-kg box-shaped containers the level of instrumentation that would have previously required 10 times the weight and size. Their satellites were called *UoSAT* (University of Surrey satellite). An early example of the piggyback opportunity was *Ariane V59* (its 59th flight): while its main cargo was *SPOT 3*, it had the space and weight available to launch no less than six microsatellites (of these, two made by Surrey): *Stella* (a 48-kg French laser target), *PoSAT* (a 50-kg Portuguese imaging satellite), *Healthsat* (an American store-and-forward satellite), *Kitsat 2* (Korean imaging satellite), *Eyesat* (12 kg) and *Itamsat* (12 kg).

SSTL's work in small satellites began in 1979. The idea came from Professor Martin Sweeting who tried to persuade the university that it was possible to send satellites into space on a shoestring budget. He managed to raise £120,000 from industry (aerospace companies, the Post Office, the Royal Aircraft Establishment and the Radio Society of Great Britain), which the University matched with

Figure 3.17. SSTL: preparing minisatellite.

£150,000 and allocated him a small laboratory. The original team comprised five university academics and technicians, with the assistance of five students. They built their clean room over a weekend out of plywood, glass and polythene, using a kitchen fan to filter out dust particles. The aim of the first satellite was to do experiments for radio hams and the transmission of radio waves through the Earth's atmosphere.

The first *UoSat* was eventually launched on 6 October 1981 piggyback with the solar mesosphere *Explorer* on a *Delta* from the Vandenberg base in California. SSTL persuaded NASA to let them fly it for free because it was an educational project. *UoSAT 1* was only 50 kg, a cuboid 1 m tall and 50 cm at the side. The SSTL people could not afford to travel to the launch so they listened to a commentary from Vandenberg Air Force Base over the telephone. Once it arrived in orbit, they learned on the job about how to control a satellite, sleeping under their consoles to awake in the small hours of the morning for the overhead passes. They could tune in at 145.825 MHz.

Once in orbit, *UoSAT* quickly proved its worth. Not only did the amateur radio community benefit, but so did scientists and technologists. It transmitted messages using a speech synthesizer. It sent back pictures of Earth with its TV camera.

Figure 3.18. SSTL: *UoSAT*.

Instruments detected electromagnetic storms in the atmosphere and auroral activity over the poles. Then disaster struck on 4 April 1982 when the data beacons were accidentally switched on together and the system no longer accepted commands. It took until September to troubleshoot the problem and restore the satellite to working order. *UoSAT 1* eventually burned up in October 1989.

When America's *Landsat 4* failed, NASA gave SSTL the opportunity to fly a *UoSAT* piggyback on its replacement mission – provided the university could build it in five months! This it did and *UoSAT 2* duly made it into orbit with *Landsat 5* on 1 March 1984. *UoSat 2*, a 50-kg cuboid, carried an electron spectrometer for magnetospheric studies, a space dust detector, Earth imager and voice synthesizer.

From *UoSAT 3* onward, SSTL developed a standardized bus. Different experiments could be located in set areas of each spacecraft. The standardized approach meant a reduction in the period between when an order was placed and when a satellite was launched to as little as 10 months and within a budget of less than £3m. The centre formed itself into a company (SSTL) in 1985, the profits generated being ploughed back into research, teaching and applications. SSTL's work, while cutting-edge, also has a strong garden shed workshop approach. Rather than develop something new, SSTL would much rather adapt a component from the local radio store for its purpose, or even, as it did recently, go scavenging on the Airbus scrap heap for leftover parts that might suit [57].

Two *UoSAT*s were flown into orbit on *Ariane V35* on 22 January 1990, which carried into orbit France's second Earth observations satellite *SPOT 2*. Because there was no other major customer with *SPOT 2*, there was room for several small satellites to be carried. *UoSAT 3* and *4* were twin 45-kg satellites $345 \times 345 \times 600$ mm put in an 800-km polar orbit, the mission cost being £700,000. *UoSAT 3*, also known as *Healthsat 1*, carried a digital store-and-forward communications transponder to test disaster alert systems, experiments to measure cosmic rays and radiation levels and new gallium arsenide solar cells. It worked perfectly – indeed, in 2001 it was pressed into service by amateur radio enthusiasts in the state of Gujarat when the state was hit by an earthquake. At a time when phone lines were down, communications through *UoSAT 3* provided much the most effective means of coordinating relief efforts. *UoSAT 4* carried an Earth imaging camera, solar cell experiments and a data-processing experiment. *UoSAT 4* broke down only 30 hours into its mission despite having multiple redundancy in its transmission systems, but that was unusual.

Ariane V52 put into orbit two SSTL satellites alongside the main payload, the French–American *Topex-Poseidon*. These were *S80/T* and *Kitsat 1*. *S80T* was a £1m CNES-funded research project into the use of satellites for hand-held communications, the entire project being designed, built, completed and launched within a year. *Kitsat* was the first of two for Korea and was a small satellite with an Earth imaging system with 4-m resolution, a store-and-forward digital communications system and a cosmic ray satellite.

Ariane V59, mentioned earlier, put into orbit two Surrey satellites: a Portuguese mini-satellite called *PoSat* and *Healthsat 2*. *PoSAT* was built by seven Portuguese engineers working alongside their British colleagues in order to gain experience in building satellites, whereas *Healthsat 2* carried store-and-forward communications systems to develop health services in Third World countries. *PoSAT* sent back fine pictures of the Earth with details of roads, bridges and weather patterns. *UoSat 12* was a 350-kg satellite launched on a Russian *Dnepr* rocket on 21 April 1999. It carried 10-m and 30-m multispectral imagers, a wide-angle camera and digital communications transponders.

By 2000 SSTL had built 20 satellites in 20 years and was then manufacturing several more. One hundred and sixty people were now employed, and the centre had grown so much that it had to move in to a new £3m building. PhDs are awarded in small satellite design and it was now considered one of the university's greatest success stories. Some of its earliest satellites are still operational, years after the end of their design lifetimes, and the centre has won a series of contracts for the provision of future satellites and equipment for a range of different missions. Its current order book is €50m. It even has its own mission control centre, able to operate 11 satellites at a time with just a single operator.

Countries as diverse as China, Thailand, Algeria, Portugal, Pakistan, South Africa, Chile, Singapore, Turkey and Malaysia all sent their engineers to study SSTL's methods to develop their own small satellites. SSTL even sold a satellite to the US Air Force as a training tool for its space cadets. It won a £3.3m contract to build four 65-kg microsatellites for the US Air Force, called *Picosat*,

Figure 3.19. SSTL – staff.

the first such contract awarded outside the United States. *Picosat* was duly launched from the new range on Kodiak Island in Alaska. SSTL has now built several small military satellites, *Cerise* and *Clementine*, for the French department of defence and *STRV* for Britain's own Defence Evaluation and Research Agency (DERA). *STRV* was a 50-kg cube developed by RAE to test solar cells, radiation and electrical anomalies. Each carried 23 American, British and Canadian experiments in platform technology, autonomy, communications, surveillance and the space environment in an orbit of 600 km to 36,000 km.

These small satellites have become ever more capable. As an example of their versatility, *SNAP 1* was flown in June 2000. This must have been the most advanced nanosatellite in the world at the time. *SNAP*, standing for Surrey Nanosatellite Applications Programme, accompanied a Russian *Nadezhda* navigation satellite into orbit in June 2000 alongside China's first microsatellite *Qinghua 1*, also built in Surrey. *SNAP 1* weighed only 6.5 kg and its engine only 450 g, yet it was able to verify its capacity for inspecting other satellites by chasing *Qinghua* in orbit, changing its orbit with its tiny pencil-size engine and closing in on *Nadezhda* to take its picture, much to the consternation of the Russians, who weren't used to this kind of thing.

In 2000 SSTL began to develop the Disaster Monitoring Constellation (DMC) a network of five 100-kg microsatellites to orbit the Earth at 686 km with imaging systems to watch for and observe natural disasters. Five countries came together for the €60m project – Thailand, China, Algeria, Nigeria and Britain – to be launched on a *Dnepr* rocket. Ideally, these microsatellites, to be launched by Russian rockets

(a)   (b)

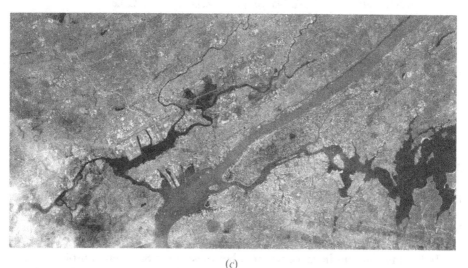

(c)

Figure 3.20. (*a*) SSTL – *SNAP*; (*b*) SSTL – ground control; (*c*) SSTL – image of New York.

over 2002–4, could pave the way for a global environmental monitoring system based on small satellites.

Following SSTL's lead, several other countries in Europe began to develop microsatellites. In Germany microsatellites were built by OHB-System (*Safir*) and the Technical University of Berlin (*TUBSAT*). In Italy microsatellites were developed called *CESAR*, *JUNO* and *COSMO* and the Italian Space Agency began a microsatellite programme called *Mita* and *Prima*. Almost all sought out piggyback launches over the next few years, mainly on Russian rockets and *Ariane 5*.

SSTL small satellites

6 October 1981	UoSat 1	Delta 2	SSTL
1 March 1984	UoSat 2	Delta 2	SSTL
22 January 1990	UoSat 3	Ariane 40	SSTL
	UoSat 4		SSTL
17 July 1991	UoSat 5	Ariane 40	SSTL
11 August 1992	S80/T	Ariane 42P	CNES/France
	Kitsat 1		Korea
26 September 1993	Kitsat 2	Ariane 40	Korea
	PoSAT		Portugal
	Healthsat 2		
7 July 1995	Cerise	Ariane 40	CNES/France
31 August 1995	FaSAT Alpha	Tsyklon 3	Chile
10 July 1998	FaSAT Bravo	Zenit 2	Chile
1998	ThaiPhut		Thailand
3 December 1999	Clementine	Ariane 40	CNES/France
21 April 1999	UoSat 12	Dnepr	SSTL/Singapore
20 June 2000	SNAP 1	Cosmos 3M	SSTL
	Qinghua 1		China
26 September 2000	Tiungsat	Dnepr	Malaysia
30 September 2001	Picosat	Athena	US Air Force

3.25 ITALY'S NATIONAL SPACE PROGRAMME

By the turn of the century, Italy had become the rising star of ESA. France remained, as ever, the leader, with Germany in second place, but with restricted budgets was slow to play a leadership role. Italy was a country producing new ideas, projects and investment [58].

Italy's involvement in space research began with the San Marco project, which was a model of lateral thinking. The San Marco idea came from Professor Luigi Broglio in the Italian Space Commission in 1961, with the Centro Ricerche Aerospaziali and the Centre for Space Research in the University of Rome. His idea was that Italy should develop a joint project with the United States for the launching of small satellite payloads into equatorial orbit from Kenya using the *Scout* launcher. The Kenyan coast offered two advantages: direct ascent to equatorial orbit and a downrange path over the ocean, away from inhabited areas.

Italy would build the launch platform, train launch crews and fire the American rocket, on the understanding that Italian satellites could reach orbit on at least some of these missions. A memorandum was duly signed between the Italian foreign minister and Vice President Lyndon Johnson in 1961. Italian launch crews were soon in the States training on the *Scout* as the platform was built off the coast of Kenya, on the equator.

Their first training launch was to put an Italian satellite *San Marco 1* into orbit launched from the *Scout*, using the *Scout*'s home port of Wallops Island, Virginia (the Kenyan platform was not yet ready). This was done in late 1964, almost a full year ahead of the first French satellite. This was a small satellite 66 cm across to measure air density and radio interference. The Italians have since claimed that they were 'the first Europeans to launch a satellite'. Strictly speaking, this is true but granted that it was an American rocket from an American base the French would probably dispute the claim!

Meanwhile, progress was being made with the new launch site in Formosa Bay, 5 km off the coast of Kenya at 2.56°S, 40.12°E. The launch site comprised a launch centre in two parts: a launch platform (San Marco) and a command platform (Santa Rita) 500 m away from one another, both platforms on stilts. They were linked to tracking facilities in Nairobi and Mombasa. The original platform was an oil rig, towed from Charleston, South Carolina. Each was triangular-shaped and embedded in the ocean floor. Eighty staff worked on the site during launches (there was a permanent facility on shore). For launches, the *Scout* rocket was brought along in a barge and then erected onto a triangular-shaped railing whence it was pointed in the launch direction and fired. The location meant that the satellites sent aloft from Kenya could be put in almost equatorial orbit – 3° being the norm. Construction took three years, 1964–7.

The first launch of an Italian satellite duly took place in April 1967, launched by Italian engineers using the American *Scout* rocket. It was the first-ever seaborne launch. *San Marco 2* measured atmospheric density while *San Marco 4* sampled the auroral area around the equator. *San Marco 5* examined charged particles, electric fields and solar fluxes. The British *Ariel 5* was also launched from the San Marco platform.

On 13 December 1970 the Italians launched the first-ever X-ray satellite from Formosa Bay, also the first American satellite to be launched outside the United States. Its scientific name was *Explorer 42*, but it was named *Uhuru* or 'freedom' in honour of Kenya's independence. The satellite was a joint project between NASA and the University of Rome.

The San Marco platform was last used in 1988 and the *Scout* rocket is no longer in production. The platform was a pioneering method of launching satellites, since put to good use by the Russians and Ukrainians in the Sea Launch project. The Ukrainian *Zenit 3* rocket was adapted for launch from a sea platform (also an oil rig) in a project involving Russian rocket scientists and the American Boeing company. The rig was towed far out in the Pacific where, accompanied by its control ship, the *Zenit 3* has been used since 1999 to put large satellites into 24-hr orbit. But the idea started in Rome in 1964.

3.26 ITALIAN SPACE SCIENCE

The *San Marco* series was the first of a number of space science missions. One of the most famous of these was *SAX*, Satellite Astronomia raggi-X. Once in orbit it was named *Bepi SAX* after the astronomer Giuseppe (Bepo) Occhialini. *SAX* was an Italian–Dutch bilateral X-ray astronomy project set up by ASI in 1984 with its

Figure 3.21. Dutch–Italian *SAX*.

Dutch opposite number the NIVR. The Dutch contributed about 10% of the €600m project, principally the wide angle camera, attitude control and solar panels. *SAX* weighed 1,400 kg and carried four Italian spectrometers designed to search for a broad range of X-ray sources and bursts. Launch in the mid-1980s was delayed because of the grounding of the shuttle and *SAX* eventually made it into orbit on an *Atlas* from Cape Canaveral on 30 April 1996. *SAX* was 3.6 m high, 2.7 m in diameter with two solar panels and entered an orbit of 583 km by 603 km with an almost equatorial orbit of 3.96°. As a result, tracking was done not just from the main Italian station at Fucino but also from a ground station in Kenya. Once in orbit the satellite began to sweep the skies for X-ray binaries, pulsars, supernovae remnants and galaxies.

Italy built the second *Lageos* satellite. *Lageos* was an American programme for a geodetic satellite and involved two small satellites, each with identical specifications. They were ball-shaped, only 405-kg and 60 cm in diameter, equipped with 426 laser reflectors and made of aluminium and brass. *Lageos* is so solid that it is expected to survive re-entry in 8.4 million years from now. The concept of the satellite is that by bouncing laser beams off the reflectors it will be possible to determine the exact distance of the ground to within 1 cm, meaning that with triangulation it will be possible to detect movements of the Earth's plates. The Italian interest in *Lageos* had much to do with detecting irregularities in the Earth's crust in the seismically active zones of the Mediterranean. The American *Lageos* was launched in 1976 and the Italian one *Lageos 2* was planned for the shuttle in 1987. Due to the disaster in 1986 it was delayed and the Italian *Lageos* was eventually launched on shuttle STS-52 on 23 October 1992.

In addition to space science, Italy has been involved in small satellites. On 15 July 2000 a *Cosmos 3M* launched the Italian microsatellite called *Mita* made by Caro Gavazzi Space. On 26 September 2000 a Russian *Dnepr* launched five minisatellites into orbit, of which two were Italian: *Megsat 1* for education built by the Meggiorini group and *Unisat* built by GAUSS (Gruppo di Astrodinamica dell'Università degli Studi della Sapienza [Astrodynamics Group of the University for the Study of Knowledge]), the latter with camera, magnetometer and space debris sensor.

3.27   ITALIAN COMMUNICATIONS

The second area of work for the Italian space programme is communications. *Sirio* was an experimental communications satellite developed entirely by Italy to test out the best way of using satellites to meet domestic Italian communications needs. A secondary aim was to test the propagation of radio waves during rain, fog and snow. *Sirio 1* was launched by American *Delta* from Cape Canaveral on 25 August 1977 and was put on station at 15°W. In 1983 it was moved to 65°E to assist China in the development of its communications satellite programme. A follow-on mission, *Sirio 2*, was lost in the *Ariane V5* failure.

Italy's second communications satellite was *Italsat*. This was a €281m programme to test out the use of communications satellites for conventional domestic and mobile communications. *Italsat* was launched on the first *Ariane* for 1991, *V41*. *Italsat* carried spot beams for the development of digital communications in Sicily, Sardinia and the Italian mainland as well as 11,000 phone call circuits and transmitters for mobile communications. The *Italsat* project enabled Italy to develop its own capacity to build communications satellites. This took 10 years, but the outcome was an indigenous capability to build such satellites, placing Italy in a good position to win comsat projects in the future.

Since then Italy developed its own military communications satellite, *Sicral* (Satellite Italiana di Communicazioni Riservate ed Alarmi [Italian Classified Communications and Alarm System]), costing €700m, launched by *Ariane* on 8 February 2001. The project was in design from 1983 and aimed to provide Italian ground troops with the ability to function abroad either with the United Nations or the European Rapid Reaction Force. *Sicral* was especially designed to cope with the type of mountainous terrain that inhibited communications in the Kosovo engagement. *Sicral* was a 2.6-tonne satellite built by Alenia, FiatAvio and Telespazio. It had nine transponders and was located over the Congo. *Sicral* had a mobile spot beam that would be initially focused on Italy, Europe and the Middle East, communicating with a control centre at Vigan di Valle in Rome and up to 1,000 mobile terminals on ground, air and naval forces, helicopters and infantry

carriers [59]. After checkout it was handed over to the Italian defence ministry at the
end of the year and provided support for Italian troops sent to Afghanistan.

In addition, the Italians have experimented with small communications satellites. *Temisat* was a small, 32-kg experimental communications satellite costing
€40m designed to interrogate and relay data from environmental platforms, but it
failed after a year.

Italian satellites

15 December 1964	*San Marco 1*	*Scout*	Wallops Island
26 April 1967	*San Marco 2*	*Scout*	Formosa Bay, Kenya
24 April 1971	*San Marco 3*	*Scout*	Formosa Bay, Kenya
18 February 1974	*San Marco 4*	*Scout*	Formosa Bay, Kenya
25 March 1988	*San Marco 5*	*Scout*	Formosa Bay, Kenya
25 August 1977	*Sirio 1*	*Delta*	Cape Canaveral
September 1982	*Sirio 2*	*Ariane 1*	Launch failure
15 January 1991	*Italsat 1*	*Ariane 44L*	Kourou
31 August 1993	*Temisat 1*	*Tsyklon 3*	Plesetsk
8 August 1996	*Italsat 2*	*Ariane 44L*	Kourou
15 July 2000	*Mita*	*Cosmos 3M*	Plesetsk
26 September 2000	*Megsat 1*	*Dnepr*	Baikonour
	Unisat		

3.28   FORMING THE ITALIAN SPACE PROGRAMME

Peculiarly, much of the early Italian space programme developed in the absence of a
formal national space programme or agency. Italy's space programme was not
formally established until 1979 when the government set up the National Space
Programme (PSN), which was required to operate according to five-year plans
and directed by the National Research Council (Consiglio Nazionale delle
Ricerche). It was not for another decade that ASI was established, in 1988, with
an initial complement of 180 staff. One of the main architects of the modern Italian
space programme was Professor Carlo Buongiorno, who became director general of
ASI. He was involved in sounding rockets in the early 1960s and then the San Marco
project.

The setting up of ASI was accompanied by a doubling of the space budget. ASI
was made responsible for operating Italy's national space plan (Plano Spaziale
Nazionale), which is approved by the Interministerial Committee for Economic
Planning (CIPE). Italy took the decision to increase its space investment in the
mid-1990s. Its space budget rose to €483m in 2001, divided about 2:1 between
national and ESA programmes, the national one budgeted at €180m in 2001 and
ESA's at €303m. ASI is now the third largest contributor to ESA, giving about 16%
of its budget and has come to play a more forward role in policy-making. Of the
optional ESA budget contribution of €220m, Italy's main contributions are for the

manned programme (€83.6m), followed by observations of the Earth (€36m) and telecommunications (€31m). The mood in the Italian space industry is now upbeat, with Italy visibly to the fore in the construction of the International Space Station. The latest ASI promotional publication suggests this is only a start: *You'll Be Getting More and More Space from Us* is its title.

Italy will feature prominently in the later description of the European space programme. Italy decided to concentrate on a number of distinct areas, such as space station modules and related robotics [60]. Italy has been an important contributor to European science missions such as the *Cassini* probe to Saturn, where it built the high-gain antenna. In bilateral programmes it supported the *Helios* military surveillance programme with France (7%). Italy selected its first astronaut Dr Umberto Guidoni in 1989, and began a series of ambitious space tether experiments with the Americans on the space shuttle.

Italy's leading aerospace company is Alenia, based in Turin, best known for its role in building the *Columbus* module and the logistics modules for the International Space Station. Alenia was a coming together of Aeritalia and Selenia, which had experience in, respectively, pressurized modules and the *Olympus* communications satellite. Alenia built the main part of *Spacelab* for the American space shuttle. Its most recent achievement was building the *Artemis* experimental communications satellite. Alenia's contribution to the International Space Station covers half its $1,300\,m^3$ pressurized space, the three logistics modules, two nodes and significant parts of the *Columbus* modules and the Automated Transfer Vehicle.

One of the leading specialized companies is Galileo Avionica, formerly Officine Galileo, part of Tecnospazio and Finmeccanica, based in Campi Bisenzio, Florence and Milan which supplies both ASI and ESA with components and instruments, especially in the areas of attitude sensors, mission payloads, power supplies, telecommunications equipment and robotics. Galileo contributed key components of *Giotto, SOHO, SAX, Cassini, Cluster,* the tethered satellite experiment and *Rosetta*.

While the main aim of the Italian space programme has been to act as a driver of Italian industrial and scientific technology, a secondary aim has been to benefit the poorer southern regions of the country. For example, Alenia's latest plant is in Naples and other production facilities have been located further south.

Italy may now be ready for its first ventures outside Earth orbit. In 2001–2 Italy made a number of outline agreements with NASA for Italian participation in extensive American plans for the exploration of Mars, paralleling French interest in the *Netlander* and sample return missions. Subject to government approval, Italy aimed to invest €200m in these programmes. Italy has expressed interest in:

- subsurface sounding radar for the 2005 orbiter;
- telecom relay for the 2007 NASA/CNES sample return mission;
- science package and drill for 2007 mission landers;
- feasibility study of NASA/ASI high-resolution orbital radar mapper;
- study of nuclear propulsion system for manned flight.

Italy is host to Europe's ESRIN (European Space Research Institute) facility, located in Frascati, south of Rome. About 140 staff work in the facility, which is

Figure 3.22. *Columbus* in construction.

set in rolling hills. ESRIN handles the Earth observation data from European satellites and now houses the largest environmental data archive in Europe. ESRIN focuses on the taking in, processing, handling and distribution of Earth resources data. Its main satellite projects have been the European Earth Resources Satellite, *ERS* and *Envisat*, but it has also handled data from *SPOT*, *JERS* and *Landsat*. Incoming data are made available in real time, prepared for fast delivery and archived. Because of Italy's key role in the new European small launcher programme *Vega*, the *Vega* project team has been located there from 2000. Italy also hosted the new Space Applications Institute in Ispra, designed to integrate a number of European research efforts in the area of agriculture, fisheries, the environment, transport and security.

3.29   *VEGA* LAUNCHER

Italy's space programme is an expanding one. The country started with an involvement in light launchers, the American *Scout*, and it is to this area that the programme has now returned with *Vega*.

Figure 3.23. *Vega* launcher.

From the early 1990s Italy began to make the case for Europe to develop a light booster for scientific satellites, able to put about 700 kg into low Earth orbit. Italy already had some rocket-building experience, making major contributions to the solid rocket boosters for *Ariane 4* and *5*. An Italian launcher would mark a major presence by Italy in European rocketry and space development.

Italy has been convinced of the need of a light European launcher for a number of years. In its research, ASI originally talked both to China and the American company that made the esteemed, but now-obsolete *Scout* launcher. Italy gave some consideration to developing the *Scout* into a new lightweight launcher capable of being fired from the San Marco platform off Kenya. However, the proposal hit legal snags, high maintenance costs in San Marco and a number of counter-proposals.

Italy then went on to propose a fresh design called *Vega*, a three-stage solid-fuel rocket. The aim was to place up to 1,500 kg into an 800-km orbit. *Vega* quickly ran into opposition from France, for two reasons. First, France took the view that the small launcher market was over-catered for, mainly by former Cold War missiles. Second, French companies had a financial interest in the small launcher market anyway and saw no good reason to fund a rival.

Despite French reservations, *Vega* was eventually approved in October 2000 with an initial budget of €128m. The development phase was supported by Italy 55%, France 30%, Belgium 5% and 10% each for Switzerland, Spain and the others (e.g., the Netherlands). The prime contractors were Aérospatiale and EADS (European Aerospace and Defence Industries). *Vega* would be a 130-tonne rocket, 28 m high using solid-fuel motors with three stages. The Italians moved quickly on the project, test-firing the Zafiro engines that summer in Sardinia. *Vega* will have a solid P80 first stage to fire for 105 seconds. The second stage was first tested at the Salto di Quirra range in Sardinia in December 2000. Its first flight is planned for 2005, using the original *Ariane ELA-1* launching pad in Kourou, French Guiana.

3.30 GERMANY'S NATIONAL SPACE PROGRAMME

For the last 25 years of the 20th century, Germany ran the second largest space programme in Europe. Germany was slow to move into rocketry after the war, probably because of the evil associations of the V-2 in the public mind. Germany was formally permitted to resume its work in rocketry in August 1952, when the FPS (Forschungsinstitut für Physik der Strahlantriebe [Research Institute for the Study of the Physics of Jet Propulsion]) was established in Stuttgart. The allies made it plain that it must engage in theoretical work only, such as the theory of jets and rockets, the physics of combustion and gas radiation. The FPS was funded by the federal government in Bonn, the state government of Baden Wurttemburg and Heinkel, Messerschmitt, Dornier, Bosch and Daimler Benz. Their budget had a ceiling of €40,000 a year.

However, they appointed as their first director a person of quite practical bent, Eugen Sänger, who had spent the 10 post-war years working for the French. He had become quite frustrated there, his activities focusing on aircraft rather than rockets, with projects being frequently interrupted by the endless changes in government of the Fourth Republic. In Stuttgart he quickly drew up fresh proposals for rockets and space shuttles. His work there lasted until November 1961 when he was dismissed for doing consultancy work for Egypt's President Gamar Abdel Nasser, whom he was helping to design a rocket that might attack Israel. He then emigrated to Egypt where he continued to work on Egyptian rocket projects with several hundred other Germans, including Peenemünde veterans, in the secret Project 333. Sänger died of a heart attack in February 1964 at 58 years, his wife Irene Bredt living on to the age of 72 until 1983. Despite his dismissal by the German government, he was later posthumously rehabilitated when Germany's space shuttle project was named after him.

When he went into exile in 1961, the French and British governments signalled to the Germans that they would be happy to see a reinvolvement of Germany in practical European rocketry, suggesting a role for Germany in the third stage of the prospective European rocket. The architect of the restored German role was *Raketenflugplatz* veteran Rolf Engel. He had helped the French get started after the war and he returned to head up the new ASAT company formed by Erno in Bremen and Bölkow [61].

A key step in the re-establishment of rocketry in Germany was the setting up in 1959 of the Lampoldshausen research centre, north-east of Stuttgart, with its associated Institute for Chemical Rocket Propulsion. The founder was – no surprise – the veteran rocket designer Eugen Sänger and the centre became operational in 1962. It was there that the third stage of the *Europa* rocket was duly tested in the 1960s (though judging by the results, not enough). Eight test stands were subsequently set up there, called P1 to P8, for anything from small rocket engines (P1) to upper stages with storable fuels (P2), cryogenic engines (P3) and main stages with storable propellants (P4). In addition to testing classic rocket engines, the centre also carried out research into spaceplanes and future space transport systems. Later, the main engines for the *Ariane* – Vulcain and Aestus – were tested in Lampoldshausen. Two hundred people now work there.

Not until Dr Gerhard Stoltenberg became Minister for Scientific Research was the decision taken to move investment in space research substantially ahead through the use of parliamentary grants. Germany embarked on a programme of small scientific satellites – *Azur*, *Dial*, *Helios*, *Aerosat* and later *Firewheel*. *Helios* broke new ground by being the first spacecraft to fly toward the Sun.

Germany's first probe was *Azur*, orbited on an American *Scout* in 1969, with seven instruments to monitor the radiation belts and solar particles in a Sun-synchronous orbit to study solar particle fluxes and their effects on the Earth's radiation belts. The purpose of *Dial*, which was launched by France in 1970, was to study the ionosphere. The first *Aerosat*, also called *Aeros*, was launched 16 December 1972 on a *Scout* from Vandenberg. Weighing 127 kg, it carried five experiments for the study of electrons and ions, neutral particles, ultraviolet solar

Figure 3.24. Engine testing in Lampoldshausen.

radiation, the temperature of neutral gas and atmospheric drag. Costing €13m, it entered an initial orbit of 250–1,000 km, 97.2°, 95 min, with the perigee later raised to 600 km. *Aeros 2* studied solar ultraviolet radiation and the Earth's ionosphere.

Germany's most ambitious early probes were *Helios 1* and *2* (1975–6). These entered solar orbit, coming as close as Mercury. Each weighed 370 kg and were designed to survive heating of up to 165°C. A new 30-m tracking dish was completed in Weilheim, Germany, to follow the mission. Each carried a 32-m dipole as well as magnetometers, a photometer and radiation counters. *Helios 1* approached 48 million km from the Sun and had instruments to study the solar wind and the surface of the Sun as well as cosmic rays and the zodiacal light. *Helios 2* came even closer, 45 million km from the Sun.

Firewheel was a spacecraft designed by the Max Planck Institute to investigate the ionosphere, but it was lost in the explosion of the second flight of the *Ariane* in May 1980. The work was resumed on *AMPTE* (Active Magnetosphere Particle Trace Explorer), launched by *Delta* from Cape Canaveral on 16 August 1984 (*AMPTE* also carried a small 74-kg, 1-m, drum-shaped diameter, British subsatellite called *UKS*, which operated for five months). *AMPTE* was a German–American mission. On 27 December 1984 *AMPTE* released an artificial comet trail 49,600 km long. This shone for 20 minutes at an altitude of 100,000 km over the Pacific.

In 1976 there was a change of policy: Germany decided to focus exclusively on participation in ESA and discontinued its national space programme entirely. From 1976 to 1989 there was no national space programme or even bilateral projects. In ESA Germany played a lead role in the *Spacelab* programme and a number of other key projects. All this was done without a national space agency, such a body not being established until 1989 as DARA (Deutsche Agentur für Raumschifffahrtange-legenheiten).

The eventual formation of DARA in April 1989 heralded the reinstitution of a national space programme. However, it was badly timed. With the unification of Germany, budgets became extremely tight and the country cut back its space spending. The period of retrenchment was at its most severe in the early to mid-1990s. The ESA contribution fell from €794m in 1994 to €662m three years later in 1997. Germany effectively pulled out of a number of European projects, like *Hermes*, to the great distress of the French. Spending on the now-restored national programme also declined, from €301m in 1994 to less than half that, €138m in 1999.

Germany resumed the orbiting of its own payloads. *TUBSAT* was the name for a series of minisatellites developed by the Technical University of Berlin (hence TUB). The first was a 35-kg, store-and-forward electronic mail satellite flown piggyback with *ERS 1* on *Ariane* in July 1991. The second was carried into orbit with the Russian *Meteor 3-6* in January 1994. The third and the fourth were the most remarkable and had the distinction of being the first satellites to be orbited from a submarine. The Russians were testing out the possibility of using old Cold War missiles, launched from submarines, as a means of putting small satellites into orbit. The Technical University of Berlin offered two microsatellites, weighing 8.5 kg and 3 kg, respectively. The submarine *Novomoskovsky* duly dived under the Barents Sea. From the shore, observers could see the rocket break the surface, shake off the cold seawater and curve upward into orbit, placing the two little satellites in an orbit of 401–777 km, 78.9°. On the American side, the shuttle STS 60 launched the small 63-kg minisatellite *Bremsat* in February 1994 to measure dust and micro-meteroids. It was named after the University of Bremen where it was made.

TUBSAT's submarine launch was not the only unusual way to put satellites in space. *GFZ* was delivered to the *Mir* space station by the *Progress M-27* freighter spacecraft, where it was then hand-launched by one of *Mir*'s cosmonauts by pushing it out through the airlock! *GFZ* stands for GeoForschungsZentrum (Geological Research Centre), which is in Potsdam. In 1994 the centre commissioned the Kaiser Threde company to build a laser-reflecting geodetic satellite. Weighing only 20 kg and about 21 cm in diameter, it was tracked in two places: Potsdam and the Cuban National Centre for Seismological Investigations in Santiago de Cuba.

Germany made an important contribution to the *Galileo* spacecraft that orbited Jupiter from 1995 to 2003. *Galileo* was one of the most advanced and ambitious spacecraft of its day, its mission being to explore the Jovian system, fly past its moons and drop a probe into the atmosphere. Although a NASA project, the crucial task of Jupiter orbit insertion fell to Messerschmitt Bölkow Blohm (MBB), which was responsible for the engine. Substantial thrust was required and the Ottobrunn plant built a main engine and 13 manoeuvring engines. No less than 850 kg of propellant was loaded: monomethyl hydrazine and nitrogen tetroxide.

This was the first time such a major component on a deep space American spacecraft had gone to an external contractor, but MBB's engines were known for their power and reliability. When *Galileo* reached Jupiter on 7 December 1995, the main engine fired for 49 minutes, putting the probe into the intended orbit, the first spacecraft to orbit the largest planet in our solar system. On 14 March 1996 the engine fired for a further 24 minutes to raise the orbit clear of the planet's deadly radiation belts. Of the 925 kg of propellant loaded at the start, 105 kg was still left.

German scientific satellites

8 November 1969	*Azur*	*Scout*	Vandenberg
10 March 1970	*Dial*	*Diamant B*	Kourou
16 December 1972	*Aeros 1*	*Scout*	Vandenberg
16 July 1974	*Aeros 2*	*Scout*	Vandenberg
10 December 1974	*Helios 1*	*Titan/Centaur*	Cape Canaveral
15 January 1976	*Helios 2*	*Titan/Centaur*	Cape Canaveral
23 May 1980	*Firewheel*	*Ariane 1*	Kourou (lost)
17 July 1991	*TUBSAT 1*	*Ariane 40*	Kourou
25 January 1994	*TUBSAT 2*	*Tsyklon 3*	Plesetsk
9 February 1994	*Bremsat*	Shuttle STS-60	Cape Canaveral
7 July 1998	*TUBSAT* N1,2	*Shtil*	Barents Sea
28 April 1999	*Abrixas*	*Cosmos 3M*	Kapustin Yar
15 July 2000	*CHAMP*	*Cosmos 3M*	Plesetsk
	Rubin		

A small German astronomy satellite *Abrixas* was launched on 28 April 1999. The launching marked the reopening after 12 years of the Kapustin Yar cosmodrome, the original Russian launch site dating to 1947. It was a 470-kg satellite to make an all-sky X-ray survey. Sadly, although placed in the right orbit, the spacecraft failed after 3 hours and was lost.

On 15 July 2000, a *Cosmos 3M* launched the German *CHAMP* (Challenging Minisatellite Payload) geophysics research satellite, made by Jena Optronik and GFZ of Potsdam and a *Rubin* micropayload (also an Italian microsatellite). The aim of the 522-kg *CHAMP* was to measure the Earth's magnetic field most precisely through the use of GPS, laser reflectors, ion-drift meter, star sensor, accelerometer and magnetometer – and an American instrument was carried to measure the choppiness of the seas.

In October 1997 Germany flew a 150-kg, 1 m across *MIRKA* capsule on the Russian *Foton 11* microgravity capsule. This was a 1-m cabin flown at the front of the standard, Vostok-type re-entry cabin on its 10-day mission. It was ejected separately before retrofire, making its own re-entry before recovery on the steppes of southern Russia. The mission successfully tested heatshields, pyrometers and ablative material, and inside the *MIRKA* there were experiments into crystal growth, human cells, cancers, flies and beetles.

3.31 *ROSAT*

ROSAT was one of the most important multinational space science projects and the European end was led by Germany. *ROSAT* acquired its name because it stands for Roentgensatellit (X-ray satellite; X-rays are called Roentgen rays in German). *ROSAT* was a German–British–American project: the two main instruments were a 80-cm, German, X-ray telescope and British wide-field camera. It was box-shaped, 2.15 m by 2.4 m by 4.5 m, with solar panels. The principal contractor was Dornier. The main funders were the German Federal Ministry for Research and Technology and NASA. Observing time was allocated 50% to the United States, 38% to Germany and 12% to Britain. The spacecraft weighed 2,426 kg. *ROSAT* was controlled from the German Space Operations Centre in Oberpfaffenhofen (about 48 minutes contact per day) and tracked by the American Deep Space Network.

ROSAT was launched 1 June 1990 by a *Delta II* from Pad 17 at Cape Canaveral Air Force Station. A special fairing was fitted to the *Delta* to accommodate the astronomical satellite, dropping off 4 minutes 43 seconds into the mission. *ROSAT* separated 43 minutes into the mission for two weeks of checkout and entered a precise orbit of 580 km by 584 km, 53°.

The objective of *ROSAT* was to carry out a six-month all-sky survey of X-ray sources picking up objects of twice the magnitude of previous experiments. Objects to be studied were neutron stars, supernova remnants, black holes, quasars and galaxies. This was the first survey of its kind and *ROSAT* managed to map 80,000 such sources, of which 6,000 were studied individually. 500 were found in the Andromeda galaxy alone. *ROSAT* achieved a rich scientific haul, with the detection of X-rays from supernovae and neutron stars, also picking up X-ray transmissions from the Moon, comets and the impact of comet Shoemaker-Levy on Jupiter. Some of the instruments began to degrade after the initial survey, but *ROSAT* continued to provide useful information. By the time it had completed its work six years later, *ROSAT*'s data had contributed to a record 2,400 publications in scientific journals.

When comet Hyakutake swept into the inner solar system in March 1996, astronomers requested some viewing time to see if X-rays could be picked up from the comet. They had not expected to find anything, possibly some dim readings. *ROSAT* took nine 2,000-sec exposures of the comet with its microchannel plate detector over a period of 30 hours. To their astonishment scientists received pictures of a crescent-shaped body of emissions spreading 30,000 km into space with an intensity about a hundred times greater than their theoretical expectations. This was the first time X-rays had been picked up from a comet [62].

3.32   GERMAN COMMUNICATIONS SATELLITES

Germany developed two communications satellites, *TV-Sat* and *DFS Copernicus*, through its state, later privatized telephone company Deutsche Telekom. Germany decided to study a direct broadcasting television satellite, and in 1979 the Federal Ministry of Research and Communications awarded a €3m study to MBB, AEG Telefunken, Dornier, ERNO and SEL to examine the possibilities of a geostationary comsat able to transmit three TV programmes into every German home with a dish of 60 cm to 90 cm. The outcome was *TV-Sat*, intended to be Europe's first direct broadcast television service, sending four channels down to people with 35-cm home receivers. *TV-Sat 1* was launched by *Ariane* on 20 November 1987, but ran quickly into trouble when one of the two solar panels failed to deploy. Desperate attempts were made to use thruster firings to shake it free, but it had to be abandoned three months later. The German telecommunications company Deutsche Telekom later ordered three communications satellites in 1983 called *DFS* (Deutsche FermeldeSatellit), able to carry television, radio, telephone, voice, fax and data links.

Domestic German communications satellites

20 November 1987	*TV-Sat 1*	*Ariane 2*
5 June 1989	*DFS-1*	*Ariane 44LP*
8 August 1989	*TV-Sat 2*	*Ariane 44L*
24 July 1990	*DFS-2*	*Ariane 44L*

3.33   GERMAN SPACE INFRASTRUCTURE

Despite the low level of purely domestic space activities, Germany has an extensive space infrastructure. Direct employment in the space industry in Germany is 12,400 people, with space comprising about 11.5% of the whole aerospace sector and total product valued at €1.3bn.

German space policies are decided by the Federal Ministry for Research and Technology, with interministerial coordination being achieved by a Cabinet sub-committee on space activities chaired by the Chancellor, showing its importance. From 1989 to 1997 German space activities were coordinated by its space agency DARA, but this was then merged with the German Aerospace Centre. This is the Deutsches Zentrum für Luft und Raumfahrt (DLR) (German Centre for Air and Space Travel). DLR has 1,600 scientists and 2,900 other staff in Berlin, Braunschweig, Göttingen, Cologne, Lampoldshausen, Oberpfaffenhofen and Stuttgart. The centres in Braunschweig, Stuttgart and Göttingen, for example, host a range of institutes dealing with such issues as fluid mechanics, aeroelasticity, design aerodynamics, combustion mechanics and flight mechanics. It is worth noting that DLR's brief is not just space, but like NASA aeronautics (the first 'A' in NASA) as well as energy and transport research. DLR has a budget of €350m and its work is divided into the broad areas of telecommunications, Earth observing, materials, mechanics and guidance and control. DLR's facilities include wind tunnels, rocket test stands, simulators and astronaut training.

DLR makes an interesting comparison with the BNSC. Whereas the BNSC coordinates, facilitates, promotes and enables, DLR holds an empire of facilities together assisted by cooperating and advisory bodies. DLR says: 'Our mission is quite clear: we are Germany's centre for research and technological development in aerospace: we operate large-scale test facilities and infrastructures and we shape space activities on behalf of the federal government in Germany and Europe' (from *DLR Research Enterprise – goals and strategies*, 1999). Germany's space interests are broad, ranging from the ISS (Institute for Space Simulation) to *Ariane*, future transportation systems, applications (*Galileo* and *Envisat*) and astronomy (where, for example, Germany has an especially strong record in infrared astronomy).

German space spending in 2001 was €690m, of which €158m went to national programmes and €532m to ESA. Germany was – and remains – the second highest contributor to ESA, though the level of its funding has been falling back over the past decade (Germany's ESA contribution was higher in 1993, €592m).

The German space budget for the four years 2002–5 was set at up to €4bn in a four-year plan announced by German Federal Research Minister Edelgard Dulmahn in May 2001 [63] and, subject to the Report of the High Level Group on the German Aerospace Industry, was accepted by Chancellor Gerhard Schröder in September 2001. The programme set Germany's national programme at €150m a year, a level criticized as inadequate and unbalanced by the German Aerospace Industries Association (BDLI). Indeed, BDLI contrasted Germany's record unfavourably with France and Italy who, BDLI says, not only invest much more but also have a bigger domestic programme and are able to expand their leading positions in Europe.

Germany's main space development facilities are called the Industrie Anlagen Betriebs Gesselschaft, located in Ottobrunn, and these comprise an electromagnetic test facility, vibration test facility, acoustic chambers, thermal and vacuum room and instruments for simulating magnetic fields. Germany inherited the developments in spaceflight that had taken place in the German Democratic Republic (GDR). Instruments from the GDR had flown on many of the *Intercosmos* satellites, as well as the *Venera 15/16* Venus radar-mappers, *Phobos* and *Vega* probes. The GDR had continued the fine tradition of optical and camera equipment that was long established in the region. Multispectral cameras from the GDR had flown on *Soyuz 22*, *Salyut 6* and *7* and *Mir*. Germany absorbed in 1991 the old institute for space research, the Institut für Kosmosforschung and with it the commitments into which it had entered, including *Mars 96*, *Okean*, *Mir* and *Koronas*.

A close accord on space development was signed up by France and Germany at the 1995 Franco-German summit between Helmut Kohl and Jacques Chirac. Under a memorandum of understanding agreed at the summit, Germany agreed to join the French proposal for a European military surveillance satellite, and there was agreement for the merger of Aérospatiale and DASA into a new Franco-German conglomerate: EADS. The first part of the deal did not stick, for in 1998 Germany had second thoughts about *Helios 2* and withdrew from the project. EADS is made up, on the German side, by Dornier, Erno and MBB. It has three facilities: Friedrichshafen, which makes *Ariane 5* parts and satellites such as *Envisat*; Ottobrunn (Munich), which made *Cassini* and *SPAS*; and Bremen, which made *Spacelab*, parts of *Ariane 1–4*, *Eureca* and then moved on to *Columbus* [64].

3.34   GERMAN MISSION CONTROL (GSOC) AND EUROPEAN MISSION CONTROL (ESOC)

Germany hosts a number of ESA facilities, the most important being ESOC in Darmstadt and the European Astronauts Training Centre in Cologne. Oberpfaffenhofen, 25 km west of the Bavarian capital of Munich on the road to Lindau, is best known as Germany's mission control, GSOC. It is 300 km to the south of Darmstadt and has the Alps to the south. About 1,100 people work at mission control.

GSOC was set up in 1968 and its first missions in the 1970s were *Helios*, *Symphonie* and *AMPTE*. In the 1980s it concentrated on the *Spacelab D* missions before moving on to *ROSAT* and *Galileo*. Signals are transmitted via the nearby Weilheim station, which comprises two 15-m antennae and one 30-m dish (used for *Galileo*). The typical work of GSOC comprises monitoring satellites, sending up command sequences for 24 hours or 48 hours at a time and real-time control. In the German Space Operations Centre in Oberpfaffenhofen, Germany has the third manned mission control centre in the world (followed by Beijing) where it controlled

Figure 3.25. Mission control, Germany.

Figure 3.26. Astronaut training, Cologne.

the work of *Spacelabs D-1* and *D-2* and then the two *EuroMir 95* missions. GSOC will be ground control for *Columbus*. Oberpfaffenhofen has been expanded at the cost of €260m to provide a mission control facility for European operations on the International Space Station. In addition to mission control, the Centre also houses:

- Office for Innovations and Technology, to encourage small- and medium-sized firms to engage in space industries;
- Institute of Communication & Navigation (e.g., development of the *Galileo* system);
- Institute of Atmospheric Physics;
- Institute of Robotics & Mechanics;
- Remote Sensing Data Centre;
- Remote Sensing Technology Institute;
- Microwaves and Radar Institute.

Cologne is best known as the host centre for the European Astronaut Corps. Cologne-Porz is also a DLR centre, hosting the Institute of Aviation Medicine, the Institute for Space Simulation and the lowest temperature wind tunnel in Europe. The ISS, for example, can recreate the environment of the surface of Mars or the surface of a comet (e.g., to test the *Rosetta* lander) and individual experiments to be flown on space missions (e.g., the *Spacelab D* missions or *Mir*).

ESOC, in Darmstadt, actually dates to the ESRO days and was set up in 1967, becoming operational in 1968. Two 15-m dishes were inaugurated in Odenwald in September 1976, directly linked to Darmstadt by land lines. About 650 people work there, coming from all the ESA nationalities, controlling possibly 10 missions at a time. The main control room is not as massive as those to be found in TsUP Moscow

Figure 3.27. ESOC mission control, Darmstadt.

or mission control in Beijing, ESOC having semi-circular banks of consoles rather than forward-facing rows. ESOC is a complex of offices, computer, control and command centres, employing 270 permanent and 500 temporary staff. It is the hub of the European tracking system, which has stations in seven other locations.

European tracking stations

Redu	Belgium	Perth	Australia
Villafranca	Spain	Kiruna	Sweden
Kourou	French Guiana	Darmstadt	Germany
Malindi	Kenya	Maspalomas	Canaries

Among the projects given priority in the domestic programme are *TerraSAR*, a commercial small radar imaging programme by Astrium; *RapidEye*, a smallsat imaging programme; *Comed*, a satellite-based broadband technology programme and an astrometric programme called *Diva*. The €340m *TerraSAR* project is for a 1,023-kg spacecraft to be launched on a Russian *Dnepr* into 500-km polar orbit. It is

5 m high, 2.3 m diameter and with a 4.8 m long radar. The images will be marketed commercially by Astrium.

For the future, Germany plans to develop and fly five *SAR-Lupe* military imaging satellites between 2005 and 2007. The Germans, apparently, were sore about the reluctance of the Americans to provide timely radar-based data during the Balkan wars of the 1990s and don't plan to be left without their own good intelligence again in the future. The contract for building the *SAR-Lupe* system was signed between the federal government and the Bremen firm of OHB in December 2001. Each will weigh 700 kg and the radar will be able to pick up all military objects 1 m across or bigger. *SAR-Lupe* will be a big breakthrough for Europe, for its military systems to date have all been optical and their abilities are limited in the dark or in cloudy weather. There appears to be an understanding that the French will continue to develop the optical *Helios* system while Germany will develop radar-based systems. The launcher is expected to be the Russian *Dnepr*.

3.35 EUROCKOT

Germany's involvement in rocketry has been limited to the third stage of the *Europa* and, later, the third stage of *Ariane*. In the 1990s Germany expanded its interests in launchers when the German company Daimler-Benz Aerospace entered a strategic alliance with Khrunichev, the owners of the Russian *Rockot* launcher, to form the Eurockot Consortium. A linguistic note is that the word *Rockot* means 'roar of sound', not 'rocket' (*raket* in Russian).

Rockot is a small, 30 m tall missile, deployed in silos around Russia and Ukraine at the height of the Cold War. *Rockot* was the first of the Cold War missiles to be adapted for civilian missions. Suborbital tests were made at Baykonur in 1990 and 1991 before the first orbital mission in 1994. For commercial applications, a new third stage called *Briz K* was fitted to enable it to put 2-tonne payloads into Earth orbit. *Briz K* weighed three tonnes, twice that when fuelled. This was a versatile stage that could fly for seven hours and restart six times, dispensing different satellites into different orbits. With *Briz K*, *Rockot* can place up to 1.85 tonnes in low Earth orbit.

Rockot was originally stationed at pads in the north-western part of Baykonur cosmodrome in Kazakhstan. They were all underground silos, and this presented a number of problems. The noise circulating in the silo during the first two or three seconds of lift-off was tremendous: there were fears that it would damage the satellite payload. At the same time as the Russian Space Agency was considering this problem, the Kazakh authorities got wind of the proposal and demanded a share of the profits of every *Rockot* mission out of Baykonur, with the result that the Russians threatened to move all *Rockot* launches to Plesetsk in Northern Russia. Which is exactly what they did. Accordingly, with Eurockot, they built an above-ground, ground-based, vertical support structure. This was a procedure familiar to most space programmes the world over (e.g., the United States, Japan, India, China), but was new to the rail-based Russians.

In 1995 the Eurockot Consortium began to market *Rockot*, offering it at €7m a launch. Within five years it had built up an order book of 12 launches for €200m. So confident was Eurockot that the company bought 45 old *Rockot*s from the Russian strategic missile forces so as to build up its inventory. In 2000 Eurockot was part-bought in turn by the German company Astrium GH, a shareholder of Arianespace (51%) and Khrunichev (49%). In May 2002 *Rockot* put into orbit an American double satellite called *GRACE* to measure small changes in the Earth's gravity. Eurockot won further contracts for putting into orbit the Japanese *SERVIS* space applications platform two *Iridium* satellites, the Canadian *MOST* space telescope and the Czech *Mimosa* upper atmosphere satellite. The project seemed to have an assured future.

3.36   THE NETHERLANDS NATIONAL PROGRAMME

If we take European space activities as a whole, France, Germany and Italy may be regarded as the 'big three' (Britain would have made it four in the early days). There is then a group of about three middle-league nations: the Netherlands, Sweden and now Britain. They are reviewed next (Britain has already been reported).

Astronautics has a long and honourable history in the Netherlands. The Netherlands Institute for Aeroplane Development (NIV in Dutch) was formed in 1947, becoming, following reorganization, NIVR (Nederlands Instituut voor Vliegtuigontwikkeling en Ruimtevaart [Netherlands Agency for Aerospace Programmes]) in 1971. NIVR was and remains the Dutch representative on the ESA and the main national space agency, promotional and governmental advisory body, the equivalent of DLR, CNES or ASI.

A national Geophysics & Space Research Committee (in Dutch GROC) was established in 1960, and this encouraged an early Dutch interest in the emerging European space programme. The Netherlands joined both ESRO and ELDO at the very start, asking for and getting one of the main ground facilities ESTEC (European Space Technology Centre), in the first share-out of sites. Originally, ESTEC was located in Delft, but doubts began to creep in about the stability of the soil there. A new site in Noordwijk was selected in 1965 and opened 1967, now extended to a large block of offices and facilities with a picturesque inland lake, adjacent to the dunes of the North Sea.

Nowadays, ESTEC is ESA's largest single site and the largest establishment for the testing of satellites in Europe. Its staff are responsible for feasibility studies, project management, the development of scientific projects and the testing of spacecraft equipment. ESTEC is equipped with a large simulator, thermal chambers, electrodynamics shakers, acoustic room and radiation facility. Here satellites are put through the range of tests necessary to ensure that they will survive the space environment: vacuum, extremes of hot and cold, vibration and noise. About a thousand ESA personnel work there from all the different ESA nationalities along

Figure 3.28. ERS picture of the Netherlands.

with about 500 Dutch people. Although it is best known for the testing facility, several other important parts of ESA are located there: technical support for ESA project teams, new technologies, quality control, the space science department and project management (except *Ariane*). Manned spaceflight is the most recent directorate to be set up in ESTEC. A new hydraulic testing facility called HYDRA was introduced in 2000, *Envisat* being the first satellite subjected to its rigours.

In addition to hosting ESTEC, the Netherlands contributed ground links for the *Europa 1* launcher. Philips developed its telemetry system. The NLR provided wind tunnel facilities. When *ESRO 1* was announced, the Netherlands' bid to build the whole satellite was unsuccessful: in the event, the contract went to Italy, but the Netherlands contributed the attitude control system (Philips and Fokker).

3.37   THE IDEA OF A DUTCH EARTH SATELLITE

By the mid-1960s, the Netherlands were still unhappy about the level of ESRO and ELDO business coming their way. This they felt was due to their lack of appropriate experience and track record in the area. They took the courageous decision to try and remedy the problem on their own and prove their capabilities to the world. In 1965 Philips, Fokker and GROC made the proposal that the Dutch should build their own satellite, and over 1967–9 the Dutch government came to approve the building of a national satellite for €80m.

The first Dutch satellite was *ANS 1* (*Astronomische Nederlandse Satelliet* [Astronomical Netherlands Satellite]), built to study stellar, cosmic, ultraviolet and X-ray radiation with one ultraviolet and two X-ray telescopes. The project director was the Head of Fokker Space, Prof. W Bloemendal, and the two main companies involved in its production were Fokker and Philips and, on the experimental side, the University of Utrecht.

ANS was duly launched by an American *Scout* rocket from Vandenberg Air Force Base in California. *ANS* was a small satellite, only 133 kg in weight, 1.123 m high, 73 cm across and with 2,000 solar cells on two panels. It was put into a near-polar, 258-km by 1,173-km, 98° orbit where it carried its X-ray telescope, X-ray detectors and photometer to study hot young stars and X-rays from distant sources. It quickly found an X-ray source in the constellation of Scorpio. The satellite was designed to operate for six months, but lasted 15 months in the end. It was then turned off on government order, even though it was working perfectly well and it eventually burned up in 1977. By this stage, though, the Netherlands had proved their ability in the field of space research and the level of ESA contracts duly increased, as they had hoped.

3.38   *IRAS*, ONE OF THE GREAT SUCCESS STORIES

The success of *ANS* paved the way for Dutch interest in what became the Infra Red Astronomy Satellite *IRAS*. The idea of a successor mission to *ANS* was considered in 1973 even as *ANS* was readied for launch: it was to have been a modest scientific follow-on called *ANS-B*. The project received government financial support in 1975 (€1.45m) and the design was finalized the following year. *ANS-B* tried to build on *ANS 1* through developing infrared observations. The vision of an infrared telescope in space was the idea of R. J. van Duinen, then head of the space exploration group at Groningen University. He was aware of what balloon-based infrared observations could achieve, but was confident that much more could be done through satellite-based observations. The Americans and the British later joined the project, making it a three-sided venture and a memorandum between them was agreed in 1977, NIVR having overall responsibility. Attempts to involve the ESA did not work out. The aspect that proved the most difficult was the devising of coolant for the telescope, a challenge the Americans took on but one that delayed the project by about two years.

Figure 3.29. Pioneering space astronomy.

The aim of the *IRAS* project was to make the first infrared astronomical map of the sky. The Earth's atmosphere blocks out infrared rays, so they can only be detected from space (either sounding rockets or satellite orbits). A large amount of the energy of stars is emitted at the infrared end, especially stars during the early stage of formation. Detecting them presents a special problem, for infrared-measuring instruments create their own heat radiation. An infrared telescope must therefore be cooled close to −273°C so as to eliminate its own sources of heat that might confuse observations. The big challenge for *IRAS* and similar infrared tele-scopes is to keep them cool. They can only detect these dim objects by picking up traces of their heat in their own ultra-cool instruments – which also meant that *IRAS* had to be pointed away from the Moon, Earth and Sun all the time. Accordingly, the telescope was cooled to a temperature of −271°C, or two degrees above absolute zero (+2°K). By picking up the faint heat of objects far away, *IRAS* would be able to see stars, dust clouds, galaxies 32 million light years away, find the centre of the Milky Way and even find small planets in the solar system. Sixty per cent of the mission would be devoted to mapping the infrared sky (98% to be covered) and 40% to dedicated observing programmes.

IRAS was a stout cylinder: the bottom part held the spacecraft systems, the top part the 60-cm telescope with an angled lens. Its double wall was cooled by liquid helium as close to absolute zero as possible. The data collection demands on the mission were quite demanding, with up to 500 million bits of data collected each day. If all went well, infrared information would be collected that was a thousand times more sensitive than anything done before.

The *IRAS* project was supervised by the NIVR with Fokker as the main constructor. Britain provided the tracking system at Rutherford Appleton Laboratory. The telescope was designed and built by the Jet Propulsion Laboratory in Pasadena, California. The rest of the spacecraft was designed and built by the Dutch National Aerospace Laboratory. *IRAS* weighed just over a tonne, 1,076 kg, was 3.57 m in height and 1.06 m diameter and had power of 350 W. The instruments were a low-resolution spectrometer and a 5.5 m long, 57 cm aperture 810 Richey Chrétien telescope [65]. *IRAS* carried a telescope with a primary mirror with 62 detectors and an accuracy of 62 seconds of arc. Electrical energy came from two solar panels on its back with a width of 3.4 m able to generate 250 W.

IRAS was launched 26 January 1983 on a *Delta II* provided by the United States from Vandenberg Air Force Base in California. The *Delta* went straight into clouds, lighting up the cloud deck for observers underneath, and 240 km away observers at the Jet Propulsion Laboratory saw the second stage ignite as it headed toward orbit. Signals picked up on the 12-m dish at the Rutherford and Appleton Laboratory near Chilton in Oxfordshire took in twice daily data dumps and these indicated all was well. Orbit was 896–913 km, 103.1 min, 99.1°. Next, *IRAS* dropped its protective cover on 31 January so as to get first light and detected 4,000 infrared sources on its first day. The level of helium remaining on that date indicated sufficient for a 250-day mission, enough time to cover the whole sky.

In April 1983 *IRAS* found a comet that was later named *IRAS*–Araki–Alcock after itself and two other astronomers who found it at around the same time, Genichio Araki and G.E.D. Alcock. It was intended that *IRAS* would scan each segment of the sky at least twice so as to eliminate any moving objects (mainly asteroids), but on 26 April ground control picked up a fast-moving fuzzy object and a global alert was sent out to ground watchers. The comet later passed within 4.5 million km of the Earth.

The following month *IRAS* on its own found a second, fainter comet, which was duly and simply named Comet *IRAS* (1983f). Later, *IRAS* found that comet Tempel, supposedly tailless, in fact had a tail 32 million km long. No tail had ever been picked up on visible photographs, so this discovery was a vindication of what infrared observations could achieve. In June it found a third comet, *IRAS 2*, at a magnitude of +16.5. *IRAS* found another two. Going through *IRAS* data, scientists identified 50 stars that were candidates for having planetary systems.

On 26 August 1983 *IRAS* completed its first full sky survey and began its second one to reverify the first round of data. By this stage JPL (Jet Propulsion Laboratory) had begun to put together a new 250,000 object sky catalogue of the objects identified and mapped by *IRAS*.

IRAS's 475 litres of helium began to run low after five months. This did not happen before *IRAS* had returned a huge amount of data, detecting small infrared objects as far away as 32,000 light years and providing astonishing images of the formation of new stars and material that later will turn into stars (protostars). *IRAS* ran out of fuel on 22 November 1983 after 300 days (as predicted), by which time it had made an almost complete survey of the sky, noting 200,000 infrared objects. Ninety-five per cent of the sky was covered four times and 72% six times, giving a high rate of accuracy. At a press conference that month, American, British and Dutch investigators presented the preliminary findings. *IRAS* observed up to 25,000 galaxies. It provided detail on the centre of our own galaxy, the Milky Way. It found that one new sun-like star per year was being formed in our own galaxy. *IRAS* found graphite dust clouds in interstellar space. It had found a ring of solid material around the star Vega, possibly where planetoids were being formed, seven comets and dust bands between Mars and Jupiter. There was intense infrared radiation in Andromeda, indicative of early star formation. Within the solar system, *IRAS* found an asteroid 1983TB that was probably the cometary remnant of the Geminid meteors. In addition, three dust rings were found between the asteroid belt and Jupiter.

The catalogue was duly published in 1984 with details of 245,839 infrared sources that scientists could obtain either on paper (400 pages) or two magnetic tapes. These listed the position and brightness of each object. The project-managers claimed an accuracy of one part in 100,000, having eliminated all possible intruders, but undertook to reprocess the data to refine the accuracy even more. Production of such a detailed outcome within three years of the start of the mission may well have been a record. The following year the California Institute of Technology (CalTech), the parent body of the JPL, constructed a building to house the *IRAS* archive. There scientists would be invited to review the archive for astronomical research projects. Even while this was being done, *IRAS* had an unexpected postscript. Of the eliminated objects 1,811 were asteroids, half the then-known total, including four discovered by *IRAS* itself. This enabled a separate asteroid catalogue to be compiled [66]. The haul from *IRAS* was a rich one, with the publication of spectacular false-colour images of the infrared universe [67].

3.39 NETHERLANDS OPT OUT OF NATIONAL *IRAS* SUCCESSORS

The Dutch considered several successors to the *IRAS* mission. These included *ARTISS* (Agricultural Real Time Imaging Satellite System) for agriculture in developing countries; *IWACXS* (Imaging Wide Angle Camera for X-ray Sources); *TIXTE* (Timing and Imaging X-ray Transient Explorer); and a tropical, Earth resources satellite with Indonesia. By the time these studies had concluded it was mid-1980 and the government had set its face against further such projects, deciding against a successor to *IRAS*. Instead, it preferred modest participation in international ESA missions in their place. In a policy document the following year, NIVR warned against this approach and that the Netherlands would lose their influence in ESA.

Thankfully, there was a way out when the Italians began to put together their new X-ray satellite called *SAX* (later, *Bepi SAX*). The Netherlands were made welcome to a number of feasibility studies over 1982–3 and in the end contributed 10% toward the *SAX* mission, building the control systems and solar panels (Fokker). The Netherlands contributed the wide-field camera that was successful in locating and positioning gamma ray bursts. As we know, this mission became a success (see Italy's national space programme).

Since then the largest single Dutch space project has been the *Sciamachy* (Scanning Imaging Absorption Spectrometer for Atmospheric Cartography) experiment lofted on *Envisat* in 2002. The only exclusively national project in the pipeline is the experimental small satellite called *Sloshat-Flevo*, a 115-kg, cube-shaped satellite to be released by the shuttle to study the sloshing of liquids in space (FLEVO stands for Facility for Liquid Experimentation and Verification in Orbit). To carry out these studies, it carries an 87-litre tank with 35-kg fuel, and instruments will watch how the fluids behave under different conditions. *Sloshat* was developed by the National Aerospace Laboratory, Fokker and the Belgian company Verhaert.

When ESA was formed in 1973–5, the Netherlands was a founder member, contributing to *Ariane*, *Telecom*, *Spacelab* and *MAROTS*. The Netherlands has subsequently participated in the major scientific programmes, and in the space station will take part in *Columbus*, the robotic arm and the data management system.

During the 1980s Dutch policy swung back from national programmes toward participation in European programmes, and on average 80% of the Dutch space budget went to ESA. The pendulum has since swung back a little: in 2001 the Dutch contributed €68m to ESA (historically between 3% and 4% of the total ESA budget) and €40m to national programmes (total €108m). About a thousand people are directly involved in the space industry in the Netherlands.

Several bodies govern space policy in the Netherlands: NIVR (the lead body), SRON (Stichting Ruimteonderzoek Nederland [Space Research Organization Netherlands], which funds scientific projects) and the Netherlands Industrial Space Organization (NISO), which brings together the industrial organizations in the field, all coordinated by the Interdepartmental Committee on Space Research and Technology. SRON has a budget for research in the areas of astronomy, astrophysics and Earth observations. There is a society, the Nederlandse Vereniging voor Ruimtevaart [the Netherlands Association for Astronautics], founded 1951, with a regular magazine *Ruimtevaart*.

The main agency for technological development is the NLR (Nationaal Lucht en Ruimtevaartlaboratorium [National Aerospace Laboratory]), which dates to 1919. It is a government-funded service, designed to aid both the public and private sector in space and aircraft development. Spaceflight is only one of its areas of work (the others include, for example, civil aviation, air traffic, avionics and military aviation) and here NLR is active in improving access to satellite remote-sensing through small remote terminals, developing the European Robotic Arm and testing the effects of heat on spaceplane surfaces.

About a hundred Dutch companies are involved, either directly or indirectly, in the space industry. The leading Dutch aerospace company is of course Fokker and there was shock in the aviation world when it went into receivership in 1996. Fokker Space, originally a division within Fokker, but later a separate company, has

Figure 3.30. Dutch astronaut Wubbo Ockels.

been an important contractor for space instrumentation and equipment, including components of the *Ariane 5*, solar panels (e.g., *IRAS*), the European Robotic Arm for the International Space Station, parts for the *Ariane 4* (thrust frame) and *5* (igniters for the Vulcain engine). Several Dutch companies and institutes are involved in the *Columbus* programme (especially the microgravity facility planned) and the Automated Transfer Vehicle. The Netherlands have contributed instruments toward *ERS 2* and *Envisat* and many other European projects.

The Netherlands is one of the few small countries to have managed to get astronauts selected for the European team. In 1977 NIVR was charged with the recruitment of the Netherlands' contribution to the first ESA astronaut selection, 192 people applying. Wubbo Ockels was chosen for the final ESA selection. Later, André Kuipers was chosen for a *Soyuz* taxi flight to the international space station.

Dutch or Dutch-led satellites

30 August 1974	*ANS* (Astronomical Netherlands Satellite)	*Scout*
25 January 1983	*IRAS* (with Britain and the United States)	*Delta*

3.40 SWEDEN

Swedish space research has strong theoretical roots (Swedish Interplanetary Society), a practical basis (ionospheric observations began in Kiruna in 1948) and an amateur role (rockets were fired from forest clearings in 1962). The first systematic rocket

Figure 3.31. ESRANGE balloons, rockets.

firings were made from a derelict farm called Kronogard over 1962–4, and it was the original engineers from this group who later generated and sustained the Swedish space effort thereafter [68].

Sweden's northerly location meant that it was ideally suited for research into the magnetosphere, ionosphere and aurorae. Sweden joined ESRO in 1962 and part of the arrangement was that Sweden opened a rocket range for ESRO appropriately called ESRANGE in 1966 (Sweden bought it back from ESRO in 1971). This was located in low, rolling, often snowy hills in Kiruna, in northern Sweden at 67°N. This was a town that had suffered from the decline of ore-mining and urgently needed diversification. Not only was the rocket range located there but so too were the Geophysical Institute, scientific radar systems, tracking systems and data-handling centres. Over 800 sounding rockets have now been fired from there in campaigns involving the different European countries, the United States and the Soviet Union.

These early activities led over time to the Swedish Space Corporation, Svenska Rymbola, founded in July 1972. The Corporation played an important role in directing Swedish space activities around a number of core areas that were suited to Sweden's location and economic situation [69]. Sweden joined ESA at its formation and decided to participate in the *L3S Ariane* rocket (Flygmotor built the combustion chambers, Saab the inertial guidance computer) while other companies became involved in *OTS* and *Meteosat*. Sweden opted to stay out of *Spacelab*, fearing it would not offer long-term opportunities. The Swedish Space Corporation is based in Solna, a suburb of Stockholm, but space activities happen in Kiruna.

3.41 A SWEDISH EARTH SATELLITE

Granted Sweden's early involvement in European space activities, it was no surprise that the idea of an early Swedish satellite came quickly onto the agenda. Sweden made its first satellite study in 1970. Ten years later the government decision was taken that Sweden should develop its own Earth satellite, modelled on NASA's atmospheric *Explorer*s. Swedish instruments had already flown on Soviet satellites (*Prognoz* and *Intercosmos 16*). Appropriately for a Nordic country the first satellite was named *Viking* and it had payloads for the study of electric and magnetic fields, particle and magnetospheric waves and the northern lights. Costs were kept down by the use of already-proven instruments and equipment. The Boeing company was invited to work with Saab to build the satellite. The costs of the mission were kept down by the use of already tested components and units.

   Viking was in the shape of a flat octagon, 550 kg in weight, 1.9 m long and 0.5 m high, with four 40-m wire booms for electrical field experiments. The €26m satellite was launched piggyback on *Ariane*, going into an 811–13,536-km, 98.7° orbit on 22 February 1986 on *V16* with *SPOT 1*. The satellite was designed to work for eight months, but in fact it worked for well over a year until May 1987 when the power supply gave out. *Viking* sent back over 20,000 images of the northern lights to the ground station in Kiruna, including images of what the northern lights look like from above.

Figure 3.32. *Freja* team at Jiuquan.

Sweden's second satellite was launched by the Chinese from Jiuquan launch site in October 1992, piggyback on the *Fanhui Shi Weixing 1-4* recoverable spacecraft. *Freja*, named after the Viking goddess of fertility, was a 259-kg spacecraft based on the *Viking* model with seven experiments to continue the study of aurorae, electrical and magnetic fields, particles, plasma, electrons and the magnetosphere, as well as carrying an experimental store-and-forward communications experiment called *Mailstar*. A small American upper stage was used to place it in a high orbit from where its seven experiments could study the northern hemisphere's aurorae, electrical and magnetic fields, particles, plasma, electrons and magnetosphere. Sweden paid China €5.05m for the lift. It also provided one of the first opportunities for Westerners to visit Jiuquan. The visiting Swedish team got a telephone line to their colleagues back home and gave an excited live commentary of the launching.

The *Astrid* satellites were scaled-down versions of *Freja* and named after the author Astrid Lindgren (the instruments were in turn named after the characters in her stories). *Astrid 1* weighed only 28 kg, cost €1m and took a piggyback ride on a *Cosmos 3M* from Plesetsk on 24 January 1995, whence its three experiments went on to study magnetospheric particles and aurorae. Its successor *Astrid 2* was slightly larger at 35 kg and carried four experiments. It was designed for long periods of unattended operation investigating the magnetic fields of the upper ionosphere, neutral and charged particles. *Astrid 2* was launched 10 December 1998, also on a *Cosmos 3M*.

Odin was a 250-kg atmospheric research and astronomy satellite launched from Svobodny cosmodrome in Siberia into its intended 98.7° polar orbit on 20 February 2001, using a converted *Topol Start-1* mobile launcher. *Odin* was 2 m tall, 1.1 m wide and with its solar panels spanned 3.8 m. It looked like a box, with a dish on top and flat-vaned solar arrays on the bottom. *Odin* was to study the atmosphere and ozone depletion and used a 1.1-m cooled telescope to study star formation as well. It followed in the tradition of *Viking* and *Freja* of versatile, small and low-cost spacecraft. *Odin*'s instruments made a detailed examination of comet Ikeya-Zhang when it visited the solar system in April 2002 and was able to calculate that the comet was releasing water at the rate of six tonnes a second. Sweden's next satellite project was to lead the European *SMART 1* Moon probe (see Chapter 10).

3.42   COMMUNICATIONS OVER LOW-DENSITY RURAL AREAS

A particular problem faced by the Nordic countries of Finland, Sweden and Norway is communications. Their relatively small, low-density populations are spread out over a vast area. The journey from northern to southern Sweden is, for example, the same distance as from Copenhagen to Rome. Such a distance offered an opportunity for the development of satellite technologies. This was something that these countries were quick to grasp and gave leading companies an opportunity to specialize in advanced communications. The first study on satellite communications was done in 1974. A feasibility study of an operational satellite for all the Nordic

Figure 3.33. *Odin.*

countries was done in 1977. This ran into many obstacles, but was salvaged as an experimental satellite in 1980 called *Tele-X*.

The idea of *Tele-X* was to develop direct broadcasting, satellite-based computer links, telephone services, mobile telex, picture telephones, small-terminal development and facsimile transmission of newspapers. *Tele-X* was a 2.3-tonne spacecraft with large solar panels measuring 20 m by 5 m. The idea was that this would pave the way for an operational system. It was mainly a Swedish project (82% of costs), but was also supported by Norway (15%) and Finland (3%). *Tele-X* was used for the transmission of public and private TV channels and for the publication of the newspaper *Aftonbladet*.

Despite the success of *Tele-X*, a united, Nordic operational system was slow to develop. Eventually, the Swedish Space Corporation took part with Luxembourg's Société Européenne des Satellites in the provision of telecommunications for the Nordic and Baltic regions through the *Sirius 1, 2* and *3* comsats. In a recent experiment the satellites were used to provide public service information for train stations, shopping malls and sports areas. In effect, *Sirius* became the operational system for the Nordic countries originally projected in the 1970s.

3.43   REMOTE-SENSING

Remote-sensing has also been a priority for Sweden, a country conscious of the fragility of its forests and waters. The first remote-sensing experiments were carried out on sounding rockets from ESRANGE in 1973. At the same time, a remote sensing station was set up in Kiruna, initially to receive American *Landsat* data, later to become the national remote-sensing centre.

Sweden was well disposed to support European projects in the area of Earth resources. When France proposed the *SPOT* programme to ESA in 1977, Sweden supported the idea (the first other country to do so). *SPOT* went ahead as a national French project, but because of its interest and support France broadened the programme to include Sweden (4%) who made the computer (Saab-Scania). A Swedish company was set up with *SPOT* to market the satellite's data (Satellitbild). In the course of time Sweden became France's largest bilateral partner, mainly in the areas of remote-sensing and telecommunications.

There was a significant expansion in Swedish space activities from 1979, when it was decided to treble the budget. The Swedish National Space Board presented a long-term plan in 1984 for the period to 1991, a plan that emphasized the importance of promoting space technology within Swedish industry. Volvo's aerospace division Volvo Flygmotor was a major subcontractor for the *Ariane*, making the combustion chamber in the *Viking* engines and for the *Ariane 5*. Instruments for scientific probes were built by IRF (Institut for Rymdfysik [Institute for Space Physics]). One of its most advanced research projects was with Germany, when Sweden spent €3.3m cooperating on the Saenger spaceplane design.

Sweden's space spending in 2001 was €71.6m, of which €55.3m went to ESA (about 2.6% of its budget) and bilateral projects, and €16.3m to the national programme. The Swedish space programme is executed by the Swedish Space Corporation (SSC), which operates under the Ministry for Industry and Trade. Policy advice is provided to the Ministry by the Rymdstryrelsen (Swedish National Space Board, also called the Swedish Board for Space Activities), set up in 1972, the body that represents Sweden on ESA.

Sweden operates a tracking and balloon-launching station at ESRANGE, which follows about 40 passes a day from satellites in polar orbit and has a 50% share in Norway's Tromsø station and its subsidiary in Svalbard (Spitzbergen). There is now an ESA station at Salmijarvi 6 km from Kiruna to follow *ERS* and *Envisat*.

Sweden has its own astronaut, but he has not been one of the lucky ones. Dr Christer Fuglesang was selected as an ESA astronaut in 1992 and trained in Moscow for the *EuroMir* missions. Ten years later he was still waiting hopefully for his first mission.

Sweden's satellites

22 February 1986	*Viking*	*Ariane 1*	Kourou
8 October 1992	*Freja*	*Long March 2C*	Jiuquan
24 January 1995	*Astrid 1*	*Cosmos 3M*	Plesetsk
19 December 1998	*Astrid 2*	*Cosmos 3M*	Plesetsk
20 February 2001	*Odin*	*Start*	Svobodny

3.44 SPAIN

These are the middle-league European national space programmes. Now we take a look at the later and smaller national programmes. Spain was not invited to join

ESA as a founder member, since the country was still ruled by General Franco. Spain already had by then a satellite in orbit, *Intasat 1*, a 20-kg ionospheric observatory built by INTA (Instituto Nacional de Técnica Aeroespacial [National Institute of Aerospace Technology]), launched piggyback on *NOAA 4* on 15 November 1974 by NASA on a *Delta*. It worked for two years sounding the ionosphere.

A domestic, Spanish, communications two-satellite system was approved in 1989, and a company was set up to manage the system and a communications satellite subsequently launched (*Hispasat*). *Hispasat 1A* was duly put in orbit by *Ariane 44LP* on 10 September 1992. This was a 2,194-kg satellite with three high-power television transponders and 10 telecommunications transponders. *Hispasat 1A* was located over the Atlantic, to cover Spain, the Balearic Islands, Canaries and Latin America with television, radio, telephone and data transmission as well as televised educational programmes. It was built by Matra Marconi in Toulouse with French and Spanish contractors. Three channels were reserved for the Spanish military. *Hispasat 1A* was followed by *1B*, *1C* and *1D*.

Spain's first entirely domestically built satellite was the educational microsatellite *UPM-Sat 1*, made by the University Polytechnic of Madrid (hence the UPM). It weighed only 47 kg, was box-shaped and rode piggyback into orbit on *Ariane 4* (V40) on 7 July 1995.

Spain's next satellite was called *Minisat*, and it was launched from the Orbital Sciences Corporation *Pegasus* system on 21 April 1997. The carrier plane for the mission was based at Torrejón Air Base in Spain, so this was the first orbital flight made from Europe. *Minisat* was built by the CASA company and was box-shaped with four solar paddles at the base. Weighing 558 kg, it entered an orbit of 560–581 km, where it operated a low-energy gamma ray emitter, extreme ultraviolet ray detector and a microgravity experiment.

Spain did later join ESA and Spain's space budget is €124m, of which €94m goes to ESA and €30m to national programmes. Spain's space programme is supervised by the Ministry for Industry and Energy but directed by the Centre for the Development of Industrial Technology (CDTI), with important parts managed by INTA, which goes back to 1942. Spain hosts a large 70-m dish for NASA's Deep Space Network, one which has been used to track *Voyager* and *Pioneer*, and provides emergency landing fields for the space shuttle in Morón and Zaragosa.

In its most ambitious venture, Spain put forward a proposal for a light rocket to be called *Capricornio*, to be launched from the Canary Islands. The idea of *Capricornio* was first mooted in 1992 by INTA as a three-stage, 15-m tall, 1-m diameter solid-fuel rocket able to send a maximum 100 kg into polar orbit up to 600 km high. With Europe's commitment to the Italian *Vega* rocket, it is uncertain if it can have much of a future.

Spain's most visible presence in space probably remains its sole astronaut, Pedro Duque. He went to Moscow in 1994–5 as back-up for the mission *EuroMir 95*, eventually making it into orbit on the shuttle in 1998 accompanying John Glenn.

Figure 3.34. Spain's *Hispasat*.

Figure 3.35. Pedro Duque on shuttle mission.

3.45 BELGIUM'S NATIONAL PROGRAMME

Belgium was a founder member of ELDO, ESRO and ESA. Most of Belgium's space budget goes to ESA (there is a national programme worth €30m), its contribution in 2001 being €140m. In recent years funding has shifted from ESA to the national programme. About 4,500 people work, either directly or indirectly, in the space industry in Belgium.

Belgium hosts the Redu tracking facility, and its main interests have been in weather satellite programmes (*Meteosat*), *Ariane* and *Envisat*. Belgium has a modern space facility in the Centre Spatial de Liège, part of the University of Liège, which has clean rooms, vacuum chambers and other test facilities. Belgium's participation in European space activities are determined by the Office for Science, Technology and Cultural Affairs.

Belgium collaborated with the Soviet Union and Russia on a number of projects, contributing instruments to *Phobos* and *Mars 96*. One of its most prominent experiments was done with Russia, during the mid-period of the *Mir* space station. Called *MIRAS* (*Mir* Infra Red Atmosphere Spectrometer) it was built by the Belgian Institute for Space Aeronomy. It was a €20m instrument affixed to the *Spektr* module to study changes in the chemical composition of the atmosphere. It had the ability to measure changes in 15 gases from an altitude of between 20 km and 120 km. *MIRAS* was 2.5 m long, weighed 235 kg and was

Figure 3.36. Belgium's *PROBA*.

attached to the outside of *Mir* by cosmonauts Anatoli Soloviev and Nikolai Budarin during a spacewalk on 21 July 1995. It was a bulky object and the cosmonauts used *Mir*'s Strela crane to fix it into position.

Belgium announced its first satellite project in 1998, *PROBA* (Project for On Board Autonomy). *PROBA* was Belgium's contribution to ESA's programme for developing new technologies, the cost to ESA being €13m. This was a small satellite, weighing only 100 kg, in the shape of an 80 cm × 60 cm × 60 cm cube – indeed, *Star Trek* fans would liken it to a Borg cube. *PROBA* carried three instruments: a high-resolution imaging spectrometer, a space radiation monitor and a debris in orbit evaluator. The cube was equipped with GPS receiver, sophisticated navigation devices and a computer system to enable it to operate with a high degree of autonomy in orbit. *PROBA* was duly launched by the Indian Polar Satellite

Figure 3.37. Dirk Frimout and Frank de Winne.

Launch Vehicle into 553–667-km orbit, 97.9°, 97 min, on 22 October 2001 from the Sriharikota range. Its small camera was soon sending back high-quality images to the ground station in Redu.

Belgium's first astronaut was Dirk Frimout, who flew as payload specialist on shuttle STS-45 in 1992. Physician Marianne Merchez was selected for the ESA astronaut squad in 1992, but retired without being allocated a flight. Later Frank de Winne was selected as an astronaut candidate and flew on *Soyuz TMA-1* to the International Space Station in October 2002.

Belgospace is the professional association for those industrial companies working in space activities, formed as a non-governmental organization in 1962. It includes Belgium's leading space companies like Verhaert and Sonaca. Several Belgium companies are involved in leading European space projects like the *Ariane 5* (Alcatel Belgium, Cegelec, SABCA), the ATV (Automatic Transfer Vehicle) and the European Robotic Arm (Spacebel).

3.46 LUXEMBOURG

Neighbouring Luxembourg is the smallest state in the European Union. It is not a member of ESA, yet despite that and its small size it is a prominent player in the European space business.

Luxembourg was one the first countries to formally use communications satellites. In some senses, this was not so improbable after all. Luxembourg was a

Figure 3.38. *Astra 1K* comsat.

country with a long tradition of international radio broadcasting (Radio Luxembourg) and, like larger countries, was entitled to apply for its own slots in 24-hr orbit from the regulatory authorities. The Luxembourg government became a minority shareholder in the Société Européenne des Satellites company (SES), which obtained a number of slots in 24-hr orbit for commercial broadcasting. SES began with the first *Astra* satellite in December 1988. This was the first of four satellites called *Astra*, broadcasting to Britain, Ireland, Germany and France, operated by a mission control at Betzdorf Castle, in rolling woods 40 km east of the town of Luxembourg, the centre of the Grand Duchy [70].

The *Astra* system was quickly into direct-to-home digital broadcasting (*Astra 1E*). This prompted a rival network to be established by Eutelsat, which launched the first of the Eutelsat *F1 Hotbird* series in August 1990. When *Astra* started, no-one had given it much of a chance in competing against the main German and French comsats. However, the channels on *Astra* were quickly leased to commercial satellite broadcasting companies and the network grew to seven geosynchronous satellites by 1998, extending direct broadcasting to eastern and central Europe, Turkey and north Africa. Castle Betzdorf expanded to 300 staff of 20 nationalities. By 2002 the eight *Astra* satellites broadcasted to 75 million homes in 22 countries. The *Astra* system seemed to have done for Luxembourg in the new century what Radio Luxembourg had done in the old.

3.47 DENMARK

Denmark set up its national Space Research Institute (Dansk Rumforkningsinstitut) in 1966, under the Ministry for Education & Research. Denmark was a founder member of ESA and in 1999 built its own first satellite. Called *Oersted*, this was a small 62-kg satellite, orbited January 1999 to study the changes in the Earth's magnetism and auroral phenomena. The satellite was named after Hans Christian Oersted (1777–1851), the founder of modern theories of electromagnetism. *Oersted* was shaped like a box 34 cm by 45 cm by 68 cm, with an enormously long (8 m) magnetic boom. It had four instruments (two magnetometers, a particle detector and star camera), five solar panels and a data transmission system, information being sent to the meteorological institute in Copenhagen. The first scientific results from the mission were presented in 2000 and in 2001 when a coloured magnetic map of the Earth was published, showing areas of high and low magnetic pressure.

The Danish Space Research Institute has a budget of over €2.6m and 40 staff. The Institute was involved in the preparation of instruments for *Granat*, International Space Observatory, *Viking* and *Cluster*. Ten per cent of its budget was allocated to a small satellite programme. Danish space activities are supervised by the Danish Space Board (DSB). The national space budget in 2001 was €29m, of which €25m went to ESA.

3.48 AUSTRIA

Austria set up its space agency in 1972 (the Osterreichische Gesselschaft für Weltraumfragen) with a €640,000 start-up budget. The Austrian Space Agency comprises representatives of government ministries, leading industrial companies (e.g., Siemens), research bodies (e.g., Joanneum Research) and the city of Graz. In its early days Austria contributed instruments to several Soviet and Russian missions, principally *Venera 13/14*, *Vega*, *Phobos* and *Interball*, the high point being a manned mission to the *Mir* space station in 1991 (see Chapter 6).

Austria became an observer in the ESA in 1975 and a full member in 1987. Austria has been involved in many of ESA's most important space programmes and has contributed instruments for *XMM/Newton*, *Envisat*, *Artemis* and *Cassini*. Austrian companies have contributed key components for the *Ariane 5* (e.g., feed lines for the cryogenic main stage, by Magna International). Austria is contributing four instruments to *Rosetta* (e.g., the microdust analysis system) and either components, instruments or experiments for *Integral*, *Mars Express* and *Metop*.

Austria's 2000 contribution to ESA was €37.9m or 1.2% of ESA's total budget, and almost all Austria's space budget goes to ESA (there is small €5m for national programmes). Austria has a Space Research Institute, which is its main research facility in the area; policy is determined by the Federal Government Advisory Commission on Space Research and Technology.

3.49　FINLAND

Finland joined ESA as an associate member in 1986 and as a full member in 1996. The budget for space spending is €39.5m (2001), but unlike many of the smaller countries spends most of it on its national programme (€25.2m, compared with €14.3m for ESA). Finland has a joint commission for cooperation with Russia and contributed instruments to the *Phobos*, *Mars 96* and *Interball* missions. Regionally, Finland has contributed to Sweden's *Freja* and *Odin* and the Scandinavian *Tele-X* communications satellite. Finland's main interests in ESA are weather (*Meteosat*) and Earth resources missions (*Envisat*). The Finnish Meteorological Institute has, through its geophysical division, contributed spacecraft instruments for Russian and American *Mars* probes and in Europe for *Rosetta*, *SOHO*, *Envisat* and *Huygens*. Finland has a national centre for space development, the Space Technology Development Centre (TEKES). Finnish space activities are organized by the Ministry of Trade and Industry, advised by the 12-strong Finnish Space Committee and the Academy of Finland and operate under a national strategy first published in 1996 and subsequently updated.

3.50　SWITZERLAND

Although a reluctant participant in international projects generally, Switzerland has been a successful participant in European space development, being a founder member of ESRO and ESA. Swiss astronaut Claude Nicollier has the distinction of having made four missions, the most of any of the European astronauts: STS-46 to deploy the Eureca platform; STS-61 the *Hubble* servicing flight, STS-75 the second tethered mission and STS-103 at the end of 1999. The Swiss company best known for its contribution to the European space industry is Zurich-based Contraves, which took part in the *ESRO 1* mission as far back as 1968 and later became known as the maker of the payload fairings for the *Ariane*. Although the payload fairing might seem a mundane item, in fact it has to be designed and built correctly or the rocket may break up. The *Ariane 5* fairing involved over 200 people and 30 companies devising materials made from light aluminium honeycomb with carbon fibre-reinforced synthetic resin maximized for efficient aerodynamic performance. Contraves has also been involved in *SOHO* (structure), *Huygens* (separation mechanism), *Hubble* (deployment mechanism), *Rosetta* (mass spectrometer) and *Envisat* (MIPAS [Mitchelson Interferometric Passive Atmosphere Sounder] trace gas detector optical system).

Switzerland's present interests in ESA are in the microgravity programme, the *Ariane 5* launcher, the *Artemis* experimental communications satellite and the *Integral* X-ray astronomy satellite (Geneva Observatory will host its archive). Switzerland contributed €74m in 2001, 2.8% of the ESA total (there is a small national programme of €2.2m). Policy is decided by the federal government on the advice of the 20-strong Federal Space Affairs Commission.

3.51 NORWAY

Norway is an example of how a small country can make a contribution to European space development that is disproportionate to its size [71]. Like Sweden, Norway had a number of advantages because of its location, especially when it came to tracking satellites in polar orbits. Norway set up a satellite tracking station in Tromsø in 1966. A station was subsequently set up on the icy island of Spitzbergen (or Svalbard) at Longyearbyen at 78°N. Similarly, Norway was in a good position to provide a base for space research from northerly latitudes. Norway opened the Andoya rocket range in 1962. It is located at 69°N, just inside the Arctic Circle, on an island off the coast, the site set dramatically against cliffs. In its first 30 years of operation, 437 rockets were fired from there, mainly to understand the atmosphere, ionosphere and magnetosphere at high latitudes.

As with Sweden, space-based communications have been a particular interest for Norway. Norway first leased communications satellite lines in 1976. An Earth station was built on the forested hillside of Eik, near Stavanger and then another at Nittedal, near Oslo. Norway got its own comsat in 1997 with the launch of *Thor II*, named after explorer Thor Heyerdahl of the *Kon Tiki* Expeditions. This was a *Hughes 376* model comsat for the Telenor Corporation designed to provide direct television to homes in Norway, the Scandinavian countries and the Baltic. *Thor III* followed in June 1998. A sign of Norway's commitment to telecommunications was that in May 2001 all the 350 citizens of the village of Modalen, near Bergen, became the first to be equipped with satellite-based, broadband Internet and television connexions (fibre-optic cable is too difficult to lay and the homes too scattered).

Norway joined ESA as an observer in 1981 and as a full member in 1987, contributing 0.7% of its budget. Norway's space budget in 2001 was €28.7m, of which €21.1m went to ESA and €7.6m to national programmes. Norway has contributed to *Cluster*, *Cassini* and remote-sensing programmes. A National Space Centre (Norsk Romsenter) was set up when Norway became a full member and the country's space budget was doubled. The National Space Centre is responsible to the Ministry for Industry and frames Norway's space programme and participation in ESA. Norwegian companies have been involved in remote-sensing, electronics and communications satellite equipment, navigation equipment and have subcontracted components of the *Ariane 5*. Norway is the fourth largest contributor to *Inmarsat*, which reflects the importance of maritime communications and shipping.

3.52 IRELAND

Ireland is an example of a country that has benefited substantially from membership of ESA. Although Ireland was not one of the countries that proposed ESA, the country joined the ESA convention on the final day that was open for signature, thus becoming at the last possible moment a founder member.

Ireland's space budget in 2001 was €6.7m, all going as a contribution to ESA (there is no national programme). Altogether Ireland won 340 ESA contracts from

1975 to 2000. While its neighbour Britain eschewed the *Ariane* project, 22 Irish companies won subcontracting work for the *Ariane 1* worth €4.5m and other contracts brought in over €4.5m over 1976–83 – far more than the cost of Irish membership, which was €2.4m over the same period. A good example of Irish participation in the International Space Station was the WebACT software, developed for the ISS to link experiments on the ground and in orbit through the Internet – a contract won from ESA by the Irish company Skytek, whose international partners are the NLR in the Netherlands and the Moscow Institute for Maths & Physics. Another successful company is Captec in Dublin which developed the software used to direct the *XMM* telescope, *SOHO*, *ISO*, *Hipparcos* and *Huygens*. Two other successful Irish companies are Devtec and Rendor, the former made components for the *Vulcain* engine for the *Ariane* and the latter made the valves on the *Ariane 4* to tolerances of two microns (there are 25,400 microns to an inch).

Ireland's contribution has not been exclusively industrial or commercial. Ireland has a long history in astronomy and for the last half of the 19th century housed the world's largest telescope, Lord Rosse's huge telescope in Birr, Co. Offaly. Ireland's renewed interest in space research can be traced to the establishment by President de Valera of the Dublin Institute for Advanced Studies (DIAS), set up in 1945 and subsequently attracting some of Europe's leading astronomers and astrophysicists. DIAS contributed key components of the *Integral* optical system. Leading Irish space investigator Susan McKenna Lalor is one of Europe's few prominent women scientists: she devised instruments for *Giotto*, the Russian *Phobos* and *Mars 96* missions and was a major contributor to the *Rosetta* mission.

3.53   PORTUGAL

Portugal was a late starter and eventually joined ESA in 1995, currently contributing €2m to the ESA budget (0.1%). Portugal embarked on its first satellite project in the 1990s with *PoSat*. The purpose of *PoSat* was to stimulate interest in satellite projects in the country and build up appropriate engineering expertise. Portugal turned to SSTL to build a small 48-kg satellite based on Surrey's standard *UoSat 5* design. *PoSat* was duly launched piggyback on *Ariane* on 26 September 1983 into a 794–806-km, polar, 99° orbit. The project cost £5m. The experiments carried were for navigation, cosmic ray detection and Earth imaging (two cameras were carried). Eleven Portuguese engineers went to Surrey for training and the satellite was tracked from Sintra near Lisbon.

3.54   COUNTRIES OF EASTERN AND CENTRAL EUROPE

The countries of eastern and central Europe participated in the *Intercosmos* programme with the USSR from 1969 to 1991, including manned missions from

Figure 3.39. SSTL – team and satellite.

1976 to 1999. *Intercosmos* was specifically designed as a programme of cooperation between the socialist bloc countries. Most of their space efforts were oriented eastward, rather than westward. *Intercosmos* comprised small satellites (25 spacecraft were flown in the *Intercosmos* programme from 1969 to 1991) and gave opportunities to fly cosmonauts up to the *Salyut 6* and *7* space station, each with a package of experiments (see Chapter 6). As a result the countries of eastern and central Europe built up considerable experience in space exploration. *Intercosmos 24*, for example, called the Aktivny mission, rode a *Tsyklon* out of Plesetsk on 28 September 1989 to study low-frequency radiation waves in the Earth's magnetosphere and the Van Allen radiation belts from an orbit of 511–2,479 km, 82.6°, 116 min. It used equipment made in the GDR, Hungary, Poland, Bulgaria, Romania and Czechoslovakia. A Czechoslovak subsatellite *Magion 2* separated on 3 October 1989, to test an electric propulsion system for station keeping.

Although the programme was primarily designed for cooperation between countries of the Soviet bloc, several individual Western European countries participated in particular experiments, especially France and Germany. The *Intercosmos* programme effectively ceased with the end of the Soviet Union, and the Russian science programme suffered badly in the 1990s from lack of funding. Only two *Intercosmos*-style missions were run in the 1990s – *Interball* 1 and *2* – the lead countries being Russia and the Czech Republic. They were launched in 1995 and 1996, respectively, with 13 countries participating, including Austria, ESA in its own right, Finland, France, Greece and Sweden.

Although the countries of eastern and central Europe continue to have links with Russia, they are now interested in defining a new role within European space strictures, an imperative that will grow as they integrate with the European Union.

Hungary is possibly the country that has made the most rapid progress toward ESA membership and has formed a national space board comprising relevant government departments and experts, a scientific council on space research and a Hungarian space office. Hungary is an active participant in the *Rosetta* project to comet Wirtanen, providing a power supply unit, dust impact monitor, data management system, plasma experiment and neutral gas and ion spectrometer. It also has an involvement with *Envisat*, *Cluster* and *Netlander*. Hungary hopes to put experiments on the International Space Station, such as a dosimeter to test the radiation exposure of astronauts. It invests considerable efforts in the application of existing space-based information; for example, in the areas of remote-sensing (e.g., crop monitoring and production forecasting), meteorological research (*Meteosat*), space physics (magnetosphere and heliosphere) and life sciences. Hungary set up the Hungarian Space Office in 1992 and has a 2000 space budget of HUF380m.

Following the revolutions of 1989, ESA expressed its willingness in principle to integrate the countries of eastern and central Europe into its membership. Hungary was the first to sign an association agreement (1991) and was followed by Poland, Romania and the Czech Republic. In 1999 ESA held its first workshop in the region called *Toward a New Partnership – the European Space Agency with the Czech Republic, Hungary, Poland and Romania*. The Czech Republic is anxious to join ESA: indeed, the Czechoslovaks were contributors to several of the *Intercosmos* missions and built five small *Magion* satellites, in effect microsatellites before they became fashionable. Romania has formed a national space agency, making as its Head the first Romanian cosmonaut Dmitru Prunariu.

Discussions have begun on the process of making several of the countries full members of ESA. Hungary joined the PRODEX programme (PROgrammes de Développement d'EXpériences Scientifiques' [Programme for the Development of Scientific Experiments]), which is a scientific cooperation programme involving some of the smaller ESA member states (Switzerland, Ireland, Belgium, Norway, Austria, Denmark) and the accession states. Hungary is to be followed by the Czech Republic, Poland and Romania. PRODEX promotes cooperation in the production of instruments for science, Earth observations, microgravity, life and physical sciences. Poland is a participant in the *Galileo* satellite navigation programme and will operate a tracking station in Warsaw.

3.55   ASSESSMENT AND CONCLUSIONS

These are the national space programmes. Most countries run their space programme on a collaborative basis within ESA, committing the bulk of their space budgets to the Agency. Some countries spend all their space budget on ESA and have no national programme at all (this was the German case for many years and continues to be the case for many of the smaller countries). In the case of other countries, the balance between ESA and national programmes has fluctuated over the years, the Netherlands being a case in point.

However, most countries have also developed bilateral programmes and projects over the years as well as healthy national programmes. Bilateral and multinational programmes and projects have also been run with the big space superpowers, Russia and the United States, producing a great diversity of activities. Many of these have been very successful, projects like *IRAS*, *Bepi SAX* and *ROSAT* coming to mind.

Structurally, the space programmes of most countries have a standard format: an executive space agency, an oversight body (often located in a government department) and a ministerial responsibility (e.g., the appropriate minister for science, research or technological development). Generally, the establishment of a strong executive government agency has been the key to a successful national space programme. The French appreciated this from the very start, and it is no surprise that CNES has been the dominant national space agency in Europe. The Italian national space programme has thrived ever since the ASI was set up, making Italy very much the rising star of the European national space agencies. The coherent, well-organized programme of Swedish space activities may also be attributed to a strong sense of direction by the Swedish Space Corporation. Although Germany has not projected the image of an agency-led space programme, it would be a mistake to underestimate the leadership of Germany's space programme. The size of the German space industry makes it the second largest in Europe and the range of its facilities can only be described as impressive and extensive.

By contrast, the lack of an executive national space agency may lie at the root of the low level of British space activities compared with other European countries of similar size. The vacuum created by Britain leaving ELDO was not matched by a correspondingly large national programme. Britain did not get its own space body until 1985 and the exercise proved abortive. The BNSC, while valuable, has played a coordinating, not an executive role.

The national space agencies have proved important in enabling countries to make strategic choices about their programmes, choices and areas of interest. The European countries recognize that they cannot all set up broad-based national space programmes: they must define a number of specialities that make the best of their history, engineering expertise, skills and geography. For example, in its national programme France has concentrated on Earth observations, the Scandinavian countries on communications and the magnetosphere of high latitudes, the Italians on manned space modules and unusual experiments with space tethers. Despite the lack of political leadership, Britain has excelled in X-ray astronomy

Figure 3.40. *Odin* at Svobodny.

and developed military communications satellites. Despite, or alongside, the imperatives of integration within the ESA, national space programmes are likely to continue. ESA programmes are generally very long term, while national programmes provide flexible opportunities for the development of national goals and skills.

4

The European Space Agency (ESA) (1974–)

We left the European space programme in a state of collapse in 1973. France and Germany had pulled the plug on the *Europa* launcher and the project lay in ruins. The European Space Research Organization (ESRO) had been through traumatic periods as well, but at least ESRO hardware and satellites were in orbit, a modest outcome for the hopes of those who had gathered in Paris, London and Switzerland in the early 1960s.

4.1 ORDER FROM CHAOS

Eventually some order was to emerge from the chaos. Several strands of development came together in a positive outcome, even though amid the fog of negotiations at the time these were not readily apparent.

The first strand was the idea of a combined ELDO (European Launch Development Organization) and ESRO with a combined, stronger secretariat. The Causse report had recognized the lack of direction from the centre of both bodies and that managerial skills and authority were necessary to manage complex, modern space projects. As it was, ELDO and ESRO mirrored the institutional weaknesses of the other. The importance of coordination had been underlined by the final *Europa* failure. Ironically, the political leader who was to force the two together came from the country that had done most to undermine ELDO, or expose its contradictions, depending on one's view. This was the British Trade Minister, the mercurial Michael Heseltine. His interventions were sudden, uneven, but had a clarity and determination that others had failed to show. In summer 1972 Heseltine abruptly flew to Paris and formally proposed a merger of ELDO and ESRO, throwing the weight of at least one government behind the idea. Heseltine's visit was so sudden – it the French were on holidays. Ultimately that did not matter, for his success was in outlining the way forward.

The second strand of ideas that was to produce a united agency was a para-doxical one, but it was a definite element in Heseltine's approach. Although the united agency should have stronger managerial authority, the programmes entered into or engaged should follow national preferences and should permit opt-outs. Heseltine told the French that Britain would support a united and strong agency that built launchers, as long as Britain could opt out from this and other pro-grammes if it wished. The idea of optional programmes became a core idea for the new agency. Countries could opt into those programmes they liked (and benefit from them), but need not contribute to those that they did not like. Forming the agency began to shape up like a package deal.

The third element was probably the most problematical: a launcher. The French remained adamant that Europe must have its own launcher. The *Europa I* programme had grown to the *Europa II* and this had grown to the *Europa III*, which would be able to put a substantial communications payload into 24-hr orbit from French Guiana. Britain apart, *Europa* had been the cement that held ELDO together. This began to crumble when Germany abruptly pulled out of the *Europa III* programme in March 1972, even though it had been engineered around its requirements. The Germans said that its €125m pre-development cost was already too costly and insisted that other less expensive possibilities be studied.

Trying to salvage something from the ruins, France took Germany up on its suggestion that alternatives be studied. The Germans may have intended the phrase as an afterthought or a face-saver, but the French turned it into a lifeline. France took the Germans up on their offer, devising a new programme called the *L3S*. Here 'L' stood for launcher, '3' for the *Europa III* and 'S' for the substitute that the Germans had asked for.

4.2   *SYMPHONIE* SAGA

France wasted no time and formally proposed the *L3S* to ELDO in December 1972, adding a number of financial terms to make it more acceptable:

- France would pay 60% of development costs;
- France would also pay any budget overruns;
- The project would be managed by CNES (Centre National d'Etudes Spatiales).

France was clearly prepared to pay a very high price for a European launcher. Why? The crux issue here appears to have been *Symphonie*. This was a German–French two-communications satellite project to test out systems for the provision of phone and television channels. Aérospatiale built the first one and MBB (Messerschmidt Bölkow Blohm) the second. The *Symphonie* project arose from different projects under way in Germany and France at the same time. In France CNES had been planning *SAROS*, a geosynchronous satellite to link France with its overseas

Figure 4.1. Original *L3S* design.

departments. Germany had been planning a satellite called *Olympia* to broadcast the upcoming 1972 Munich Olympics to the world. In November 1966 the two countries agreed under the terms of the French–German treaty of 1963 to pool their efforts. It was the first joint venture between France and Germany, their only initial problem being to agree on a name for the merged project. Athos and Concerto were rejected and they settled on *Symphonie*. *Symphonie* was to have been launched on the *Europa II*, but with Europe launcherless they were obliged to pay the Americans for a launcher.

Under an international agreement between the Western industrialized nations signed in August 1964, responsibility for international satellite communications was assigned to the Intelsat Organization, which was 56% owned by the American government. Intelsat was made a private American company (Comsat) and sole production administrator or manager. Under Article 14 of the Intelsat Convention, member states agreed not to take actions that might be financially prejudicial to Intelsat. The Franco-German proposal was seen as a threat under Article 14 and to the monopoly hitherto enjoyed by American launchers and satellite makers. The French and Germans argued that, as a regional European satellite, it posed no serious threat to American interests, but their arguments cut no ice. France discretely went to the Soviet Union to try and persuade the Russians to launch *Symphonie*, but the USSR refused for fear of provoking the Americans. France and Germany had no choice but to go back to the Americans, who agreed to launch *Symphonie*, but required a formal agreement that *Symphonie* only be used for experimental and not commercial purposes.

Two *Symphonie* satellites were built, one by Aérospatiale and Thomson, the other by AEG Telefunken, MBB and Siemens. *Symphonie* was an original design, for this was the first time European companies had built communications satellites. *Symphonie 1* and *2* were duly launched by *Thor Delta* on 19 December 1974 and 27 August 1975, respectively. Each operated for almost 10 years, far beyond their design life. They were indeed used for experimental purposes, especially to demonstrate the potential of satellite communications for developing countries. *Symphonie* became the basis, in time, for the French, German and other operational European communications satellites.

Symphonie launches

19 Dec 1974	*Symphonie 1*	*Delta*	Cape Canaveral
27 Aug 1975	*Symphonie 2*	*Delta*	Cape Canaveral

The French learned a hard lesson from this experience, namely the absolute importance of an independent European launcher. Indeed, many years later several French space leaders mused whether there ever would have been an *Ariane* without the American pig-headedness over Article 14 of the Intelsat accords.

4.3 *L3S* PROPOSED

France had originally put forward the idea of a new European rocket, the *Europa IIIB*. This would have French first and second stages with *Viking* engines and a take-off thrust of 220 tonnes and a height of 36.5 m. On top would be a liquid oxygen and hydrogen third stage built by MBB of Germany, able to put payloads into 24-hr orbit. France's partners felt that this proposal was too ambitious and too expensive, particularly in the light of Europe's recent experiences. Accordingly it was scaled back, the third stage becoming a joint French–German venture, and acquired its new title, the *L3S*. The objective was to send 750 kg to geostationary orbit. The development cost was then given as €380m.

The *L3S* was debated at what became a stormy ELDO meeting in December 1972. More important, in January 1973 President Pompidou managed to persuade Chancellor Willi Brandt to come on board the *L3S* project. The Germans weighed in with 20% of the costs, the other countries contributing very small amounts. It was agreed:

● to formally cancel the *Europa III* programme (*Europa II* survived until the German and French finance ministers pulled the plug the following April);
● to begin the *L3S*, on the terms France had proposed;
● to merge ELDO and ESRO into a single agency;
● Britain would get the guidance system for the *L3S*.

Former French Prime Minister Michel Debré, who was involved at the time, later revealed that France would have gone it alone as a national project had the other countries refused. Not to develop the *L3S* would have meant leaving spaceflight to the Americans and the Russians and later to the Chinese and Japanese. 'To give up would have been shameful,' he said later. But it would have been a French venture, not a European venture, and there probably would have been no European space agency.

An important plus point was that the *L3S* was designed from the start to send commercial communications satellites to geosynchronous orbit. Its performance could be optimized around this basic requirement. By contrast, *L3S*'s rivals were launchers originally designed with another purpose in mind, but adapted for the geosynchronous mission.

That lay in the future. So far, a concrete proposal was emerging from the ashes of the *Europa* programme. But would there be an agency to run it?

4.4 FOUNDATION OF ESA

The decisions of ELDO in December and of the French and German governments in January were supposed to be formalized by a conference on 12 July 1973, which should have been the founding date of the ESA. This was actually the sixth meeting of all the ministers responsible, and its purpose was to confirm the French proposal for the *L3S* programme made the previous December. It quickly became apparent that no agreement was possible, so the Belgian chairman, rather than let it break up, adjourned the meeting until 31 July. The conference had lasted less than two hours, and there was little confidence that differences could be resolved [72].

The chairman took advantage of the interval to broker a deal that could get everyone on board the new agency. His core idea was to develop a system of incentives to get the big players linked to a set of projects. At around the same time as the *L3S* idea, Germany had been trying to find a way of involving Europe in the American post-*Apollo* programme *Spacelab*, a project in which it was keenly interested, but France less so. Britain was interested in a third project, much smaller than *Spacelab* or the *L3S*, for a Maritime Orbital Test Satellite (*MAROTS*). Could a formula be worked out to involve these three leading countries in a balanced and fair way? In a further twist, participation in the *L3S* was to be made optional, making it hard for member states to veto the idea. This was a development of the Heseltine idea and would have the merit of getting the British into the new agency. He tried to build a deal in which the French would get the *L3S*, the Germans the lead role on cooperation with the American proposal for *Spacelab* and the British the prime contractorship for *MAROTS*. Each of the three stakeholders would be associated with three founding projects. The exact financial formulae for doing so was of course another matter.

4.5   ESA IS FORMED

The Belgian chairperson brought everyone together again two weeks later, and the decision to form the European Space Agency was taken by eleven countries meeting in Brussels on 31 July 1973. The outcome was a complex, but precise formula, one designed to learn all the lessons from the ELDO and ESRO experience. It involved first of all a complex package for three founding projects: the *L3S*, *Spacelab* and *MAROTS*, but one which had the merit of getting the three big players on board. This was what was agreed:

Country	L3S (%)	Spacelab (%)	MAROTS (%)
Germany	20.12	54.1	20
Belgium	5	4.2	1
Denmark	0.5	1.5	–
Spain	2	2	1
France	63.9	10	12.5
Britain	2.47	6.3	62.2
Switzerland	1.2	1	–
The Netherlands	2	2.1	4.7
Italy	1.74	18	2.3
Sweden	1.1	–	3

This in effect defined the first tasks of the new agency as: to build a European launcher, the *L3S* (€760m), collaborate with the United States in the construction of *Spacelab* (€250m) and build *MAROTS* (€80m). France duly became the lead country for the *L3S*, Germany for *Spacelab* (followed by Italy) and Britain for the smallest of the three projects *MAROTS*. The glue that held ESA together was the fact that the three leading countries each got their own pet project approved.

As important as this deal was, ESA embodied a number of ground rules that distinguished it from ELDO and ESRO:

- The budget was set each year by its council, which comprised the Minister for Science in each member state or his or her delegate. This would provide for smoother running than the bigger set-piece conferences of ELDO.
- All member states must contribute to the general administrative budget and the science budget. These were called the mandatory programmes. The general budget covered administration, facilities, training, research, information systems and technical investment. These would be a small part of the overall budget, but they would be compulsory.
- Apart from the two mandatory budgets, member states were free to contribute, or not, to other programmes. These were called the optional programmes. These include, for example, launchers, manned flight, Earth observations and telecommunications.
- *Juste retour*: member state companies should expect to win contracts for the execution of programmes broadly commensurate with their financial input.
- For the sake of stability and good planning, long-term financial and management plans were introduced to govern budgets and programmes a number of years at a time.

This was a clever attempt to prevent a recurrence of the destabilizing features of ELDO and ESRO. The mandatory budgets were much smaller than the larger launcher and *Spacelab*-type programmes and likely to put membership under least strain. Countries were not forced into participating in programmes that they disliked or found too costly (the focus of British resentment from 1966). Optional programmes were fairer, because they confirmed a *juste retour* only to those countries that opted in to them. Optional programmes also provided an incentive for their promoters to solicit partners in other member states [73].

So much for the political and financial glue for ESA. What about the management issues that had dogged ESRO and ELDO? It was agreed that each project would have a single, strong management – either directly by ESA itself or by a delegated agency (e.g., CNES). The old system of letting member states manage a project informally between themselves, the ELDO problem, was over. Specifically, the *L3S* would be managed by CNES directly. This was a break with the principle of collective management, but CNES would pay for the management costs. In a further reinforcement of the principle of vertical management, any subcontracting would be done by CNES and the French contractors directly. This had a number of advantages. First, the French had their own management team, giving them confidence in their own project. Second, whatever anyone else thought of the French role in Europe, their management capacity was not in doubt. Third, something probably left unsaid, the French could not blame anyone else if the *L3S* went wrong. The only snag was that it gave the French quite an advantage in the share-out of *L3S* sub-projects, but most rocket-building expertise was in France anyway.

The agreement to set up ESA was made in 1973, the convention coming into force on the first day of 1976. Although Ireland did not sign the original agreement, the country joined on 31 December 1975, the last possible day to sign up under the original convention and may in that sense be considered one of the original members.

Founder countries of the ESA (1973–5)

Belgium	The Netherlands
Britain	Norway
Denmark	Spain
France	Sweden
Germany	Switzerland
Italy	Ireland

Later members of ESA

Austria	1985
Norway	1985
Portugal	2000

But would it work? Initially, the new agency seemed to be heading into the black hole that characterized ELDO. Very little progress was made in the first year, and 18 months after it was founded the agency still did not formally have a name. There was much dispute about contracts and which company would get responsibility for what. British enthusiasm dimmed when Michael Heseltine's government went out of office in March 1974. Two months later in May the *L3S* rocket was suspended by new President Valéry Giscard d'Estaing as too expensive and was sent for evaluation. French space leaders had to fight a rearguard action to save the *L3S*, and its funding was not restored until October 1974.

In the event, the diplomatic wheels went on turning and brought the new machinery into effect. The organization formally and finally acquired its name, the European Space Agency, in early 1975. The first Director General Roy Gibson took up his post on 16 April 1975.

Despite such a shaky and downright unpromising start, the organizational arrangements and highly political deals that founded ESA proved remarkably durable. They were largely unchanged 30 years later. ESA came under intense strain in 1987, due to British intervention once again, but the difference was that this time it weathered the storm.

4.6 BUILDING THE *L3S*

The real test for ESA was always going to be the new *L3S* launcher. British press opinion generally regarded the project as a hiding to nothing, consoling itself that at least British money was not being wasted. The temporary suspension of the project in mid-1974 did not help. Investment in the *L3S* had serious knock-on effects for other French space projects, as the French did not have a bottomless budget. Funding the *L3S* meant a huge investment in rocket building, infrastructure and facilities – but national space projects were slashed to pay for it. The Centre Spatiale de Toulouse (CST) even considered diversification into solar-powered home-building and cultivating soya beans. Matters were settled in a traditional way when the workers in CST came out on strike and the government backed down. Indeed, the strike put CNES in a stronger position than ever, because the settlement involved fresh funding for a domestic French communications satellite *TDF* (Télédiffusion de France).

Figure 4.2. Problems at CNES.

These hiccups concealed the fact that CNES had actually embarked on one of the major European engineering projects in the last quarter of the century. The decisions as to who did what for the *L3S* may not have appeared as epoch-making at the time, but they were in the event to determine the main managerial and industrial players in the European launcher industry for the next 40 years. Aérospatiale was the prime contractor with responsibility for the integration of all stages of the vehicle. The role of the French engine company SEP (Société Européenne de Propulsion) was to build the first and second stage engines, while the third went to Air Liquide and MBB. SEP developed what was called the *Viking* engine, based on the successful engines used on the *Diamant*. The other main companies involved were Matra, the French electronics company, Germany's Dornier and Sweden's Volvo. A crucial development was the third stage. Up to that point only one country had flown operational hydrogen-powered upper stages, the United States with *Centaur* (the Russians actually built one for their man-on-the-moon programme, but it had not yet been used). SEP had made studies of a cryogenic third stage as far back as 1962. When the *L3S* was commissioned, they had a design ready to be used [74]. In dramatic contrast to the *Europa* rocket that was led by Britain, the British share of *L3S* was only 2.47%. This covered Ferranti's participation in the inertial guidance system and Marconi in the central digital computer. Later, British Aerospace and a small company, Avica in Hemel Hempstead, became involved. The legal agreement for *Ariane* was signed on 21 September 1973 at Neuilly-sur-Seine. The total cost was then estimated at around €760m.

Preparations for the launch of *Ariane* went ahead over 1977–8. Ground tests and engine firings passed their milestones, some snags being experienced on the way, but none causing lasting delays. ESA announced that there would be four development flights before the system could be declared operational and offered free space on these missions with the proviso that success was not guaranteed. India leapt at the opportunity, asking Europe to fly an experimental communications satellite called *APPLE* on the third mission.

Even as work progressed, CNES faced another knotty problem: the name of the launcher. *L3S* was not a title calculated to win broad popular appeal. CNES began to search for a name that might win popular favour. Several were proposed:

- Phoenix, the bird that rose from the ashes;
- Lyra;
- Ganymede.

Vega was also considered, but coincided with the name of a Belgian beer, and in the interests of not giving Belgium or beer an unfair commercial advantage (or disadvantage if things went wrong) was eliminated. The names were submitted to the French minister responsible Jean Charbonnel, but he chose none of the three. Being a classicist, he chose *Ariane*. It was an ominous choice, for in classical mythology Ariane died from wounded love when she was abandoned. Was he tempting fate? The first *Ariane* was built over 1977–8 and the testing regime appeared to go successfully.

L3S launcher

Dimensions	47.6 m
Weight	203 tonnes
Capability	750 kg to 24-hr orbit

	First stage L-140	Second stage L-33	Third stage H-8
Dimensions	18.7 m × 3.8 m	11.3 m × 2.6 m	8.5 m × 2.6 m
Weight	159.55 tonnes	36.79 tonnes	9.6 tonnes
Engines	4 × Viking II	Viking IV	HM-7
Fuels	UDMH, nitrogen tetroxide	UDMH, nitrogen tetroxide	Hydrogen, oxygen
Thrust	140 tonnes	70.76 tonnes	2.5 tonnes
Burn time	138 seconds	131 seconds	563 seconds

4.7 PREPARING *ARIANE*

Since the end of the *Europa* programme in 1973 and *Diamant* two years later, Kourou had become very quiet. There was a real danger that the facilities would be reclaimed by the jungle. The *Diamant* pad had become a fireman's store.

The *Europa* launch site in French Guiana was now modified: the base was lowered and the tower made taller [75]. It was called the ELA 1 (Ensemble de Lancement Ariane 1). The launch process between the completion of rocket construction at Aérospatiale in Les Mureaux in Paris, and launch from Kourou was to take several months. First, the stages of *Ariane* were shipped by barge down the Seine to Le Havre, whence by ship across the mid-Atlantic to Kourou. In September 1979 the first *Ariane* arrived in Cayenne from Le Havre. The different parts were taken to Kourou by barge and road soon thereafter. *Ariane 1* was brought to the pad on a mobile launch platform on rails. It then reached an enclosed air-conditioned service tower that looked like a grain silo. Here the payload was added and checked out in clean room conditions. For a payload, the first *Ariane* was fitted with an Aeritalia-built instrumentation unit. Its role was to measure all the key stages of the ascent such as noise, stress, acceleration, temperature and pressure. To simulate a real satellite, it was filled with 600 kg of aluminium alloy [76].

As the launch of *Ariane* approached, critical opinion in the British and American press was sceptical. *Ariane* might, it is true, win a number of commercial payloads before the shuttle entered full commercial operation – but otherwise it was an expensive indulgence by the French in seeking an independent launcher capability. Keeping the programme going must have required some nerve. *Ariane* had acquired only three initial customers – one German communications satellite, one French comsat and the French *SPOT* satellite. There was a breakthrough in December 1977 when Intelsat placed an order for two *Intelsat IV* comsats. Knowing that the Americans were offering the shuttle for $110m, CNES pitched in with a 'promotional' offer of $100m and persuaded Intelsat against the wisdom of putting all its

Figure 4.3. *Ariane 1*, Christmas 1979.

eggs in the shuttle basket. A week later, ESA ordered a production run of 10 *Ariane 1*s.

Pre-launch preparations and the launch itself were supervised from a hardened bunker 200 m from the pad. The countdown began 38 hours before launch. The maiden flight was set for 15 December 1979 and the general aim was to fly the first *Ariane* not just before the end of the year, but also the decade. The service tower was removed after five hours. Each stage was filled with fuel, the liquid hydrogen stage just three hours before lift-off.

It was nerve-wracking. An hour before launch there was a delay, which at one stage looked so serious that it might cost a month's hold-up, pushing the launch into the new year, hardly encouraging. But the countdown resumed for a second attempt. The electrical systems were armed at 40 minutes and the automatic sequencer came in at 6 minutes. At zero, the engines lit up, smoke billowed from the launch stand for three seconds – but then two seconds to take-off the computer had information that engine pressure had fallen (erroneous in the event) and the engines cut out.

Thousands of journalists, well-wishers and people involved in the industry saw it all. However, engineers worked hard to get the rocket off the ground before Christmas. Launch was reset for 24 December, Christmas Eve. This time, the third attempt was assisted by French President Valery Giscard d'Estaing who personally pressed the firing button.

4.8 INTO THE NIGHT SKY: 24 DECEMBER 1979

With the flags of the ESA participating nations painted on its side, *Ariane 1* was successfully launched into a warm equatorial night-time sky. *Ariane 1* put a test payload called CAT (Capsule Ariane Technologique) into an orbit of 202 km by 35,753 km. There was celebration all over Kourou (they could go on an albeit late Christmas holiday break happy) and jubilation in the French government in Paris. The American commercial launcher monopoly had been broken at last. In the hangar alongside the launch site, champagne flowed at a noisy, rowdy party. It was a wild night. The third stage dumped some frozen gases on the launch site and the controllers had a snowball fight with what was left.

The first flight was called LO-1 (Lancement [Launch] 01). After several launches, the 'L' designation was replaced by 'V' for 'vol' (after the French word for 'flight') and thereafter each succeeding mission received a V designation (V will be used here throughout).

No sooner had the launching been achieved than CNES and ESA moved to create a company that could promote, market and manage the operations of *Ariane*. Originally it was called Transpace, but that sounded too American, so the more mundane Arianespace was decided upon. Formed on 26 March 1980, France was the principal country involved. It was registered under French law with an initial capital of €40m and with 50 shareholders, including Aérospatiale, Matra, Erno, Dornier, British Aerospace and major European banks. Arianespace was a consortium of 36 manufacturers in the air, space and electronic industries and 13 banks (the two key players though were CNES, which had a 32.45% shareholding, and Aéro-spatiale, later EADS [European Aerospace and Defence Industries], with 27%). Arianespace also had responsibility for marketing and procurement.

4.9   THE *ARIANE* TEST PROGRAMME: TRIALS AND TRIBULATIONS

The second mission was set for May 1980 and marked the arrival of a new Danish Director General in ESA, Danish shipbuilding engineer Erik Quistgaard. Mech-anical problems and then bad weather called off three countdowns. When *V2* even-tually flew, all went well until it reached an altitude of 50 km at 59-seconds, but then *Ariane 1* lost pressure in one of its four first stage engines and then another, with the result that the rocket had to be destroyed at 108 seconds at an altitude of 64 km. Examination of telemetry found that pressure in one of the engines had begun to fluctuate only 6 seconds after launch, again at 28 seconds and 63 seconds. All the other engines then went down at 104 seconds. The rocket crashed and a salvage boat was sent out to retrieve what was left. Wreckage came down 26 km off the coast.

Two German satellites *Firewheel* and *Amsat*, were lost. CNES was accused of rushing its fences to get *Ariane* airborne for commercial missions and taking advantage of the delays to the shuttle. The debris was trawled out at sea in an attempt to identify the cause of the failure. The engines were found on 20 June 1980. Thankfully, the offending engine was found in the sea just off Devil's Island, and analysis revealed what had gone wrong. The post-mortem traced the engine failure to tiny manufacturing imperfections in the system that injected fuel into the combustion chamber, causing vibration and instabilities in the burning process. The problem had not shown up in over 200 tests, showing just how subtle these problems could be. Repairing the engines cost €16m, forcing a 14% overrun in the year's budget for the rocket.

V3 was more successful on 20 June the following year, putting into orbit *Meteosat 2* and *APPLE*, an Indian test satellite designed to pave the way for India's first generation of communications satellites *INSAT*. The purpose of *APPLE* was to test out means of stabilizing satellites in 24-hr orbit, and was a major opportunity for the Indians to develop their skills and experience. Despite a solar panel that jammed, *APPLE* relayed television programmes and educational teleconferences in an experiment that lasted just over two years. *Meteosat 2* itself provided weather pictures of the Earth from 0° longitude every 25 minutes from a 40-cm telescope in the visible, infrared and water vapour bands. However, its trans-ponder broke down and *Meteosat 1* was pressed back into service for this purpose.

4.10   *ARIANE* OPERATIONAL

CNES was now confident that the rocket could be declared operational. On 20 December 1981, two years after the first launch, *V4* climbed into the night-time tropical sky of French Guiana, and 16 minutes later *Marisat* was on its way to geosynchronous orbit. By this stage confidence in *Ariane* was growing. Even though the American shuttle had now begun its test flights, *Ariane* had 25 payloads in its order book worth €500m, and in the next month January 1982 it was declared operational.

Ariane V5 was the first money-earning commercial payload and was launched on 10 September 1982. Tucked into the nosecone fairing were the €40m *Sirio 2* and the €30m *Marecs B* marine communication satellite. The dawn launch appeared to go perfectly, but 560 seconds into the flight, nearly halfway into the third stage burn and long past the normal danger points, the turbopump lost pressure from 32 atmospheres to zero and the stage failed. By this time CNES engineers could well be described as worried, for *Ariane* had failed twice in its first five attempts – a reliability rate of only 60%, which was well below the 98% aimed at and the level necessary to instil confidence in the satellite-maker. In Britain – with only a minor involvement in *Ariane* – the press ran stories that *Ariane* would be ditched in favour of cheaper, sooner-available American rockets. The investigation attributed the failure to insufficient lubrication of the turbopump during ground testing and subsequent gear failure.

Figure 4.4. *Ariane* engines.

So there was a do-or-die atmosphere in Kourou on 16 June 1983 when *Ariane V6* moved swiftly off the tower, placing the European Communications Satellite *ECS 1* accurately in orbit. The *V6* launch began a run of success for *Ariane*. *V7* headed into the night sky on 18 October with an *Intelsat VI* comsat, *V8* the following March. *V12* broke new ground in February 1985, for it was the first *Ariane* to carry two non-European satellites *Arabsat* and *Brazilsat*.

As the launchings began to go smoothly the order books lengthened, with 27 satellites booked by early 1984, valued at €938m (40% of which was for non-European customers and half of the world market at that time). After the 10th mission, ESA handed responsibility for *Ariane* over to the new company Arianespace. Compared with the old days of ELDO, how the tide had now turned!

Fixing *V6* did not bring all failures to an end and the engineers were to be reminded from time to time of the fragility of their creations. *Ariane V15* was lost on 13 September 1985. The feed valve on the third stage failed at ignition and the rocket crashed into the Atlantic, losing *Spacenet 3* and *ECS-3*. Embarrassingly, the launch was witnessed by a stony-faced President Mitterand. Hydrogen leaked from the main injector valve, freezing over the entire area and preventing ignition.

Six months later *V16* went off perfectly on 21 February 1986. By this stage *Ariane* had the world launcher market largely to itself, since the American space shuttle had tragically just been lost at Cape Canaveral in a terrible fireball. *V16* put into orbit the first European Earth resources satellite *SPOT 1* and the first Swedish satellite *Viking*. Three hours after launch the small Swedish contingent in the Jupiter launch centre relaxed when *Viking* was kicked into its own 811–13,536-km, 98.7° orbit to study the ionosphere and magnetosphere and report back to the control centre in Kiruna, north Sweden.

4.11   THE PROBLEM OF THE THIRD STAGE – AGAIN

V18 suffered a third stage engine ignition failure on 31 May 1986, 4 minutes and 30 seconds into its mission. Although the ignition command went in, the ignition process shut down as soon as it began. The rocket had to be destroyed and the failure forced an intensive €80m ignition systems procedures review and reconstruction. A German official was appointed by ESA to investigate. The igniter system fell quickly under suspicion, and it was determined that the igniter had shocked the hydrogen feed line, preventing sufficient hydrogen from reaching the combustion chamber. The margin of error was actually very small, but very small was still too much.

V18 provoked a reconsideration of the cryogenic third stage. This was the third time the third stage had gone wrong (*V5*, *V15*, *V18*) and the *Ariane* was grounded for 15 frustrating months for a searching re-evaluation [77]. Insurers put their rates up by 25%, in effect betting on a failure each fourth flight. In the worried SEP plant at Vernon near Paris, engineers were put on three shifts, 24 hours a day, to locate and

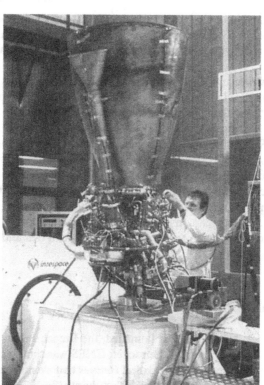

Figure 4.5. *Ariane* hydrogen upper stage.

fix the problem. They ignited the fuel under different conditions, using photographic frames at 5,000/sec to track the problem.

The new ignition system involved improved flows from valves and twin lighters pointing in different directions, a trebling of the electrical charge used and a re-sequencing of ignition to ensure more fuel was available to burn and putting more oxygen into the burn. Forty-one simulated altitude tests were carried out to check the modifications. By this stage *Ariane* had launched 18 times, but been lost four times, a higher failure rate than the programme promoters would have liked. Once again the programme was delayed while the problems were fixed. Arianespace was quick to keep customers and the public up to date with the nature of the problem and its responses. Despite the hiatus, *Ariane*'s order book actually lengthened to 46 satellites. Half of these were European, a quarter were international projects and a quarter were American.

A feature of this and previous *Ariane* investigations was to appoint commissions of inquiry immediately and to bring in outside experts. A preliminary report was normally released after a couple of weeks, explaining progress so far and indicating the lines of investigation. An interim report and then a final report would follow, explaining the most likely cause of failure and the measures required to rectify them. This open approach kept customers informed – and as a result kept them as customers.

The countdown resumed in August 1987 after a long grounding. The Launch Director then announced a further delay so that he could put his engineers back on a five-day week and get them in good shape for the launch. The checks – and the relaxed approach – obviously worked, with *V19* putting up two satellites just before midnight on 16 September 1987. The accuracy in placing the satellites into their proper orbit was 0.005%. Two months later *V20* orbited Europe's first direct broadcast satellite, the Franco-German *TVSat-1*. Confidence returned at last: just as well, for Arianespace now had a record 41 satellites on its order book worth €3bn. Early the following year *V21* orbited *Spacenet* and *Telecom 1C*, the first of eight launches planned for that year. In mid-1988 Arianespace reported profits up 50% to €70m and ordered another 40 *Ariane 4*s to keep up with demand.

4.12   UPGRADING *ARIANE 3* AND *2*

Even before the first *Ariane* was ever launched, the French space agency CNES had been giving attention to upgrading the rocket. CNES was thinking so far ahead that the first ideas of a mini space shuttle called *Hermes* had even been floated. The idea of a follow-on version to the *L3S* had been first mooted in 1978, before *Ariane 1* had even flown. The first phase of a development plan for a new *Ariane* was approved in July 1979 and the design process took place over 1980–2. *Ariane 3* anticipated the day when there would be a need for a launcher to bring over 2 tonnes to geosynchronous orbit. Alternatively, with a new shroud and deployment system, *Ariane 3* could put *two* smaller satellites into geosynchronous orbit. The country-by-country budget for *Ariane 3* was led by France (63.9%), followed by Germany (20.1%) and Belgium (5%). Smaller contributions came from Denmark (0.5%), Sweden (1.1%), Switzerland (1.2%), the Netherlands and Spain (2% each) and Britain (2.5%). The main contractors were Aérospatiale (prime contractor), SEP (propulsion), Dornier (second stage).

Despite the numbering, *Ariane 3* was the successor to the *Ariane 1* (*L3S*). It was a much upgraded version of the *L3S*, with two strap-on boosters, improved engines and a lengthened third stage. *Ariane 2*, which came after *Ariane 3*, was a subordinate version with the second and third modifications, but without the strap-ons [78]. To keep costs down CNES required the improvements to respect already tested technologies (no new test series were budgeted), to fit existing launch and handling facilities and not to require any retooling on the *Ariane* production line.

Ariane 3 had nine specific improvements over *Ariane 1* [79]. These were:

- new payload fairing;
- lengthened payload deployment system;

- longer third stage tanks with 10.5 tonnes of propellants, 30% more (compared with 8 tonnes on *Ariane 1*);
- third stage combustion pressure up from 30 to 35 atmospheres;
- longer engine nozzle (20 cm) on third stage with improved thrust;
- pressure in second stage engine up from 53.5 to 58.5 atmospheres;
- new inter-tank structure to support solid-fuel boosters;
- two 70-tonne, solid propellant strap-on boosters (not on *Ariane 2*);
- pressure up in first stage engine from 53.5 to 58.5 atmospheres.

Ariane 3 would fly a different ascent profile from *Ariane 1*. First, solid-fuel strap-on boosters firing for 20–30 seconds would give it an extra kick soon after lift-off. These were made by Difesa e Spazio of Italy. They were then dropped, meaning that they were no longer carried as deadweight. Second, extra fuel meant that *Ariane 3* could reach 250 km, rather than 200 km, before coasting into orbit. As a result, orbital injection was much further downrange – so tracking equipment was moved from Brazil to a new tracking station in the Ivory Coast. Once it reached orbit, the new fairing enabled two satellites to be deployed at once. The carbon fibre shroud was called SYLDA (Système de Lancement Double Ariane). Better engine performance was achieved by improved pressures and a different fuel mixture. *Ariane 1* used pure UDMH, while *Ariane 3* used three-quarters UDMH and one-quarter hydrazine hydrate. Thrust was up 9%.

Ariane 3

Dimensions	49.13 m tall × 3.8 m diameter			
Weight	234 tonnes			
Capability	2.58 tonnes to 24-hr orbit			

	First stage	Strap-ons	Second stage	Third stage
Engines	4 × Viking 5	2 × P7	Viking 4	HM-7B
Fuels	UDMH, nitrogen tetroxide	Solid	UDMH, nitrogen tetroxide	Hydrogen, oxygen
Thrust	267.1 tonnes	144 tonnes	78.5 tonnes	6 tonnes
Burn time	142 seconds	30 seconds	140 seconds	735 seconds

Ariane 2

Dimensions	49.13 m tall × 3.8 m diameter		
Weight	215 tonnes		
Capability	2 tonnes into 24-hr orbit		

	First stage	Second stage	Third stage
Engines	Viking 5	Viking 4	HM-7B
Fuels	UDMH and nitrogen tetroxide	UDMH and nitrogen tetroxide	Hydrogen and oxygen
Thrust	267.1 tonnes	78.5 tonnes	6 tonnes

Ariane 3 was introduced on the 10th flight of *Ariane* (*V10*), on 4 August 1984. ESA took a calculated risk, one which saved €60m, not having a test flight and going directly to a commercial launch, carrying *ECS-2* for Eutelsat and *Telecom 1A* for the French national television network. The second *Ariane 3* carried *Spacenet 2* and the European Maritime communications satellite *Marecs B-2* into orbit on 10 November. *Ariane 3* introduced the new system able to lift two satellites at a time, called SYLDA. European confidence paid off and the launching was watched by 250 observers, including some from China, a country which that year had just joined the exclusive club of countries with the ability to send spacecraft to 24-hr orbit. The last *Ariane 3* was *V32* on 12 July 1989, which put the *Olympus* communications test satellite into orbit.

When *Ariane 3* began to fly, the United States had the major share of the world launcher market. By the end of 1984 this had begun to change. By then *Ariane* had an order book of 30 satellites worth €1bn, compared with the United States' 76 contracts for €2.5bn. *Ariane* launches were initially much more expensive than American ones (€40m compared with €15m), but the Americans came under government pressure to charge more economic fees. Europe was able to win launches when *Ariane* became seen as a reliable launcher and its insurance rates gradually became lower than the Americans. The Arianespace company included a number of banks and they were able to offer favourable loans to companies wishing to fly *Ariane*. *Ariane* was also able by its two-in-one launching system to clear its order book more quickly. *Ariane* also offered its own insurance: 13% premium for full compensation or 11% for an *Ariane* relaunch. By the mid-1980s the Americans were feeling the pressure and acknowledged Europe's combination of technological prowess and shrewd marketing (80).

The destruction of the space shuttle *Challenger* on 28 January 1986 had a major impact on the *Ariane* programme. The Americans had been moving all their commercial payloads onto the space shuttle and had run down their production of expendable launch vehicles. This had the virtue of enabling the shuttle to be utilized to the full, albeit with the puzzling logic of risking manned spacecraft to launch unmanned satellites. When the shuttle was grounded in the aftermath of the disaster these payloads had no home, a grim situation compounded by the Americans suffering a series of losses of their expendable *Titan* and *Delta* rockets in the months that followed. The Americans did a U-turn, deciding not to use the shuttle ever again to launch commercial payloads and to rebuild their expendable launcher fleet – but this would take time. In the meantime there was a large number of commercial payloads in search of a home – which only *Ariane* could provide.

Ariane first launches

Ariane 1	24 December 1979	*Ariane 4*	15 June 1986
Ariane 3	4 August 1984	*Ariane 5*	4 June 1996
Ariane 2	31 May 1986		

Figure 4.6. *Ariane 3.*

4.13   LAST OF THE FIRST: THE *ARIANE 4*

In January 1982 ESA approved the construction of the *Ariane 4* rocket and allocated an initial development budget of €207m. The aim was to produce a flexible rocket that could compete with the top end of the world's launcher market. To give an idea of the growth of the *Ariane* rocket, *Ariane 1* was 207 tonnes and could orbit a payload of 1.7 tonnes, *Ariane 4* weighed 470 tonnes and could orbit 4.2 tonnes [81].

That *Ariane 4* was intended to be seriously competitive was apparent when the development budget was matched with the building of a new launch pad, ELA 2. The new pad was designed to service the *Ariane 4* and to achieve the then unprece-

Figure 4.7. ELA 1 and 2.

dented (outside the USSR) launch rate for a single large rocket of eight a year. It was also designed should the unthinkable happen and ELA 1 be wiped out in a catastrophic launch explosion. ELA 2 comprises a launch preparation area and a pad. The preparation area consists of an 80-m tall assembly hall and the rocket was assembled there. This marked a change with ELA 1, where assembly was done at the pad. Once checked out the rocket is brought on a double railway track down to the pad, a journey that takes one hour. The weather must be absolutely calm, for there is a slight danger of the unfuelled rocket tumbling over during a strong wind. On the pad it is surrounded by a 74-m tall handling platform and an umbilical tower, that houses all the electrical and fuel connections, which are released at launch. The solid rocket motors and payload are added at the pad itself. A turntable is provided to enable a faulty rocket to be brought back and a new one to be brought down to the pad. The new launch tower was 26 m taller to suit the *Ariane 4*. ELA 2's first mission was with the *Ariane 3*, Mission *V17* on 28 March 1986, with the American *G-star 2* and *Brazilsat 2*. With the conclusion of the *Ariane 4* programme in 2003, ELA 2 is expected to be dismantled.

The big problem for the designers was how to generate substantially greater lift-off thrust using the existing *Ariane 1, 2* and *3* technologies? Originally, they proposed to add a fifth engine on a larger first stage, but this would have led to a radical redesign of the whole first stage. Instead, the first stage of the *Ariane 3* was lengthened, with four Viking engines feeding off tanks holding 210 tonnes of propellant, much increased from 145 tonnes; a new shroud was provided; more strap-on rockets were added, either liquid or solid fuel; and the second and third stages were the same as *Ariane 3*, emphasizing some continuity.

In effect, the CNES engineers opted to modify and improve the *Ariane 3* design and make it available in a number of different variants using six different combinations of strap-on rockets, each able to lift slightly greater weights:

- no strap-ons (the 40 version);
- two solid-fuel strap-ons (42P);
- four solid-fuel strap-ons (44P);
- two liquid-fuel strap-ons (42L);
- four liquid-fuel strap-ons (44L);
- two liquid-fuel and two solid-fuel strap-ons (44LP) (L for liquid, P for solid fuel, or 'poudre' in French).

The liquid-fuel strap-on boosters were an unusual arrangement, only China having a comparable system. On the 44L version there would be four boosters each generating 67 tonnes of thrust, firing at lift-off at a 10° outward angle. Fuelled by nitric acid and UDMH they would burn for 138 seconds, dropping off when *Ariane* reached five times the speed of sound. These were the only entirely new feature of *Ariane 4*, but they were based on existing and proven rocket engines. The solid-fuel rockets were improvements over those of *Ariane 3*. The weight of propellant was increased from

7.5 tonnes to 9.5 tonnes and burn time increased by between 29.4 seconds and 40.6 seconds.

The first stage engine burned for 206 seconds instead of 140 seconds, done by stretching the fuel tank of the first stage. The L220 main stage was a much stretched version of *Ariane 3*, the length increasing by 6.7 m and the propellant capacity from 144 tonnes to 227 tonnes. The *Viking* engines were designed to fire for 205 seconds rather than 138 seconds. The main body of the rocket comprised three stainless steel tanks: one for oxidizer, one for fuel and a smaller one for water coolant. The solid rocket boosters would fire 4.2 seconds after the ignition of the main engines and would burn for 89 seconds during ascent, before dropping away. Liquid-fuelled strap-ons stayed on longer, 149 seconds, until *Ariane 4* had reached an altitude of 37.5 km. On top of the third stage was a larger vehicle equipment stage with a computer for sequencing, guidance, control, tracking, telemetry and explosives for self-destruction. The payload capacity of *Ariane 4* was raised to 4.3 tonnes into geostationary transfer orbit, with costs per kilo down 55% compared with *Ariane 1*.

A new dual-launch fairing was introduced called SPELDA (Structure Porteuse pour Lancements Double Ariane) made by British Aerospace. This carried the two satellites, one on top of the other, enabling the two to be cleanly separated one after the other. SPELDA was much lighter than its predecessor SYLDA and the nose fairing was available in a number of different variations: normal, lengthened and dual-launch (able to launch two satellites). To improve the guidance system much more accurate laser-based gyros were fitted.

So although *Ariane 4* represented considerable growth, it was based on conservative and evolutionary, not revolutionary, principles of design. The *Ariane 4* programme won the support of Belgium, Denmark, Spain, Ireland, Italy, the Netherlands, Germany, Britain, Sweden and Switzerland as well as France itself. The main contractors were Aérospatiale (first and second stage), MBB Erno (liquid-fuel rockets), SEP (engines), MATRA (equipment bay), Air Liquid (third stage tanks and insulation), BPD Snia (solids) and British Aerospace and Contraves (fairing). Development costs were €650m, including the new ELA 2 launch site.

The first *Ariane 4* flew on 15 June 1988. For the first mission CNES opted for firing the most powerful version, the 44LP, a sign of its confidence in the system. The ground controllers roared their approval as *Ariane 4* headed out over the sea. On 44LP lift-off, eight engines light up almost simultaneously (four main engines, two solid strap-ons and two liquid strap-ons). The solids fired for 50 seconds and then dropped off, making the launcher lighter. The liquid strap-ons fell off at 143 seconds, again lightening the vehicle. All went perfectly, the maiden launch putting into orbit *Meteosat P2, Amsat IIIC/Oscar 13* and *Panamsat 1*. It was also the first test of the new SPELDA fairing, allowing two satellites to be stacked one above the other. *Ariane 4* was about 15% larger than *Ariane 3*, somewhere between America's *Delta 2* and *Titan 3*.

Figure 4.8. *Ariane 4.*

Ariane 4 (44LP version)

	First stage	Strap-ons	Second stage	Third stage
Dimensions	57.1 m tall × 3.8 m diameter			
Weight	420 tonnes			
Capability	3.7 tonnes into 24-hr orbit			
	L-220	PAL, PAP	L-33	H-10-3
Dimensions	10 m × 1.4 m	4.7 m × 0.8 m	2.06 m × 0.66 m	
Weight	14.72 tonnes	2.93 tonnes	0.71 tonnes	
Engines	4 × Viking SC	4 × Viking 6, 4 × P9.5	Viking 4B	HM-7B
Fuels	UH25 and nitrogen tetroxide	CTPB and UH24, nitrogen tetroxide	UH25 and nitrogen tetroxide	Hydrogen and oxygen
Thrust	306 tonnes	152 tonnes, 132 tonnes	81 tonnes	6.4 tonnes
Burn time	205 seconds	89 seconds	125 seconds	780 seconds

By the time the first mission reached orbit, *Ariane* held 50% of the commercial launcher market. This was a development flight, and in December 1988 *Ariane 4*'s second flight was its first commercial one, putting Britain's *Skynet 4B* and Luxembourg's *Astra* satellites into orbit. By this stage *Ariane* had ten launch successes in a row, which Arianespace marked by ordering the unheard-of total of 50 new rockets to add to the 21 already in production. The production line would be ratcheted up to eight vehicles a year. About 12,000 jobs were involved. For SEP, for example, the order meant the construction of no less than 346 *Viking* engines. The order of 50 was unheard of in rocketry – no wonder the Americans now appreciated the seriousness of Europe's intent – and the order was then extended by half to 75.

4.14   COMMERCIAL SUCCESS

With *Ariane 4*, Europe hit a rich market of commercial opportunities, the 1990s seeing strong demand for commercial communications satellites. With *Ariane 4* Arianespace was able to offer a variety of funding packages from €50m to €130m, depending on the cargo, the weight and the performance required. Generally the period between signing a contract and launch was in the order of not less than 36 months. In addition, customers would be charged transport, integration and insurance charges, with a view to a free reflight in the event of a launch failure.

However, *Ariane 4* did have its share of failures too. When *V36* took off on 22 February 1990 a problem developed in the main D engine only 6 seconds into the flight. Thrust fell from 58 atmospheres to 30. The rocket began to tilt and the engines badly scorched the launch tower. The computer commanded Engine C to gimbal to compensate for the loss of thrust in engine D. As *Ariane* climbed, the other engines were commanded to gimbal too, but they had reached the extreme limits of their gimbal by 90 seconds. Despite their best efforts the whole rocket began to pitch over. It was now 12.5 km downrange. By 101 seconds the whole stack had become unstable and it exploded 9 km high over Kourou. The launch teams were told to go indoors and close the windows, in case toxic fumes were blown from seaward back inland. Two satellites were lost – the Japanese *Superbird BS2X* and the French *TDF-2*.

After the explosion *Ariane* flights were immediately suspended and a seven-person board commissioned to report within a month. Some 350 pieces of debris were recovered (including the tubing of engine D) in the subsequent exhaustive search of the coast for what had gone wrong. This was tough-going, for much of it had fallen into mangrove swamps. The review board reported quickly and deter-mined that a foreign object – probably a rag – had blocked the fuel pipe. An investigation team identified the culprit as a foreign object getting into a 44-mm water line. The review board recommended 44 changes that should take place to prevent a recurrence.

Progress was resumed with *V37*, which lifted into the skies over night-time French Guiana on 24 July 1990. To make sure there were no more foreign objects, Arianespace even fitted miniature videos into the fuel pipes. *Ariane* then had an unbroken run of 26 successes, to the point that the media rarely reported *Ariane*'s launchings any more. *V38* orbited Britain's *Skynet 4C*. *V50* saw the intro-duction of an improved third stage called the H10+. This had a new tank, 32 cm longer and 26 kg lighter than before, able to carry 340 kg more fuel, improving payload by 110 kg and burn time by 20 seconds.

The year 1994 was set to be busy with 10 launches planned. Things went wrong at the very beginning. On 24 January *Ariane 44LP* on *V63* failed at 6 minutes and 43 seconds during the third stage burn – again – and the payload of *Eutelsat 2F5* and *Turksat* crashed into the Atlantic off west Africa. A fault in the cooling system was blamed. Insurers picked up the bill for €400m. The setback had remarkably little impact on *Ariane*'s order book, which actually lengthened after the crash. The problem did not take long to resolve and *Ariane* was airborne again in June (*V64*), July (*V65*), August (*V66*) and September (*V67*).

Business was back to normal again – or so everyone thought. The troublesome third stage cut out prematurely minutes following launch of *V70* on 30 November 1994 and *Panamsat 3* was lost, crashing 1,100 km off west Africa. Pressure in the gas generator became stuck at 70% of normal, far too little to get the payload to orbit. This was attributed to a leak in the oxygen feeding circuit and the review board recommended 21 changes to prevent a recurrence. *Ariane* was grounded again, its seventh failure. The makers of *Panamsat* announced they would fly a replacement – but on a Russian *Proton* rocket, not an *Ariane*. Insurance premia on the *Ariane* rose from 17.5% to 20%.

Arianespace worked hard to get *Ariane* aloft again, anxious not to lose time or business. It took them four months, with *V71* flying smoothly on 28 March 1995 carrying aloft two comsats, *Hot Bird 1* and *Brazilsat B-2*; and *ERS 2* reached orbit on 20 April 1995 on *V72*. It was the last *Ariane 4* failure.

ARIANE 1   ARIANE 2   ARIANE 3   ARIANE 4   ARIANE 5   ARIANE 5 Evolution

Figure 4.9. *Ariane* launch family.

4.15   *ARIANE 5*

The first ideas for *Ariane 5* as a heavyweight launcher began to emerge in 1977. Studies of the demand for geostationary satellites indicated that by the end of the 1990s *Ariane 4* would no longer be equal to the task. A more powerful, probably hydrogen-fuelled launcher would be required, one capable of lifting 5.5-tonne payloads into geosynchronous orbits. Such a proposal was presented within CNES in 1979, even before *Ariane 1* had flown. *Ariane 5* would have a liquid-hydrogen-powered first stage, marking a significant break with *Ariane 1, 2, 3* and *4* (though there was as yet no mention of large solid rocket boosters) [82]. Putting a manned spaceplane into orbit was pencilled in as another possible objective of the programme. Technical studies went ahead.

In May 1982 Aérospatiale first publicly put forward the *Ariane 5* concept, able to put up to 15 tonnes into low Earth orbit. The idea of using powerful solid rocket boosters on the side of the new rocket, on the same lines as the American space shuttle, was first mooted. In January 1983 ESA responded formally to the concept, approving a €13m space transportation systems and a long-term preparatory programme to look at the options for Europe after *Ariane 4*. The ESA council in June 1984 agreed to begin initial work on a new, large cryogenic engine, the HM-60.

Development of a new engine was one of the most critical stages in the whole project [83]. The development programme for the new *Vulcain* engine alone cost €850m, involving over 200 tests in Vernon (70 km west of Paris), Lampoldshausen and Ottobrunn in Germany. The 1,100-kg engine was required to deliver 800 kN of thrust at pressures of 100 atmospheres for 10 minutes, with temperatures reaching up to 3,200°C. From 1985 SEP began work on the turbopumps, gas generators and thrust chambers. During the 10-min burn, the engine would consume 26 tonnes of liquid hydrogen and 132 tonnes of liquid oxygen. SEP carried out 500 ground tests, with burn times of up to 900 seconds, one and a half times that required during a standard mission. To verify its performance, SEP set up a mini mission control and a monitoring system noting 50,000 points of performance on 600 channels every second.

Ariane 5 was formally adopted as an ESA programme at the Rome council meeting in January 1985. It was a massive programme: 6,000 people involved full-time, 6,000 work contracts, 1,100 industrial firms, 70 million production hours. A version of *Ariane 5* would, it was agreed, be able to put into orbit a manned European spaceplane *Hermes*. In July 1986 *Ariane 5* was upgraded so as to ensure it could put 5.2 tonnes in geostationary transfer orbit and to improve safety margins for *Hermes*. The intention of *Ariane 5* was to offer a 20% cost reduction per kilo compared with *Ariane 4*. Studies were also carried out into making parts of *Ariane 5* reusable, but despite showing some early promise this came to nothing.

4.16 A RADICALLY NEW SYSTEM

Although a continuation of the *Ariane* family and with the same family name, *Ariane 5* was a radically new launcher. *Ariane 2, 3* and *4* had all built on what had gone before, but *5* owed little to them. The original designs for *Ariane 5* showed it to be essentially a larger version of the *Ariane 4*. By autumn 1986 quite different illustrations were published of the designers' thinking. They showed a central core being powered by two large, solid rocket boosters, rather like the bottom of the then-grounded American shuttle. *Ariane 5*'s increased lift was achieved by the use of hydrogen fuel, welded instead of bolted booster casings and a much more powerful *Vulcain* engine with a larger throat, richer mixture, and higher chamber pressure.

The costs of *Ariane 5* were agreed as follows:

Ariane 5 programme participants (%)

France	46.2	Sweden	2
Germany	22	Switzerland	2
Italy	15	Norway	0.6
Belgium	6	Austria	0.4
Spain	3	Denmark	0.2
The Netherlands	2.1	Ireland	0.2

Figure 4.10. *Ariane 5* solid rocket boosters.

Britain did not participate, saying that Europe's needs could be best met by *Ariane 4*. Thus, when *Ariane 5* flew, the flags of the countries above were on it, but not the Union Jack. The development cost was €718m, with a 20% contingency allowance. Several years later the ESA council allocated a further €1,026m to fund upgrades to the launcher. The final development costs, by the time of the 1996 launch, were quoted as high as €8bn.

Final approval for the *Ariane 5* was confirmed at the famous, or infamous Hague council meeting of November 1987, when it was supported by 12 member states, but clearly opposed by Britain. First, approval of the programme was considered essential if Europe was to launch large dual payloads to 24-hr orbit in the late

1990s. Second, *Ariane 5* was seen as vital for any European manned space programme capability. Because of both these factors a key element in the programme was the target of 98% reliability for *Ariane 5*, compared to 90% for its predecessors.

4.17 TECHNICAL CHARACTERISTICS

What of its technical characteristics? *Ariane 5* was 5.4 m in diameter, 51.37 m tall in its highest configuration and weighed 746 tonnes at launch. The big main stage was 30 m tall and burned for 580 seconds, powered by a single 1,300-kg Vulcain HM-60 engine. The HM-60 dated to the early 1980s when the first theoretical studies were done by SEP. HM stood for 'hydrogen motor' and '60' because it is 60 tonnes in weight. The HM-60 was comparable in size with the Space Shuttle Main Engine (SSME) – but of more conservative design. SEP went for an engine that used a separate generator to drive the turbopumps, while the SSME pre-burned the propellant to do so. The SEP approach was less efficient, but quicker and cheaper to develop. The HM-60 engine could not be throttled, unlike the shuttle. It was able to achieve a pressure of 100 atmospheres and a thrust of 91 tonnes – much less than the SSME, but still a significant breakthrough for Europe. SEP's early start on the HM-60 made a big difference, as it takes about 10 years to develop and certify a new engine on this scale, overcome the problems of turbopumps and sort out seals and bearings. The management of the HM-60 programme emphasized simplicity, the ability to inspect and check systems easily and reliability, over performance. Even if the HM-60 was not a revolutionary design, the scale of the motor was much larger than anything Europe had done up to that point. Even though it was an expensive project, ways had to be found to keep costs down.

It was still a monster. The main stage held 130 tonnes of liquid oxygen and 25 tonnes of liquid hydrogen in a 15-tonne shell. The Vulcain engine was required to burn for 10 minutes with turbopumps spinning at 33,500 r.p.m. Only when it reaches full thrust does the signal go for solid rocket booster ignition.

There was huge power in the strap-on side boosters: the solid rocket boosters each weighed 39 tonnes empty, 237 tonnes fuelled, generated 750 tonnes of thrust, were 3 m in diameter and 30 m high. They would fire for 132 seconds before dropping away at between 55 km and 70 km altitude. For the first two minutes of flight, until it reached 60 km, 90% of the rocket's thrust would be provided by the solid fuel boosters. The solid rocket boosters used lightly welded joints instead of the American system of rubber 0-rings that failed so disastrously for *Challenger*.

The upper stage was quite short, 4.5 m long, 5.4 m diameter, weighed 6 tonnes, carried 1.7 tonnes of monomethyl hydrazine (MMH), 3.5 tonnes of nitrogen tetroxide and had a thrust of two tonnes. The Aestus upper stage was designed to separate at 140 km, fire for 1,100 seconds and was restartable.

Figure 4.11. *Ariane 5* bulkhead.

Ariane 5

Dimensions	54.8 m tall × 11.4 m diameter
Weight	746 tonnes
Capability	6.8 tonnes into 24-hr orbit

	First stage EAP H155	Strap-on EPC	Second stage EPS
Dimensions	30.7 m × 5.4 m	4.7 m × 0.8 m	4.5 m × 5.4 m
Weight	167.5 tonnes	237 tonnes	1,190 kg (dry)
Engines	Vulcain HM-60	2 MPS	Aestus
Fuels	Hydrogen and oxygen	Solid	MMH and UDMH
Thrust	104 tonnes	1,340 tonnes	2,000 tonnes
Burn time	580 seconds	132 seconds	1,100 seconds

Aérospatiale was the chief contractor, responsible for the first stage and the solid-fuel boosters. Matra Marconi made the equipment bay, CASA and Saab Ericsson the payload adaptors, Oerlikon-Contraves the top shroud. Daimler Benz in Bremen made the storable propellant stage. The fuel tanks were made by DASA, CASA and the Zeppelin company in Friedrichshafen. The solid rocket motors were made by Europropulsion, a consortium of Fiat and SEP. The first stage *Vulcain* main engine was made by SEP and 17 subcontractors at Vernon, 70 km west of Paris. MAN Technologie of Germany had responsibility for the case segments of the solid rocket boosters, the skirt of the main stage, the joint of the *Vulcain* engine and tanks for the upper stage and became one of the main contractors outside France.

Ariane 5 had a major non-French contractor. This was DASA (Deutsche Aerospace), headquartered in Munich but with facilities in Friedrichshafen and Leopoldshafen. The attitude control system was called SCA (Système de Contrôle d'Attitude), and it was responsible for guiding *Ariane 5* from booster separation to satellite ejection. DASA was responsible for the upper stage and the overall attitude control system. There was a new fairing called SPELTRA, (*Structure Porteuse Externe de Lancements Triples Arianes*), made by Dornier of 7-m tall, 5.4-m diameter carbon fibre. This was the reinforced top of the *Ariane* where one or two satellites were housed. Despite its size and strength, it weighed only 850 kg.

Ariane 5 had a relatively smooth development history. The first version of the new *Vulcain* engine was tested in July 1990 and the solid-fuel booster in February 1993. Its new main engines were successfully test-fired on 17 November 1994 and again on 8 December. A series of testing milestones were passed relatively smoothly over 1993–5, one exception being a tragic incident in May 1995 when two technicians died from nitrogen poisoning when a drainage plug was not sealed properly.

4.18   NEW PAD AT KOUROU

Ariane 5 required a new launch pad and a substantial expansion of Kourou. The new pad was called ELA 3. ELA 3 was approved in 1988 for the *Ariane 5* launcher and aimed to handle up to eight launches a year. ELA 3 was a massive piece of construction, cutting a swathe through the coastal plain. The much larger scale of *Ariane 5* required canalization of the river Kourou (to take the barges with the centre stage), building of a new national road, making an artificial lake and a new hydroelectric power plant. This needed 2,100 ha, 40 km of new roads, a 7-km railway, 4 million m^3 of earthworks, 80,000 m^3 of concrete and 7,000 tonnes of metal.

The year 1992 saw the completion of a propellant plant for *Ariane 5*. This covered 300 ha and comprised buildings, mixers, casting pits and control facilities. The purpose of the plant was to mix and pour the fuels used for the solid rocket segments of *Ariane* (a small amount would be made in Italy). As a result, the solid rocket fuel is made on site (although a small segment is transported from Italy). The company responsible is the French–Italian Regulus, which can make enough solid fuel for up to 10 *Ariane 5*s per year. About 120 local people are employed in the fuelling. There are *Ariane 5* tracking stations in the small hills of Montagne des Pères 10 km away and at Montabo, Cayenne, 60 km downrange.

Once the two main stages arrive in the Centre, they go to the 58-m-high Launcher Integration Building. The solid-fuel boosters, having been fuelled at the solid propellant plant, are then attached vertically on top of the mobile launch table. The topless rocket, without payload, is then moved by rail to the Final Assembly Building, which is 83 m high. Here, in clean room conditions and an air-conditioned environment, encapsulation takes place. ELA 3 provided for a faster processing time and aimed to cut launch campaigns down to 21 days. On ELA 1 and ELA 2, the rockets were prepared on their pads, out of doors, surrounded by a high service tower. With ELA 3 the rocket is stacked and prepared in three buildings – the Bâtiment d'Intégration Propulseur, the Bâtiment d'Intégration Lanceur and the Bâtiment d'Assemblage The upper stage is fuelled before the entire *Ariane 5* is brought to the pad on a stand that crawls along a wide-gauge railway. The rail crawler brings the completed *Ariane 5* down to the pad 2,800 m distant where it is fuelled from tanks below the mobile launch table – a fast system based on the Ukrainian *Zenit*. All the systems are checked out, the final fuels are loaded, the launcher is armed and the batteries are powered up. The structures at the pad are actually quite limited, enough to hold the rocket for fuelling and lift-off. Beside the pad are four lightning towers and a water tower holding 1,500 m^3 for coolant.

A new launch control centre was built for *Ariane 5*, called Jupiter 2, 15 km distant and able to withstand a launcher explosion. A typical countdown lasts 6 hours and has a high degree of computer control, especially to supervise the fuelling. The control room has two tiers – an inner, small control room of four ranks for consoles and behind a glass screen a large auditorium with seating for journalists and visitors.

A giant, new processing facility was completed in Kourou in April 2001, called blandly S5. The general purpose of the €64m facility was to provide much improved handling facilities for the payloads of *Ariane 5*, but specifically to handle large payloads such as *Envisat* and satellites for the International Space Station (e.g., Automated Transfer Vehicle [ATV]). Payloads and components assembled in many different locations could now be assembled on one single floor. S5 has a floor area of 3,800 m² and can handle four geosynchronous comsats and eight medium-size ones at a time, separated by screens as required. The set of buildings, cut into the French Guiana jungle, comprises non-hazardous checkout areas, two hazardous processing halls and transfer corridors, combining the necessary mixture of supervision and security [84].

4.19 FIRST FLIGHT OF *ARIANE 5*

Arianespace set 29 November 1995 as the date for a first launch. This slipped by six months – not bad for an entirely new rocket system – and *Ariane 5* counted down on 4 June 1996 in front of the world's press. By the time of the launch, ESA, CNES and Arianespace were confident. The first mission was given a new title V501. Two demonstration flights were planned before *Ariane* entered a commercial series. The first would carry the four *Cluster* scientific satellites; the second an atmospheric re-entry demonstrator. By this stage *Ariane 5* had cost €6.1bn. This was about 20.9% over-budget over an 11-yr period, but not bad when compared with many other rocket programmes [85].

There was an hour's delay due to cloudy weather. Lift-off took place at 9.35 a.m. Grey and white clouds billowed around the stand as *Ariane 5* rose to the cheers of scientists and engineers. It rose serenely. Then, all of a sudden, half a minute into the flight, it could be seen to tilt, veer off-course and was promptly destroyed by the range safety officer. A huge fireball erupted in the sky, a massive umbrella shape of deadly gases spewing out and raining down on the launch site where journalists, VIPs and visitors scurried to take cover under trees and cars and buildings and hid until the clouds cleared. It was one of the most televised and spectacular television images of the year. At least, unlike *Challenger*, no-one was hurt. The black box was recovered from the nearby mangrove swamp. As for the *Cluster* scientists, they were devastated. They had worked 10 years to put their cargoes onto this mission. When pressed for radio and television interviews, some stammered, some broke down while others sobbed inconsolably in the corner. Classicists added to the gloom by reminding everyone that Ariane, after whom the rocket was named, had a brief moment of glory rescuing Theseus from the Minotaur – but then came to a sticky end.

Immediate analysis pinpointed the start of the disaster to the 30-sec point in the launch. The solid rocket boosters could be suddenly seen to gimbal 6.6° down, followed by the *Vulcain* main engine. At this stage *Ariane* was at an altitude of 3,400 m, 2,000 m downrange and travelling at Mach 0.73. By 31.3 seconds, the rocket had rotated to an angle of attack of 30° and the upper stage ruptured. The rocket went to a 90° angle and the destruct command went in. Last telemetry was received at 34 seconds. What caused this sudden, erratic, manoeuvring behaviour?

ESA put a brave face on the disaster and expressed confidence that ultimately *Ariane* would be a success. A board of investigation was commissioned to report by 15 July 1996. The cause of the accident was at first a mystery. The board had to piece together the following jigsaw of information. There had been a hold at 7 minutes in the countdown to await better weather (not considered significant when the investigation began). The following was the sequence:

- the back-up inertial guidance system failed at 36.7 seconds;
- the main guidance system failed immediately thereafter at 36.75 seconds;
- having lost their guidance, the nozzles swivelled to their extreme position (after 39 seconds). *Ariane* steered off-course and had to be destroyed (after 41 seconds).

The board of enquiry reported on 19 July 1996. What the investigative group found was this (and this description is a simplification). In the event, the hold in the countdown had proved to be a key factor. To steer *Ariane 5* through the ascent, it used a software programme, which is normal. *Ariane 5* used elements drawn from the *Ariane 4* software programme, into which was pre-programmed the slow launch ascent profile of *Ariane 4*. This code also had a procedure for updating the launch profile with respect to the movement of the Earth every time there was a countdown hold and, had the hold not taken place, would never have come into play. Once the hold took place, the system triggered the *Ariane 4* software. The *Ariane 5* software was not protected against an inappropriate intervention from the *Ariane 4* software.

Once *Ariane 5* was airborne late, it became reliant on the *Ariane 4* software. This now found that the *Ariane 5* vehicle was not following the proper ascent profile for an *Ariane 4*, which by definition it could not. When the new *Ariane 5* exceeded the old *Ariane 4* limits at 36 seconds the system crashed. Deprived of commands the engines had gimballed wildly. Had the countdown delay not taken place this launch might well have been successful, but an accident would have been triggered sooner or later. The board simulated the launch using these assumptions and replicated the accident immediately. It issued instructions that, in future, software was to be purpose-built and double-checked and that redundant systems from previous launchers was not to be used for convenience or to save costs. It was, like most launch accidents, an embarrassing and avoidable disaster, the kind of thing that always looks obvious in retrospect. *Ariane* was supposed to achieve a reliability rate of 98.5%, which means only two accidents in 150 launches, so it was already living on borrowed time.

The redesign of *Ariane 5* added a further 5% to the budget (€208m) and lost the programme about one year. ESA decided that there must be *two successful* qualification flights before going to a commercial mission. *Ariane 4* meantime continued its work, with a run of successful missions over the summer.

4.20   REBUILDING CONFIDENCE

For a payload the second flight *Ariane 502* was to carry a 4.7-tonne, mock double comsat, with two microsatellites. These large models, built by Kayser Threde, were to be subjected to the same experience as real satellites and accordingly were rigged with sensors to measure acceleration, shock, noise and vibration. The microsatellites were *AMSAT* and *TEAMSAT*. *AMSAT* was a 550-kg radio amateur satellite to be attached to one of the comsat models and left in an orbit of 500 km by 35,000 km. *TEAMSAT* was to be deployed from the other comsat model and was a 350-kg payload with five student experiments devised by European universities under the supervision of former ESA astronaut Wubbo Ockels. These experiments were to investigate telemetry systems, atomic oxygen in Earth orbit, orbital debris, navigation and tether systems with subsatellites.

After 15 months of redesign and modifications, all was ready for a new *Ariane 5* attempt on 30 October 1997. The countdown had some hairy moments, sufficient to add to the already high sense of tension. The countdown was relayed throughout ESA centres across Europe. The console lights went red at $T-48$ seconds, due to a minor electrical problem that took 30 minutes to clear. After some delays due to telemetry problems and poor weather, all was ready at 10.43 a.m. Within 0.35 seconds of ignition of the P230 solid rocket boosters *Ariane 5* was on its way, on pictures broadcast throughout the French and European news broadcasting networks. In the European relay centres the television transmission broke down at $T-13$ seconds – though the radio commentary could still be heard. Hearts almost stopped as *Ariane* passed the 30-sec mark. At 90 seconds, the television picture came back and observers could see *Ariane* reassuringly sailing on majestically, climbing ever higher, releasing its solid rocket boosters at 2 minutes 2 seconds at 60 km. The first stage engine burned out at 9 minutes. After a short period of coast the *Aestus* upper stage ignited. Here there was a problem for *Aestus* cut out at 18 minutes and 33 seconds, just 6.5 seconds early, leaving the mock comsat and microsat payload in a lower orbit than that originally intended. Leaving this flaw aside, Jupiter control felt it could declare the mission a success at the 30-min point. ESA breathed a sigh of relief and the champagne was uncorked. It was the 101st *Ariane* mission. They had done it at last. Europe had its new rocket.

Despite this a number of loose ends required investigation after the mission. Analysis of post-flight data found high roll rates at the end of the main engine burn. Exhaustive simulations found that this was due to rough surfaces on the turbine exhaust areas, which encouraged the rocket to start spiralling and also led to the premature shutdown. Had a real satellite been launched it would have been left in a less-than-perfect orbit. ESA was not confident of going straight to a commercial launch and its original planned *W2* comsat was switched to *Ariane 4*. Instead, Kayser Threde was commissioned to make a dummy satellite *Maqsat 3* of the same dimensions as *W2*, a feat that it pulled off in a record 11 weeks. Also onboard would be an *Apollo*-style capsule to be recovered, the purpose being to test out capsules that Europe itself might one day use for manned missions to and from the International Space Station.

Figure 4.12. *Ariane 5* take-off.

The third and final test flight was set for the autumn of the following year. On 21 October 1998 the engineers and visitors still watched anxiously as the hydrogen-powered stage headed into the clear-blue South American skies. The only snag was that the parachute of one of the spent solid rocket boosters failed, so it could not be recovered. Once in orbit *Ariane 503* released its first payload *Maqsat* into a 1,027 km by 35,863 km geostationary transfer orbit.

Ariane now proceeded to the trickiest and most challenging part of the flight – the mission involving a half-scale *Apollo*-type capsule called ARD (Atmospheric Reentry Demonstrator). ARD was released 12 minutes after launch at 209 km to follow a trajectory 830 km high. Using its own seven 400-kN thrusters and navigating by means of GPS (Global Positioning System), ARD manoeuvred through a high-speed re-entry, coming in at 120 km at a speed of 27,130 km/hr. Parachutes opened 30 km high over the Pacific, the cabin descending gently 4.9 km from the Marquesa islands south of Hawaii. ARD followed a profile similar to the early unmanned *Apollo* missions (*4* and *6*). The recovery of ARD following a flawless launch left mission controllers not just relieved but delighted as well. Maybe the jinx had been shaken free at last. Confidence in *Ariane* had now picked up again. The order book stood at 43 satellites worth about €5bn, with the total number ordered since 1980 passing the 200 mark. ARD went on display at the next Paris air show, upside down so that viewers could admire the way in which its heat shield had withstood re-entry.

4.21 *ARIANE 5* OPERATIONAL

Ariane 5's first operational mission came on 10 December 1999, when it put into orbit Europe's most ambitious ever telescope project. *XMM* (X-ray Multi Mirror), the satellite in question, was a long, black telescope in the shape of a camera lens. *Ariane 5* gave the nervous scientists no cause for worry and put *XMM* into its planned eccentric initial orbit of 825.6 km by 113,946 km, within 1.5 km of perfect accuracy. On 21 March 2001 *Ariane 5* launched through dark skies to put up two commercial satellite payloads – the digital *Asiastar* and India's *INSAT 3* (*Ariane 505*). This was a vindication of the *raison d'être* of *Ariane 5*: to launch two large commercial comsats at a time. ESA quickly pointed out that double launches reduced costs to clients, made the *Ariane 5* more internationally competitive and brought more profit. At the time Arianespace had 39 satellites on its order book and 50% of the world launcher market.

With *V135*, *Ariane 5* put into 24-hr orbit the heaviest ever commercial payload *Panamsat 1R* and then three small satellites, four payloads altogether, two being for the British defence research agency DERA (Defence Evaluation & Research Agency, subsequently renamed QinetiQ).

4.22 KOUROU TODAY

By the new century the Kourou site had expanded considerably, covering 96,000 ha with 1,400 permanent staff and many more present at the culmination of launch campaigns. The launch centre had made a considerable impact on the local economy, contributing to the building of bridges, roads, schools, hotels and restaurants. Nowadays, Kourou is called the CSG (Centre Spatial Guyanais). It is owned by CNES, made available to ESA and Arianespace and 70% paid for by France as part of its contribution to ESA. With the completion of ELA 3, the Centre will probably retain the same size for the moment, though at one stage there were plans for the building of a shuttle runway for the *Hermes* spaceplane.

Figure 4.13. Modern Kourou.

What of the CSG nowadays? The space centre accounts for 50% of the local economy. The population has risen seven times since it became a space centre. Both the French government and CNES have put considerable efforts in local development plans, knowing that maintaining political stability and well-being is essential to the space programme. The department (French Guiana) has, even despite the space programme, high rates of local unemployment. The government has tried to respond to this by building up indigenous industry that employs local people. Efforts are made to preserve traditional crafts, fishing and hunting. The solid rocket fuels are mixed there, instead of being imported from Russia. The latest project is to develop local water-bottling facilities, instead of importing water from the Antilles. There are plans for sugar factories and brick-making plants in a drive for import substitution. Very few tourists or visitors come to French Guiana – yet as a tourist destination it could offer an unusually exotic combination of virgin jungle, rocket launches and a horrific penal colony. CSG is very conscious of preserving the local environment and has installed water and air meters and makes inventories of the local bird, fish and wild life.

What of the Iles de Salut? They have been given national heritage status. Devil's Island, the smallest (14 ha) has been closed and left to its memories. Ile Saint Joseph (20 ha) has been given over to its wildlife. Ile Royale, much the largest (62 ha), is a tracking station for the ascending *Ariane*. A ferry service has opened to Kourou and the guards' barracks have become a hotel. The dormitories for prisoners have now been converted to accommodate between 150 and 200 tourists. Those buildings that could be salvaged have been listed and refurbished, like the hospital and the church. The prison gardens have been renovated into a botanical park. Some parts of the old colony were so far beyond repair, like the workshops, the asylum and death row, that the jungle has been allowed to reclaim them. Capt. Dreyfus's cell has been rebuilt – and was reopened by his grandson in 1995.

4.23   COMMENTS AND CONCLUSIONS

Few could have imagined when the *L3S* was drawn up that they were witnessing the birth of one of the most successful rockets in history. If there is one name synonymous with Europe in space, it is *Ariane*. Its creation was inauspicious, set against a background of quarrelling, divergent objectives and organizational chaos.

Ariane owed much to French foresightedness, presenting the proposal at a time of extreme discouragement in the development of European launchers. CNES has been single-minded in pressing the programme through and was given consistent support from the French government. Failures were analysed carefully and the lessons applied thoroughly. Proposals for the next stage of development were presented and approved in a steady pipeline. CNES applied the lesson arising from ELDO (namely, the importance of a controlling, integrated management). For all its success the *Ariane* design was actually quite conservative, each step being a logical, often a small step beyond what had been achieved before, keeping costs down, time lines on schedule, avoiding overruns and building confidence.

Just as *Ariane* showed that Europe could learn the technical lessons arising from its earlier experience in rocketry, the new ESA was an attempt to apply the political lessons. The period 1973 to 1975, which marked its establishment, must have offered little confidence for ultimate success. However, the politicians, diplomats, civil servants and negotiators, especially in July 1973 the Belgians, crafted a clever formula that would avoid conflict (opting out), provide stability by setting down a small proportion of mandatory programmes (administration and science), give the three big countries leadership of their pet projects (*Ariane, Spacelab, MAROTS*), incentivize working together (opting in) and ensure there were clear lines of management (either ESA directly, or delegation to CNES). The formula was broader and deeper than the many formulas that had failed ELDO and ESA in the past.

In delivering a launcher programme, ESA managed to deliver the ELDO part of its heritage. But what about the ELDO legacy: the scientific programme. Could ESA do this too? This is now examined by Chapter 5.

5

Scientific and applications programmes

Of the two precursor organizations to ESA (European Space Agency), ESRO (European Space Research Organization) had been much more successful than ELDO (European Launch Development Organization). In the last year before the collapse of ESRO and ELDO, the former had seen several satellite projects come to fruition. Several were still in the pipeline and these continued to progress despite the many political changes surrounding the establishment of ESA: *COS* and *GEOS*. So were several collaborative missions with the Americans, like *IUE* and *AMPTE*. With this experience behind it Europe was then set for a dramatic mission to a comet in the mid-1980s, where it made a spectacular mark on space science. Here we review European activities in space science, communications and applications, in that order.

5.1 EUROPEAN SCIENTIFIC MISSIONS

In the 1970s ESA carried out scientific missions on a case-by-case basis and on the basis of what it inherited from ESRO. Missions had been proposed by scientists, researchers and industry, had been peer-reviewed, eliminated, approved or amended and then selected as finances and priorities permitted. ESRO had seen seven satellites put into orbit and more in preparation.

First of these was *COS B*. This was Europe's first astronomy satellite. The 280-kg *COS B* had the distinction of recording the first X-ray emissions from a quasar OX 169, finding that their pattern changed every six hours. First analysis of the data revealed that black holes had a huge mass, up to a million times greater than the Sun. Despite a design life of only one year and with spare propellant for three years, *COS B* operated for seven, until 1983.

GEOS was the ninth European satellite and the first to reach geosynchronous orbit. It was the first-ever *scientific* satellite intended for geostationary orbit, the first with a high data relay rate (over 100 kb.p.s.) and the first ESA satellite with a

Figure 5.1. *COS B* ready for launch.

hydrazine-based attitude control system. Weighing 524 kg it was drum-shaped and had nine instruments for studying the magnetosphere and electrical fields. The prime contractor was British Aerospace, assisted by 15 companies in 10 member states. One of the most important features was a new solid propellant apogee booster motor 1.1 m long and 305 kg in weight made by SBIA Viscosa and SEP in France and flown for the first time.

When it was launched on 19 April 1977 there was a malfunction on the *Delta* upper stage, leaving it far short of its intended orbit, only 11,040 km instead of

36,000 km, and spinning at 2 r.p.m. rather than 90 r.p.m. The mission was salvaged and, with the instruments working properly, declared operational in May even though the orbit was imperfect. *GEOS* was controlled by Darmstadt and sent data back at the rate of 20,000 characters or 10 pages a second.

With *GEOS 1* only a partial success, the back-up model was taken out of storage at ESTEC (European Space Technology Centre) at Noordwijk, the Netherlands, shipped to the United States and launched in 1978. *GEOS 2* measured electrical, particle and magnetic fields and the magnetosphere for two years until the end of its mission in July 1980. Germany and Switzerland funded a follow-on mission so it was taken out of retirement for further work until its fuel supply ran out in 1983 and then fired into a graveyard orbit out of harm's way.

Exosat was one of ESA's first major science projects and its first X-ray observatory. Nine institutes from six countries were involved, the main companies being Germany's MBB (Messerschmidt Bölkow Blohm) and Krupp, France's Aérospatiale, Spain's CASA, Belgium's ETCA, MSDA of the UK and Selenia of Italy. Originally, it was to ride an *Ariane* to orbit, but was switched due to the loss of *Ariane V5* and the resultant delays. *Exosat* was launched on a *Delta* from Vandenberg Air Force Base in California on 26 May 1983 into an eccentric orbit of 340 km by 192,000 km, 72.5°, with the object of detecting X-ray sources in our Galaxy and others further afield. It was a smaller spacecraft than its contemporary *IRAS*, 3.3 m tall and weighing 510 kg, but made over 2,000 observations of objects, which might be neutron stars and black holes. *Exosat* operated by a mixture of direct observations for up to 80 hours at a time of particular objects and by lunar occultation (watching objects disappear behind and then reappear from behind the Moon). This was a technique that permitted objects to be located and mapped with considerable accuracy. *Exosat* operated for three years. One of its most remarkable discoveries was of a white dwarf star rotating every 11 minutes in the gravitational grip of a neutron star in the constellation of Sagittarius, 20,000 light years distant.

5.2 EARLY COLLABORATIVE MISSIONS

Turning now to the early collaborative missions, *IUE* (International Ultraviolet Explorer) was originally an ESRO, NASA (National Aeronautics and Space Administration) and UK Science Council project. Although this was not a big telescope, it had a high observational efficiency and could make long, 14-hr exposures of stars as dim as 17th magnitude. NASA supplied, built and operated *IUE*, ESA supplying the solar arrays and Britain the television cameras and image-processing software. *IUE* used a *Delta* to go into a highly eccentric orbit of 46,080 km by 251,200 km on 26 January 1978. *IUE* took the first high-resolution ultraviolet spectrum of a star in another galaxy, the first ultraviolet impression of a supernova, the emission lines of faint sources and pictures of globular clusters 15,000 light years away. *IUE* weighed 320 kg and had a 50-cm telescope. In its first three years *IUE* collected 12,000 ultraviolet spectra, providing fresh information about solar winds, quasars, supernovae and X-ray binaries.

AMPTE (Active Magnetospheric Particle Trace Explorer) was a three-part project between the United States and Germany, with Britain joining later. *AMPTE* aimed to study how the solar wind entered and behaved in the Earth's magnetosphere and its tail, by releasing trace ions of lithium and barium into the magnetospheric stream to identify the pattern. There were three satellites: the American one (closest to Earth), the German one (releasing the tracers) and its British subsatellite called *UKS*. The subsatellite was 1 m diameter, 74 kg, with two booms to study natural magnetic particles, fields and waves. The resulting traces were extremely tenuous, with 1 kg able to make a cloud 30,000 km in diameter. A cloud of barium atoms was released high over the Pacific on 27 December 1984, making an artificial comet, which solar winds then drove away from the Sun's direction.

ISEE 1 and *2* (International Sun Earth Explorer) were collaborative missions between NASA and ESA, with NASA responsible for *ISEE 1* and ESA for *ISEE 2*. The two were launched together on a *Delta* on 22 October 1977, with a separation manoeuvre being performed the second day. The larger *ISEE 1*, weighing 320 kg, entered an orbit of 280 km by 138,124 km while *ISEE 2*, weighing 158 kg, flew from an almost identical 279 km to 138,330 km. Their aim was to observe the 11-year solar cycle that had begun in June 1976 and watch its effects on the magnetosphere, ionosphere and upper atmosphere. They were successors to the series of highly successful Interplanetary Monitoring Platform series of satellites flown by NASA from the 1960s. *ISEE 1* and *2* involved 117 principal investigators from 10 countries.

ISEE 3 was a 460-kg spacecraft, but carried as much as 94 kg of propellant in case the *Delta*'s injection was inaccurate, but the opposite was the case. *ISEE 3* was put into what is called a halo orbit on the inner side of the Earth from the Sun on 12 August 1978. The halo orbit was also known as the L1 libration point – a position in space 1.5 million km on the sunny side of the Earth where solar and earthly gravitational forces balance one another and where the satellite moves in the form of a circle, one perfectly positioned to study solar phenomena as they begin to interact with the Earth. Here it explored the reaction between the incoming solar wind and the Earth's geomagnetic environment between 510,000 km and 1.6 million km from Earth. Of *ISEE*'s 13 scientific instruments, three came from Europe (France, Germany and the Netherlands). The first was to investigate protons in outer space, developed by the Space Research Laboratory of Utrecht and Imperial College London; the second to study the composition of solar particles (Max Planck Institute); and the third an experiment to examine magnetic field lines (Paris Observatory, Meudon).

5.3 THE IDEA OF A COMET MISSION

ISEE's mission took place at the time of closest approach to the inner solar system of comet Halley. This was the best known of all the comets, circling the solar system every 76 years. Its appearance had been first catalogued by Chinese astronomers, it was known in Julius Caesar's time and was clearly shown in the Bayeux tapestry when it appeared before William the Conqueror's invasion of England. It had last been seen over the skies of Europe in 1910.

Originally, the United States proposed to lead a project to intercept comet Halley, with ESA taking responsibility for a solar electric propulsion spacecraft [86]. However, the United States were not able to get approval for a probe (funds for space missions were starved during the Carter presidency). When the 1981 NASA budget was published the previous summer, 1980, funding was so tight that NASA was unable to continue its participation. Instead, as comet Halley began to come closer to the inner solar system, a mission was devised for *ISEE 3* by Dr Robert Farquhar of NASA's Goddard Centre [87]. *ISEE* was commanded in 1982 to make a series of course adjustments to intercept a small comet in advance of Halley's arrival. *ISEE 3* looped five times around the moon, at one time zooming to only 194 km over the lunar surface, generating more and more momentum on each fly-by and sufficient to hurtle it out of the Earth–Moon system toward comet Giacobini Zinner. This comet had been discovered by Giacobini in Nice in 1900 and rediscovered by Zinner in 1913, orbiting the Sun every six years or so and associated with the Draconid meteor shower.

ISEE 3, now renamed *ICE* (International Cometary Explorer), crossed the tail of Giacobini Zinner on 11 September 1985, the first spacecraft to intercept a comet [88]. The fly-by distance was 7,900 km, sufficiently near to the nucleus to obtain useful information and sufficiently far into the tail to learn about its properties. Of the 13 instruments on board, 7 were suitable for cometary observations. These were able to pick up the bow shock of the comet, the comet's electronic ions, the region of turbulence between the solar wind and the comet's own magnetic forces and the levels of cometary dust (less than feared, giving hope to later comet interceptors). Although *ISEE 3* did not have a camera, it was able to pick up useful data about the comet's tail [89].

5.4   *GIOTTO*

So the United States for financial reasons could not get to close quarters with comet Halley. But what about Europe? In the late 1970s, ESA priorities in this area were determined by the Solar System Working Group. In 1978 the Working Group selected a lunar polar mission as ESA's prime objective, deeming a comet mission too difficult. A stellar-observing mission, which later emerged as *Hipparcos*, was also strongly in the running. In an unusual development, ESA's Science Advisory Committee overturned the recommendation and went for the much riskier comet Halley mission, a project that had to be put together at much shorter notice. Its members argued that the Moon and the stars could wait, but not Halley [90]. In July 1980 Europe decided to fly its own mission and asked British Aerospace to be the prime contractor. This was brave as ESA was going through the debris from V2 at the time. ESA allocated €80m to the project for a 750-kg probe with seven experiments for a 1,000-km/hr close fly-by at a speed of 70 km/sec. Time being pressing, the *GEOS* spacecraft was used as a model. The probe was named after Giotto di

Bondone (1267–1337) who drew the 1301 comet for his fresco in the arena chapel in Padua.

Although the Americans were not now going to Halley, two other countries beside Europe were. Japan launched two probes, *Sakigake* and *Suisei*, although both studied the comet from some distance (7 million km and 200,000 km, respectively). The Soviet Union launched two *Vega* probes (*Vega* standing for VEnus and HAlley), with a substantial level of collaboration from France. *Vega 1* later intercepted Halley at 12,000 km and *Vega 2* at 7,000 km. They found the nucleus of the comet to be pear-shaped, emitting clouds of dust and gas.

Giotto was built in Bristol over 1982–3 and brought to Toulouse in France in spring 1984. There it underwent an environmental test programme where it was tested for vacuum, heat, noise and vibration for up to 15 days at a time before being shipped to French Guiana. *Giotto* left Europe for French Guiana on 27 December 1984. It was an ambitious schedule, one which did not allow for delays. *Giotto* was launched by the second last *Ariane 1*, a doubly nervous moment for there was no back-up spacecraft. *Ariane* had never before launched a deep space mission and the window was limited to two weeks (or face a 76-year delay thereafter). At the last moment there was a 6-min hold when a recorder at a downstream tracking station went down. Eventually, *Giotto* had a flawless launch from Kourou on 3 July 1985 (*V14*), watched by many political and scientific visitors. *Giotto* took off into cloudy skies and was quickly lost to view. The Mage apogee boost, which took place at the next perigee, was another nervous moment. Only when that burn was safely completed could the team breathe easily. In fact it was so accurate that the first course correction was cancelled. *Giotto* soon began its 700-million-km chase across the sky. By November all its instruments had been checked out and were in working order, the probe having used up only 2.1 kg of its 66-kg propellant.

Giotto had a simple shape, a short cylindrical drum 1.81 m in diameter, 2.96 m tall, girdled by solar cells, blunt on one end with a 1.47-m-wide receiving dish on the other, topped by a tripod strut. Ten scientific instruments were carried.

Instruments carried by *Giotto*

Camera	Plasma analysers (two)
Neutral mass spectrometer	Magnetometer
Ion mass spectrometer	Energetic particles detector
Dust mass spectrometer	Optical probe
Dust impact dectector	

Giotto was designed to intercept the comet bottom-end first, with the antenna still pointing back to Earth. On the bottom, it had a 1-mm-thick aluminium alloy shield to absorb the cometary particles and a dust detector to count the hits. The energy particles detector was called EPONA (Energy Particles ONset Administrator), developed by Susan McKenna Lalor, Denis O'Sullivan and Alex Thompson.

The *Giotto* probe was controlled from Europe's own mission control, ESOC (European Space Operations Centre) in Darmstadt, Germany with the help of two Earth receiving stations: Weilheim in Germany and Parkes in Australia, the latter

Figure 5.2. *Giotto* ready for launch.

the famous dish in the sheep paddock subsequently immortalized in the movie *The Dish*, the station that took the signals from the first men walking on the Moon. Darmstadt held a communications session with the probe each day. The camera was checked out by taking photographs of the Earth, then the star Vega and the planet Jupiter. Darmstadt began to test out the procedures for the comet encounter.

By September 1985 *Giotto* was more than 13 million km from Earth. The 10 experiments were checked out and switched on. *Giotto* was spinning at 15 r.p.m. exactly as planned, and the solar cells were generating 143 W. There was a brief moment of worry in January 1986 when *Giotto* was temporarily lost by Parkes, but it was quickly recovered with the help of the NASA Deep Space Network. In February 1986 *Giotto* recorded a strong solar flare two days before it caused electrical disturbances on Earth. On 6 and 9 March 1986 *Vega 1* and *2* made their closest approach to the comet. ESA representatives were at the Institute of Space Research in Moscow at the time and they flew straight away to Frankfurt with the *Vega* coordinates in order to settle the final path of *Giotto*, then only three days away from Halley and closing fast. And 'fast' was not an overstatement: the speed of interception was 68.7 km/sec.

5.5   AIMED AT THE HEART OF THE COMET

Giotto was aimed at the heart of the comet. At midnight on 13–14 March 1986 *Giotto* intercepted Halley after a journey of 700 million km, some 150 million km away from Earth. An hour before interception *Giotto* recorded the first dust impact on its bumper shield. Giotto shot in toward Halley at 42 km/sec, all its instruments now turned on. Multicoloured images flowed in to mission control in Darmstadt showing the pear-shaped nucleus in red, yellow, purple and green. There were 30 more in the next half hour. At 6 minutes dust penetrated the bumper shield. Five minutes to go and the pictures became clearer, plasma activity was intense, the rates of dust impact were now into the hundreds. At closest approach, 2 minutes from closest fly-by, pictures from *Giotto* began to wobble and 4 seconds later were lost. Ground controllers feared the probe had been knocked out – in fact the dust particles coating the probe had temporarily knocked the dish antenna out of alignment and permanently damaged the camera. With 7 seconds to go, 770 km out, the data began to fail and at 3 seconds to go transmissions were lost [91].

The probe came to exactly 596 km from the nucleus, so close that it was coated with up to 12 kg of cometary dust. Intermittent signals resumed 19 seconds later and Darmstadt recovered control after 31 minutes. At one stage dust particles were striking the probe at 120 a second, so the engineers who made the shield had done their job properly. The spacecraft was damaged and battered, but still operational. The software had been corrupted by the electrical activity of the comet and the camera had been lost.

Between them the *Vega* and *Giotto* images gave us the first-ever close-up of a comet. Halley was a pear-shaped object, 8 km by 8 km by 16 km across, with a mass of 100,000 million tonnes, rotating every 52.9 hours or so. Surface temperature

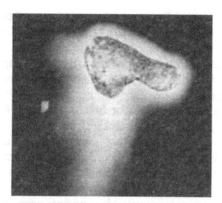

Figure 5.3. Comet Halley close up.

ranged from 300°K to 400°K. It appeared to have hills, valleys and even a crater. Most of the surface was inactive, but about 10% of the comet comprised seven jets spewing out 1,200 tonnes of dust every minute. The dust was made up of 80% water, 10% carbon monoxide and the balance carbon dioxide, methane, ammonia and cyanide. Compared with the scientists' expectations, the comet had a darker coat of dust, higher levels of energy and lower density than they had imagined.

After the successful interception *Giotto*'s instruments were switched off and the ESA scientists set to work analysing the treasure trove of data. *Giotto* continued in a solar orbit of 96 million km to 192 million km. By early 1990 ground controllers had begun to consider a follow-on mission for the probe, namely a visit to a second comet. Five cometary possibilities were considered. ESA allocated an additional €10.9m. *Giotto* was awakened from its slumbers by a powerful radio signal from Darmstadt routed via the NASA Deep Space Network station in Madrid on 19 February 1990 to prepare for a course correction to steer it in the direction of a second comet by the name of Grigg Skellerup for 10 July 1992. A weak signal acknowledging the command was received back on Earth 2 hours later – but *Giotto* was still alive. The high-gain antenna link was restored on 24–25 July and there was a full systems check the next day. To send *Giotto* in the direction of another comet required making use of the Earth for a gravitational slingshot manoeuvre. *Giotto* came within 23,000 km of Earth on 2 July 1990 when the manoeuvre was made, paving the way for what was called the *Giotto* Extended Mission. After a small course correction *Giotto* was then put back into hibernation for almost two years.

On 4 May 1992 *Giotto* was woken again. It was at a much further distance from Earth compared with its Halley interception, 219 million km distant. *Giotto* would have less solar power available. To conserve its energy only the minimum levels of instructions were sent up to the probe. Within two days ground controllers expressed their satisfaction with the status of the spacecraft. This was a small comet, with a nucleus of only 230 m, less than a 20th the size of Halley, a brightness of only 15 magnitudes and a barely visible tail. It was one of a hundred Jovian comets, so-called because they orbit out to the distance of Jupiter: Grigg Skellerup circled the Sun every 5.09 years. The interception pattern was different, *Giotto* catching up with Grigg Skellerup rather than driving head on into it.

On 10 July 1992 *Giotto* intercepted comet Grigg Skellerup at a distance of only 200 km from its core. At a distance of 1.4 million astronomical units (AU) or 160 million km out from Earth, *Giotto* made another close pass, but this time coming in at an oblique angle and taking only four hits, compared with 4,000 from Halley. *Giotto* first detected the comet at a distance of 600,000 km, 12 hours before interception, when it first registered cometary ions. The dust coma was detected at 20,000 km and the bow shock was first recorded at 15,000 km. About 300 scientists gathered in ESOC for the encounter and they were more than pleased with the data returned [92]. This was much less of a media event than the Halley interception, not least because there was no camera available (attempts were made to revive it, but it must have been knocked out during the Halley encounter). Despite that, *Giotto* was able to tell the scientists that Grigg Skellerup was a different kind of comet with a much more dispersed gas cloud 200,000 km in diameter.

The spacecraft continued to be in excellent shape. Some consideration was given to a third mission, but the fuel remaining at this stage was down to 4 kg out of the 69 kg when it started [93]. After such heroic endeavours it probably deserved a resting place in a museum. Europe had taken on a daring interplanetary mission and executed it with knifelike precision with a robust spacecraft that survived one of the closest encounters imaginable.

5.6 *HIPPARCOS*: MAPPING THE STARS

Hipparcos (High Precision Parallax Collecting Satellite) was a stellar-mapping satellite and named after the famous Greek astronomer who lived from 190 to 120 BC and who accurately measured the distance from the Earth to the Moon for the first time. The idea of a European astrometric satellite – one devoted to precisely measuring the position of the stars in the heavens – dated to 1967 when Professor Lacroute proposed the notion to the French space agency CNES (Centre National d'Etudes Spatiales). Thirteen years later, the concept was adopted by the ESA [94]. Developed by Matra of France, the aim of the project was to precisely map and catalogue the location of between 120,000 and 300,000 stars to an accuracy a hundred times greater than hitherto known. The accuracy of *Hipparcos* was intended to be 0.002 arc seconds down to stars of magnitude +9.

Costing €400m it was launched by *Ariane V33* into the night sky of 8 August 1989, but its apogee motor completely failed to ignite to circularize its orbit at 35,880 km. The bad news was relayed to the scientists just as they were getting onto their plane to fly from French Guiana back to Europe and the atmosphere in the plane was later described as 'subdued'. Three times over the succeeding week ESOC in Darmstadt tried to relight the motor, but it stubbornly refused to respond. The mission faced disaster.

Hipparcos had electrical power for only a few hours a day. The main problem was that the Van Allen radiation belts through which it was passing were degrading its solar panels every day. Mission controllers tried to put together a rescue plan that involved the use of the small hydrazine station-keeping thrusters to raise its orbit.

Hipparcos used its control thrusters to raise its orbit as much as possible. On 8 September 1989 it managed to raise its perigee by 500 km. The solar panels were deployed on 12 September and the star-mapping equipment was turned on on 26 September. ESOC worked hard to try and retrieve what it could from the mission, acknowledging that the orbit provided limited opportunities, a degradation in mapping accuracy of around 15% and that the radiation belts would eventually affect the solar cells. An additional tracking station in Perth, Australia helped by taking in the data. Ultimately, ground control was unable to get *Hipparcos* above the radiation belts – six attempts to relight the motor failed – and the radiation eventually degraded its systems and the spacecraft was abandoned in August 1993. The *Hipparcos* catalogue was published in 1997. The scientists involved in the mission tried to interest ESA in a replacement mission for €160m, but they generated insufficient support within ESA.

In the end and despite all the difficulties, *Hipparcos* measured 120,000 stars to an accuracy of 1–2 milli-arc seconds [95]. A further million were identified, but with less accuracy. In each case *Hipparcos* provided details of their positions, distances, velocities through space, whether they were binary or not, colour and variability. In 2000 ESA released what was called the *Tycho-2 Catalogue*: the 2.5 million brightest stars in the sky, with details of their position, motion, brightness and colour, all thanks to *Hipparcos*. This was a remarkably positive outcome, a bright contrast to what had seemed a hopeless situations in the hours following the flawed launch.

5.7 INFRARED SPACE OBSERVATORY (*ISO*)

ISO was a follow-up to the pioneering work of *IRAS* (Infra Red Astronomy Satellite). ESA signed for the €230m project in summer 1990 with a view to improving our knowledge of the infrared universe by a thousand times. It was designed for an 18-month lifetime to observe everything from planets and gaseous clouds to the weakest of heat sources in the universe with a 60-cm telescope. It would carry a camera, imaging photopolarimeter and two spectrometers into an orbit of 1,000 km to 70,000 km, spending two-thirds of each day outside the radiation belts.

ISO was launched on 17 November 1995 on *Ariane 44P*, settling in an orbit of 1,038 km by 70,578 km, 5.2°, 24 hr. Made by Aérospatiale, DASA (Deutsche Aerospace) and Fokker, *ISO* was 5.3 m high and 2.3 m wide. The orbit was such that *ISO* offered about 13 hours observing time a day – all done in real time since there was no tape recorder – and this was made available to the worldwide astronomical community and NASA. Mission control was in Villafranca, Spain. As was the case with *IRAS*, a significant part of the payload was supercool helium, 2,250 litres, in order to keep the temperature down to 2°K for an intended duration of 18 to 24 months. It was an expensive project costing over €750m. *ISO* was primarily aimed at galaxies, star formations and intergalactic clouds and dust – but it was also directed toward objects closer to home such as Saturn's moon Titan, comets, asteroids and the giant planets.

The initial results from *ISO* were impressive [96]. Among its early targets were the colliding galaxies of Antennae, 60 million light years away, the M51 whirlpool galaxy, dust clouds in the Milky Way, expelled stellar debris in the Southern Cross and supernova remnants. Later it imaged disk areas around stars where prototypical solar systems might be in formation. By penetrating dust boundaries *ISO* found 100,000 stars at the centre of our own Galaxy, many of them red giants.

ISO operated until 16 May 1998 when its helium became exhausted, and it was effectively switched off by the ESA ground station in Villafranca, Spain. By then a total of 26,000 observations had been made, the archiving of which would take till 2006. Its last image was of a giant star in Canis Major, showing a disk of matter circling it. *ISO* was put into an orbit that will make it burn up in the atmosphere in less than 30 years. The *ISO* data archive went online in August 1999, and astronomers are encouraged to download the data. It quickly attracted 1,100 registered users.

Although the primary focus of *ISO* was on stars, it obtained a considerable volume of data on asteroids, even though this comprised only 0.7% of total observing time, less than 100 hours. This enabled 173 separate investigations of asteroids to be carried out, with detailed analysis of the asteroids Vesta and Hygiea.

5.8 NEW PLANNING FRAMEWORK

The first round of ESA science missions had been selected on merit through competition or inherited from ESRO. In 1983 ESA decided to put its space science missions on a more long-term basis. The Agency appointed a team of scientists under Johan Bleeker and made a call for mission concepts: one that was widely supported (70 were received) and then analysed over a three-day meeting in Venice, Italy. In the end, the committee produced a report called *Space Science Horizon 2000*, often referred to as *Horizon 2000*. This adopted the principle of 'cornerstone' missions, projects that will advance space science substantially in distinct areas over a period of many years. It was hoped that over the 20-year period Europe would fly four cornerstone missions and could fly up to seven others. The cost of the programme was estimated to be €200m a year by the end of its time [97]. Although only supposed to run to 2000, in fact it ran far beyond this date.

The principles of *Horizon 2000* were that European space science should be of the highest scientific standard, mix larger and smaller projects, provide continuity, have a high technological content, balance the different European agencies and remain within a modest financial envelope. The cornerstones must also fulfil the criteria of stimulating scientific and engineering achievement, as well as placing Europe in a leading position in these fields. The four cornerstones adopted for *Horizon 2000* were an X-ray mission observatory (*XMM/Newton*), a submillimetre observatory (*FIRST*), a solar terrestrial observatory (European Southern Observatory) and a multi-point space plasma physics mission (*Cluster*). Others were added later, like *Integral* and *Rosetta*.

Several other mission possibilities were also discussed; for example, a Mars probe (*Kepler*), participation in a Saturn mission (*Cassini*) and a lunar orbiter. Also in the melting pot for consideration were an ultraviolet observatory (*Lyman*), a very long baseline inferometry mission (*Quasat*), a gamma ray observatory (*GRASP*) and a number of candidates for an asteroid mission (*Agora*, *Vesta*). These missions would be organized on the traditional basis as costs permitted, and in the event ESA was able to approve the *Cassini* mission, but not the others.

Horizon 2000, with the cornerstone concept, was later succeeded by *Horizon 2000+* (chap. 8), which continued the idea. For the first planning period *Horizon 2000* took the form of missions that were both exclusively European, like *XMM*, and others that were international in nature though with a high European component. These were missions like the spectacularly successful space astronomy mission *Hubble*, the solar probe *Ulysses*, *Galileo* to Jupiter, *Cassini* to Saturn, the *SOHO* solar observatory and *Cluster*. Each is reviewed in turn.

5.9 HUBBLE

There have been six great telescopes in our history, from Galileo's first in the 16th century to the Rosse Leviathan in Ireland in the 19th century. The Hubble Space Telescope continued in this noble tradition, although there was a stage in its development when this adjective would not have been used. The idea of a large space telescope able to see deep into the universe from above the perturbing effects of the atmosphere went back a long way – it had been proposed by Oberth long ago. NASA had carried out studies in the 1960s. Several small spaceborne observatories had been launched from 1966 onward – *OAO 1* (Orbiting Astronomical Observatory) – and these held out promise for a much larger observatory.

Hubble began life as the Large Space Telescope in 1976 and was set to have a mirror of 2.4 m diameter. The aim of the telescope was to make a quantum leap forward in astronomical observations from Earth, imaging objects 20 billion light years distant and 50 times fainter than could be seen from the ground, operating across the ultraviolet, infrared and visible wavelengths. The telescope was designed to be repaired in orbit. It was a big object, 14.3 m long and 4.7 m diameter.

NASA invited Europe to take a share in the telescope and contribute 15% of the costs and benefit from 15% of observing time. This was approved by ESA in October 1976, the European financial contribution set at €80m. Europe won the contract for a number of important components, such as the solar array and the photon detector assembly for the faint object camera (British Aerospace). Although a small object, the 318-kg faint object camera was one of the most crucial components on board, being designed to pick out the most dim objects under study. The camera was built by Dornier in Germany and Matra in France, with British Aerospace making the photon detector assembly therein. The detector was designed to be sensitive enough

Figure 5.4. *Hubble* space telescope.

to pick up a single candle light at 40,000 km. By contrast, the two solar arrays were the largest attachment to the spacecraft and were required to produce 5 kW of electricity. Each had 24,380 cells and the overall span was $33\,m^2$. A feature of the wings is that they had to be rolled up into two 20-cm-diameter drums.

Although *Hubble* was ultimately more than a triumph, it had a most troubled history. In 1983 Congressional investigators found that the programme had slipped its time schedule badly, costs had overrun and it was badly managed. *Hubble*'s launch was delayed by the *Challenger* accident and did not reach orbit, courtesy of the space shuttle *Discovery*, until April 1990. At once there were problems getting data back and transmitting images. These were sorted out and *Hubble*'s first images reached Earth in May 1990. To the horror of ground control, the images could not be focused. *Hubble* had what was called a 'spherical abnormality', which meant that the rays could not be focussed on one point. Instead, there was a perpetually blurred image. NASA's critics demanded blood for what was clearly a monumental foul-up.

NASA managed to design a series of computer corrective mechanisms to enable scientists to make sense of the anomalous mirror. Ultimately, Hubble required a visit by astronauts to repair and replace the defective equipment. The assignment went to shuttle mission STS-61. ESA astronaut Claude Nicollier was assigned to the mission, in recognition of Europe's 15% contribution to the project and its role in making the solar arrays and the faint object camera. The shuttle *Endeavour* blasted off into a moonlit sky over Cape Canaveral on 2 December 1993 for one of the most challenging space repair missions ever. Two days later the shuttle was alongside the orbiting telescope and Claude Nicollier used the remote arm to berth it in the shuttle's cargo bay. On a spacewalk, Kathy Thornton and Tom Akers rolled up the European solar arrays and replaced them with a new set. The new arrays had several improvements, the main one being a device to prevent the arrays from flexing in orbit in response to the changing extremes of heat and cold. Later, they installed the corrective optics device to compensate for the spherical aberration. After almost two weeks' work they released *Hubble* to fly back to its operating altitude. Repairing *Hubble* was added to the list of NASA's most remarkable achievements. Within days, the pictures coming from Hubble showed that it had all been worthwhile.

Hubble was repaired a second time by STS-82 and the crew of *Discovery* who left Earth on 11 February 1997. The repair did not affect the two European components on board – the solar wings and the faint object camera. Once again, the repair work was done smoothly and successfully in the course of five spacewalks. The third repair mission was STS-103, *Discovery* again, launched 16 December 1999. On board was Claude Nicollier again, making his fourth shuttle flight, and Frenchman Jean-François Clervoy, his third. It was Clervoy who brought the *Hubble* into the shuttle bay using the robot arm on the third day of the mission. Nicollier took part in the second spacewalk of the mission, lasting a staggeringly long 8 hours and 10 minutes, during which the astronauts replaced *Hubble*'s computer and fine guidance sensor. Clervoy released the *Hubble* on the seventh day while they were flying over the Coral Sea in the Pacific. The repair mission was another triumph.

The solar arrays were in sufficiently good order not to require replacement on STS-103, but a replacement was scheduled for the next mission, the fourth. This was STS-109 flown by *Columbia* in March 2002. The European faint object camera was replaced by a new one called the Advanced Camera for Surveys, while the solar panels were replaced by a newer lighter pair with 20 times greater generation capacity.

The main achievement of Hubble has been to look back in time further than any astronomers had originally thought possible. The astonishing quality of *Hubble*'s images were not just of value to astronomers, for the pictures of colliding galaxies, star nurseries and gas clouds (one set was aptly called 'the pillars of creation') made a profound impression far outside the scientific community.

5.10 *ULYSSES*

Ulysses started life as the International Solar Polar Mission (ISPM). This was a relatively small spacecraft, weighing 350 kg, with a daring mission – to swing out to the Jovian system, loop around the planet at a distance of 425,000 km and use its gravity to hurl it back inwards towards the Sun. Once in the vicinity of the Sun ISPM would then look down on the north pole of the Sun, circle behind it and then complete its mission with an examination of the Sun's south pole. This was the first out-of-ecliptic mission: hitherto all spacecraft launched from Earth into solar orbit stayed within the narrow band of the Earth's orbit, ±7.5°.

When conceived in 1977, ISPM was originally to be a double mission, with one European and one American spacecraft, each flying out to the Sun simultaneously, one over its south pole, the other over the north. Launch would be by the shuttle in 1983 when its powerful upper stage would send each probe on a 15-month journey. For the European spacecraft, Dornier was appointed prime contractor with Thomson, FIAR, Fokker, SENER, Ericsson, Contraves and British Aerospace. There were 120 investigators in 47 laboratories in 12 countries. *Ulysses* carried nine instruments weighing 55 kg overall. ISPM acquired the title *Ulysses* in July 1984 in honour of both Homer's ancient explorer and the explorer who went behind the Sun in Canto 26 of Dante's *Inferno*.

Ulysses was in the shape of a rectangular box, 3 m long, 2.14 m tall and 3.3 m wide with a large 1.6-m receiving dish, nuclear radio-isotope power source and at the other end a boom with a magnetometer, X-ray and gamma burst detectors, and radio and plasma wave detectors. Other experiments were designed to measure cosmic rays, Jovian magnetic fields, solar wind, protons, electrons and ions, solar flares and radio bursts and Galactic radiation.

Budgetary difficulties forced the cancellation of the American probe in 1981. This was a shock for the Europeans, who operated within more stable, longer term planning frameworks. Thankfully, the Americans by no means withdrew from the project and still covered about 50% of the cost, with NASA to supply the launcher (still the shuttle), tracking services and the nuclear fuel, with ESA contributing €150m to flight operations. Five of the nine experiments on *Ulysses* had American principal investigators. ESA decided to go ahead anyway.

Eventually the launch of *Ulysses* was set for May 1986 on a *Centaur* rocket from the payload bay of the shuttle. When the shuttle was lost that January the mission was postponed indefinitely, the second major setback to the project. *Ulysses* was shipped back to Europe, going into storage in Dornier's plant in Friedrichshafen. The experiments were even returned to their principal investigators.

When the shuttle returned to flight in September 1988, *Ulysses*' hibernation came to an end. New parts were installed and the probe was shipped back to the Cape. An Air France 747 flew *Ulysses* into Cape Canaveral Air Force Station in May 1990. Eventually the launch was reset for 1990, but on a much less powerful upper stage.

Ulysses was eventually launched by the shuttle on 6 October 1990, some 13 years after the mission was first planned. *Discovery*, the shuttle involved under the command of Dick Richards, deployed *Ulysses* only hours after reaching orbit. The upper stage fired and *Ulysses* left Earth at the greatest speed ever, 15.4 km/sec, passing the orbit of the Moon in only 8 hours with sufficient energy to reach Uranus. On 6 November 1990 *Ulysses*' 7.5-m axial boom was deployed. In July 1991 *Ulysses* fine-tuned its path to set it precisely on course for Jupiter. It met the bowshock of the Jovian system on 31 January 1992, some 113 radii out. Its signals sent out a stream of readings to Earth 667 million km distant, taking 37 minutes to arrive.

Figure 5.5. *Ulysses* flew to the Sun.

Ulysses passed at great speed through the Jovian system on 8 February 1992. Closest approach was 380,000 km over the cloud tops. Data were relayed about the magnetic field, solar wind, ions, radio and plasma waves, cosmic waves and solar particles. On Earth the data were straight away burned onto compact disks for storage and analysis. The big worry was whether and how *Ulysses* might survive the Jovian radiation belts, but it came through unscathed and then began to swing out of the planetary ecliptic on its looping trajectory back towards the Sun. *Ulysses* was only the fifth spacecraft to Jupiter, following *Pioneer 10* and *11* and *Voyager 1* and *2* and preceding *Galileo* and *Cassini*.

Next, *Ulysses* headed for the south pole of the Sun into regions of space never before explored, describing a giant loop out of the ecliptic. It passed its furthest point south 80.2° on 13 September 1994, surprising scientists by failing to detect a southern solar magnetic pole. *Ulysses* was still in perfect working order, transmitting at 60 million bits of data daily. The Sun, it turned out, was a much more complex body than the scientists had thought. *Ulysses* found very long period electromagnetic waves of about 10–20 hours, high latitude bursts of particles and clouds of ionized gas. Data were relayed back at 2,024 bits/sec with additional recorded information sent back from two tape recorders. At one stage *Ulysses'* signals took 50 minutes to reach Earth [98].

Ulysses then crossed into the equatorial zones of the Sun in spring 1995, heading northwards. When *Ulysses* was planned, the objective was to fly to the Sun during the period of maximum solar activity, but because of delays *Ulysses* had now arrived during the solar minimum – but this did not reduce the scientific value of the mission.

Ulysses reached its maximum northerly latitude, 80.2°N on 31 July 1995. Three years later, by 1998, it had circled out to the orbit of the planet Jupiter and then headed back towards the Sun's south pole for a second pass around the Sun. This time things would be different for the Sun was well past its quiet period and heading for its time of maximum solar activity.

In 1995 ESA added €14m to its existing €168m budget for *Ulysses* to provide for fly-bys during the 2001 solar maximum. All this time *Ulysses* was tracked by a joint ESA/NASA team at the Jet Propulsion Laboratory (JPL) in California for 8 hours a day as it orbited the Sun up to 1.34 AU out. *Ulysses*' equipment was still working remarkably smoothly 10 years after its launch, with the number of anomalies very small. *Ulysses* flew through the tail of comet Hyakutake in May 1999, finding that the tail extended far across the solar system, at least 150 km in length.

In its extended mission *Ulysses* flew over the south pole of the Sun on 8 September 2000 and the north pole again on 3 September 2001. It continued to report back on the chaotic and blustery solar wind of high latitudes. Another key discovery by *Ulysses* was that there were really two solar winds, a fast lane and a slow lane. The probe also identified heavy-grained interstellar dust [99].

5.11   *CASSINI*

Cassini was the last of the great solar system explorers. During the 1970s and 1980s the United States built ever bigger, larger, heavier and more capable spacecraft to explore the solar system. Typical were the two *Viking* missions to Mars (1976), *Voyager* to the outer planets (1977), *Galileo* to the Jovian system (1989) and *Magellan* to Venus (1990). Although their achievements were stupendous and their discoveries awesome they were extremely expensive missions taking almost 20 years from conception to completion. In the 1990s NASA administrator Dan Goldin took the view that they were too expensive, used up substantial resources for over a dozen years each and meant that interplanetary missions could be flown only every seven years or so. They were also too risky, for everything hinged on a successful launch and flight to target, a point underlined when America lost *Mars Observer* in 1992 and the Russians *Mars 96* a few years later. Hence the 'faster, cheaper, better' philosophy in which smaller, smarter probes would be launched more frequently producing more results sooner, generating popular support for planetary exploration. And if things went wrong the next mission would soon be ready anyway.

By the time 'faster, cheaper, better' had come in, plans for a big and expensive American space mission to the Saturnian system were well developed. A mission to Saturn had been approved by NASA's solar system exploration committee in 1983, with a particular aim of targeting the moon Titan, the only orbiting moon in the solar system with a dense atmosphere – 1.5 bar – though at a cool enough 94°K. Coincidentally, in 1983 Europe had begun a six-year feasibility study of a Saturn mission, the strategists being Dr Daniel Gautier of the University of Paris (Meudon) and Dr Wing Ipp of the Max Planck Institute.

The American Saturn orbiter quickly ran into budgetary problems. Following a joint study in 1984 it was agreed the following year to merge the two projects. While NASA would supply the 1,540-kg Saturn orbiter, ESA would supply the 190-kg Titan lander. The main spacecraft had acquired a name *Cassini*, as had the lander

Huygens, named after the celebrated Dutch astronomer Christian Huygens (1629–85). Cassini was the French–Italian astronomer who discovered several Saturnian moons and studied the planet's ring system. *Cassini* would explore the Saturnian system for four years while *Huygens* would deploy at 180 km a parachute for a 2–3-hr descent by a probe to the surface.

Costs in the project continued to rise reaching €1.6bn. NASA responded by dropping some experiments, deciding not to do scientific experiments before Saturn arrival and spreading out the development period (which in the long run often proves to be a false economy). Despite that, it was the largest, heaviest, most expensive and most ambitious interplanetary spacecraft ever launched, with a probable final cost of €4bn. There were 12 main scientific instruments from 17 countries.

The contractor for the building of *Huygens* went to Aérospatiale in Cannes, the first time the company built an interplanetary spacecraft. The heat shield was made at its Aquitaine factory. The spacecraft's thermal protection systems were the responsibility of Dornier, the subsidiary of Daimler Benz, in Munich. The parachutes were made by Martin Baker of Uxbridge and the software by Logica in Britain.

One of the big problems faced by the designers of *Huygens* was that so little was known about the atmosphere of Titan. Indeed, it took the Russians many missions to Venus to get the size of parachute right so as to achieve a successful descent there. The *Huygens* project was at least able to benefit from the experience of *Galileo's* Jupiter probe that also faced many unknowns. Titan is the second largest moon in the solar system – only the Jovian Ganymede is larger – and larger than the planet Mercury. Some astronomers believe that its present atmosphere is like the early primordial atmosphere of Earth many eons ago. The designers had no idea as to the nature of the surface: solid or liquid, or maybe something in-between.

Huygens was designed to enter the atmosphere at 20,000 km/hr. Its 2.8-m-diameter heat shield would slow the probe for parachute deployment, a 2-hr descent and a 2-hr surface transmission. The first drop tests of *Huygens* models were made from a balloon over Sweden in 1995 [100]. Other tests were done to test the ability of the spacecraft to survive what are certain to be very cold conditions. The six *Huygens* experiments are shown below:

Huygens experiments

Aerosol collector and pyrolyser	France
Atmospheric structure instrument	Italy
Descent imager and spectral radiometer	United States
Doppler wind experiment	Germany
Gas chromatograph mass spectrometer	United States
Surface science package	Britain

All depended on a successful launch, and this just had to go right. There were many environmental protests at the use by the spacecraft of radioactive fuel, but the

Figure 5.6. *Huygens* heads for moon Titan.

countdown eventually reached zero at 4.43 a.m. on 15 October 1997, some 14 years after the mission had been approved. In the event the *Titan* launcher performed perfectly. Saturn was visible high above as the *Titan 4B* rode a pillar of yellow–orange flame into a cold night sky over Cape Canaveral. The view was crystal clear and viewers could see the solid rocket boosters peeling away 2 minutes and 24 seconds into the mission. Over west Africa the *Centaur* upper stage ignited for 7 minutes, placing *Cassini* outbound within 0.004° of the planned pathway.

Cassini's journey to Saturn was one of the most complex trajectories ever planned, all designed to save fuel and maximize the size of the payload sent. Although destined for the outer solar system *Cassini* first headed inwards, passing Venus in April 1998 and again in June 1999 in order to gain energy before swinging outwards. On the way back out through the inner solar system *Cassini* zoomed past the Earth at 520 km in August 1999. The Americans pointed out somewhat mischievously that if anything went wrong and the nuclear-laden *Cassini* then crashed on Earth, the likely impact point was somewhere near one of Saddam Hussein's palaces in Baghdad. With the assistance of Earth's gravity *Cassini* gained momentum to swing through the asteroid belt (early 2000), passing Jupiter at a distance of 8 million km (December 2000), en route to arrival in the Saturnian system on 1 July 2004. The transit through the Jovian system brought mixed news: positively, stunning pictures were obtained of Jupiter's clouds, storms and moons and these could be matched with those of the orbiting *Galileo*; negatively, problems were found in *Huygens'* data relay system, which was likely to adversely affect the information it could transmit during its descent. The settings for the bandwidth for transmission from *Huygens* to *Cassini* were set too narrowly, with the result that some data might be lost. It seems that the designers had paid insufficient attention to the fact that both spacecraft would be moving at some velocity during the descent. This meant that, due to Doppler shift, the frequencies would move along the wave band – a basic issue for satellite communications – and would eventually be lost, rather like tuning an FM receiver to a shifting signal on a car radio. At least this was realized before the landing was attempted.

Controllers had some time to sort out the problem. A *Huygens* recovery task force was formed. By 2001, 10 options had been presented to anticipate the problem, though an investigation was highly critical of the manner in which designers had permitted the problem to arise in the first place. ESA finally adopted the revised plan in autumn 2001, involving software changes, reconfiguring the orbital altitude of *Cassini*, warming up the transmitter in advance and delaying the landing.

Once it arrives *Cassini* will burn its engine for 96 minutes to enter the first of 34 Saturnian orbits, making a complete tour around the planet every 147 days. On 25 December 2004 the European lander *Huygens* will be released to make its final descent three weeks later on 14 January 2005. *Huygens* will intersect the atmosphere 1,270 km above the planet's surface at Mach 20. The heat shield will be exposed to a temperature of 12,000°C as it barrels into the Titanian atmosphere. The probe will be stressed to 16 G as it abruptly decelerates. When the accelerometer detects a speed of Mach 1.5, which is expected at 180 km altitude, then a mortar will deploy a 2.5-m drogue parachute. In a few minutes the probe will be down to Mach 0.6 and 140 km. The instrument covers will now be blown away and the vital part of the mission, measuring Titan's atmosphere, will now begin. Its main parachute will deploy. The 335-kg probe will take 150 minutes to come down using its 8.3-m-diameter main parachute, an altimeter measuring the altitude as it descends and other instruments measuring atmospheric composition, aerosols, lightning, wind speeds and temperatures. No-one knows what *Huygens* will find, assuming it survives its perilous journey – but it is designed to float in case it splashes down! It is possible that the probe will survive on the surface, though there is only a limited amount of battery power and transmissions from there will last a couple of hours at most. The probe carries a device to measure whether the surface is solid, liquid or something gooey in-between.

5.12 *SOHO*

SOHO (Solar Heliosphere Observatory) was a joint American–European project. It was launched by an American *Atlas* on 2 December 1995, and became one of the great solar observatories, also writing a chapter in the folklore history of recovered lost spacecraft.

SOHO had three prime objectives: to see how the solar wind interacted with the Earth's magnetic field, to investigate the outer layers of the Sun and to probe its interior structure. *SOHO* was placed in an orbit 1.5 million km away from Earth at the L1 (libration) point where it could best study the Sun, its corona, radiation and solar wind. Such a position was ideal for long-term, uninterrupted observations. Weighing 1,864 kg, *SOHO* was a box-shaped spacecraft 3.6 m in height, 3 m in diameter, with solar panels able to generate 1,400 kW. It had a payload of 12 instruments for studying the Sun, made in Britain, France, the United States, Germany, Finland and Switzerland, and a total of 39 institutes were involved. Funding came from NASA (€500m) and ESA (€550m). Prime contractor was Matra Marconi of France in Toulouse and Portsmouth.

SOHO took several weeks to reach its L1 point and arrived there in March 1996. The images sent back of the Sun were spectacular, and popular participation in the project was stimulated by posting these images live on the Internet. *SOHO* became one of the best-known unmanned spacecraft, its images of the Sun attracting over 5 million Internet requests per month. One of *SOHO*'s early findings was that big ejections from the Sun were preceded by multiple magnetic loops in the inner corona. In effect, this was the Sun reorganizing its magnetic field – but knowing that this would precede a solar storm meant that an alert could be given in good time.

Ground control for *SOHO* was the Goddard Centre. Here, on 24 June 1998 a ground controller accidentally turned the €1bn satellite to face away from the Earth and contact with the probe was lost. There was widespread mourning in the scientific community involved in the project because of the success of the project until this stage. In a desperate attempt to resume contact with the spacecraft, NASA called into action its 305-m radio telescope in Puerto Rico. This telescope had been famously used to search for messages from alien civilizations and there was something ironic about using it to try to recontact a small spacecraft within the solar system.

On 23 July 1998 the radar did manage to find *SOHO* at a distance of 1.5 million km and ascertained that it was spinning. On 3 August it picked up a 10-sec burst of signals from this direction, indicating that it was still alive. A lengthy downlink was commanded on 8 August, the batteries were recharged and by 18 September they were able to declare the satellite recovered. As a result of being out of orientation, some parts of *SOHO* had frozen over and others had been roasted, but ground control nursed it back to life. Despite enduring an experience in which its equipment had been subjected to thermal extremes, full high-quality images were being returned once more by October. The recovery of *SOHO* became one of the great epics of space recovery, and ground controllers will tell its tale late into the night for many years to come.

The ground controllers were put to the test again at the end of the year when the gyros failed, probably damaged by heat and cold when out of control. The controllers managed to send up suitable software commands to control the spacecraft without the benefit of gyros. This solution worked much better than imagined, and by autumn 2001 *SOHO* still had a fuel reserve of 123 kg, enough for another 10 years. The solar arrays had degraded a little, but were still working at 90% efficiency.

Results from *SOHO* were presented to an ESA conference in Tenerife in 2000. *SOHO* found that visible radiation from the Sun had risen 0.5% over 1995–2000, as the Sun moved to solar maximum. *SOHO* data were then factored into climate records since 1700 to try and ascertain if the Sun played a more important role in global warming than people had imagined. *SOHO*'s ability to watch solar eruptions led to a direct improvement in solar storm forecasting from 27% accuracy to 85% accuracy [101]. *SOHO*'s instruments were able, because of the Sun's rotation, to get early notice of storms sweeping around from the Sun's far side, and as a result warnings of solar disturbances could be given an extra week's notice.

Astonishingly, *SOHO* managed to find 500 comets! These were mainly Sun-grazing comets or comet fragments close to the Sun, many of which are drawn into the Sun and then incinerated. Hitherto, they had been so close to the Sun as to be unobservable. *SOHO*'s position was perfect for observing other comets, like Hale-Bopp and Wirtanen, and especially their tails as they approached the Sun.

5.13   CLUSTER

Cluster was essentially a European project, though like many others we have studied it had a strong international dimension. Several of these were under the aegis of the Inter-Agency Solar Terrestrial Physics Programme (IASTPP). Hardly one of the best known, international, scientific collaborative programmes, it was nevertheless one of the most fruitful. IASTPP was set up in 1977 by NASA, ESA, the Japanese Institute for Space and Astronautical Research (ISAS) and the Space Research Institute (IKI) in Moscow and was funded by the participating countries to support investigations of the Sun–Earth environment. *Cluster* was technically a IASTPP project, although it was popularly seen as a European project. The other projects in the series were *Geotail* (Japan/United States, 1992), *Polar* (United States, 1996) and *SOHO* (Europe/United States, 1995).

Cluster was Europe's main Sun-observing mission of the very late 1990s. The aim was to place four identical spacecraft in slightly different orbits between the Earth and the Sun so as to provide precise measurements of the solar wind and its interaction with the Earth's magnetosphere. Four spacecraft were duly built at a cost of €550m and installed on Europe's *Ariane 5* for its maiden voyage on 4 June 1996. They were shaped like flat drums, 2.9 m in diameter, 1.3 m high and weighing about 530 kg each. Instruments were designed to measure electric fields and waves, electrons, ions, plasma and particles as the four *Clusters* crossed in and out of the magnetic field, bowshock, magnetosphere and magnetic tail. Seventy laboratories were involved from all over Europe. Each of the *Clusters* had 11 identical instruments, making 44 in all.

Grown men who had contributed the previous 10 years of their lives to the project cried their eyes out when *Ariane 5* exploded only a few seconds into its maiden voyage. The atmosphere among the scientists was later described, benignly, as 'suicidal'. However, ESA director Roger Bonnet brought the period of mourning to an abrupt and premature end and convened a special meeting in

Figure 5.7. *Soyuz*, launcher of *Cluster*.

London in July 1996 to revive the project. They must pull themselves together and try again.

By November ESA had chosen the quickest and cheapest option available: to rebuild the ground spares and launch them on two Russian rockets, an option costing an additional €210m. The Russian Space Agency was able to offer its trusty *Soyuz* rocket, which had already made in its many versions no less than 1,650 previous missions. The entire process was speeded up because the company that made the *Soyuz*, the Progress works of Samara, had entered an international marketing alliance with the French company Starsem, signed in August 1996.

Starsem was able to sell launcher space on the *Soyuz* at a cost of 40% less than that of the American *Delta 2*.

The *Soyuz* rocket would require some adaptation if it was to launch the *Cluster* probes into their precise and unusual orbits. Here, an upper stage was available from the Soviet Mars programme. A new powerful and versatile upper stage, *Fregat*, was adapted from the manoeuvring stage used for the Phobos missions of 1988. Even then *Soyuz Fregat* was much less powerful than *Ariane* and two launchings had to be used. The shroud on top of *Soyuz* was smaller, so some of the booms on *Cluster* had to be shortened. *Fregat* had 5,350 kg of fuel and could be restarted up to 20 times.

There could be no more mistakes and two demonstration missions were organized on *Soyuz* rockets. On 9 February 2000 *Fregat* made a series of high-altitude burns, before concluding with an unrelated experiment to send rubber inflatable cones through re-entry for a series of tests of new ways of returning to the Earth (these were made by Lavotchkin of Moscow and Daimler Chrysler of Bremen). On 20 March *Fregat* was launched again for a series of engine burns and mock separation tests.

These missions paved the way for the real thing. On 16 July and 9 August 2000, respectively, the *Soyuz Fregat* launched the two sets of *Clusters*, putting them after a series of complex manoeuvres into their final orbits of 19,000 km to 119,000 km. They were put initially into a parking orbit, and then *Fregat* relighted. Six further engine burns were required to settle them into their final orbit. They then used their own propellants to trim their position. By 24 August all four satellites had reached their assigned points in the sky.

Following a competition open to the public they were then renamed *Rumba*, *Salsa*, *Samba* and *Tango* to reflect the way in which they would dance in formation in their orbits facing the Sun, sometimes only 100 km apart and then moving to

Figure 5.8. *Cluster* reaches orbit.

20,000 km away from one another. Scientific results were coming in before the year was out. The four satellites measured the big solar storm of 9 November 2000, the spacecraft drifting in and out of the magnetosphere as the storm reached the Earth.

5.14   *XMM/NEWTON*

We conclude the examination of European space science missions with the astronomy missions, *XMM* and *Integral*.

XMM was Europe's multi-mirror X-ray telescope and the largest scientific spacecraft developed in Europe at that stage. It was a big object, coloured plain dull black, shaped like a single lens reflex camera and broadly equivalent to the *Chandra* telescope just launched by the Americans from the shuttle. The aim was to take a total of 30,000 spectra over 10 years, determine more precisely the nature of black holes, better understand X-ray bursts and determine the origin of cosmic rays. It was also aimed at what were called vampire stars – binaries where one of the two would suck in and consume the matter from its companion [102].

XMM was made under the guidance of Dornier in Friedrichshafen, with 35

Figure 5.9. *XMM/Newton.*

European companies involved from 14 countries. It used a series of nested mirrors, which meant that even though its aperture was only a third that of *Chandra* its gathering power was three times greater. *XMM* had 174 mirrors covering a total area of $120\,m^2$, three X-ray mirror assemblies with a focal length of 7.5 m, making it in theory at least between 5 and 15 times more sensitive than NASA's *Chandra* and with a sharper view. It was 11 m tall, weighed 3,764 kg and cost €900m including the launch (much less than *Chandra*'s total of €3.5bn).

Launched by *Ariane 5* on 10 December 1999, *XMM* was released while it was flying over the Middle East and first signals were immediately picked up in Villa-franca, Spain and then by the European space control centre in Darmstadt, Germany. After 30 minutes the solar panels unfurled. *XMM*'s orbit took it to 7,000 km by 113,946 km, a third of the way out to the Moon, but carefully planned so that it would be outside the radiation belts for 40 hours out of each 48-hr orbit. The solar panels spanned 16 m and provided 1,600 W of power. *XMM* opened its doors for first light on 25 January 2000. One of the first *XMM* pictures released showed the red–pink swirl of the M81 liner galaxy in Ursa Major. Another showed areas of hot and cold gas, matter and stars in the Larger Magellanic Cloud 160,000 light years from Earth.

XMM was named *Newton* after its launch in honour of the great English physicist. However, the naming was poorly publicized and most people had by then got in the habit of calling it *XMM*. The instruments were checked out by March and handed over for calibration, going fully operational in July 2000. By the end of the year a special edition of *Astronomy and Astrophysics* published 56 scientific papers on the results from the calibration and performance verification phase of *XMM*. In the case of a binary pair in Taurus, *XMM* was able to pick out carbon and nitrogen that *Chandra* had been unable to see. ESA said that *XMM* would be able to cover in a day the area of sky that the first X-ray satellite *Uhuru* had covered in three years from 1970.

5.15 INTEGRAL

Integral (INTernational Gamma Ray Astrophysical Laboratory) was selected in 1993 under the *Horizon 2000* programme. It was intended as a logical gamma ray observatory successor to the American *Compton* gamma ray observatory and the Russian *Gamma*. The objective of the mission was to take detailed, high-resolution spectral images of the objects identified by *Compton* and *Gamma* and to identify their location more precisely. *Integral* would make observing powers one hundred times greater. In addition to studying known objects it would find many more phenomena that are presently unknown. *Integral* was expected to focus on violent cosmic events in the centre of our own Galaxy: neutron stars, binaries, bursts and black holes.

Integral was based on the proven 4-m by 3-m *XMM* bus, a significant cost-cutting measure, and the 2.3-tonne payload carried four instruments: cooled

Figure 5.10. *Integral* observatory.

gamma ray spectrometer (1,300 kg), high-resolution gamma ray imager, X-ray
monitor and visible light monitor. The Russian *Proton* offered the most suitable
launcher, one able to put the observatory in an extended orbit well outside the
harmful radiation belts (*Ariane* would not have reached such an extended orbit).
Most of the observations will be made as the satellite rises to and falls from apogee.
Integral was built in two sections in Alenia's plant in Turin and was then sent to
ESTEC in the Netherlands for testing in the large space simulator. The spacecraft
was designed for between three and five years' life, the data being transmitted to
ESOC in Darmstadt and Redu in Belgium and the archive being kept in Switzerland.
The mission cost €350m.

   Russia offered a free launch in exchange for access to 25% of observing time.
The *Proton* rocket counted down at Baikonour cosmodrome on 17 October 2002. It
was a clear day as the *Proton* ignited, spewing the telltale brown nitric smoke
billowing out against the autumn steppes. As it went through maximum dynamic
pressure, *Proton* shook off three successive steam clouds, one in the shape of a smoke
ring. Four hours later *Integral* settled into its planned 685-km to 153,000-km
orbit, period 72 hr, deployed its solar panels and sent back first signals to ground
control.

ESA scientific missions

9 August 1975	*COS B*	X-ray, gamma ray astronomy
20 April 1977	*GEOS 1*	Geomagnetic fields
14 July 1978	*GEOS 2*	Geomagnetic fields
26 May 1983	*Exosat*	X-ray sources
2 July 1985	*Giotto*	Comet Halley
8 August 1989	*Hipparcos*	Star mapping
6 October 1990	*Ulysses*	Sun probe
2 December 1995	*SOHO*	Solar observations
4 June 1996	*Cluster 1* (fail)	Sun–Earth system
10 December 1999	*XMM*	X-rays
16 July 2000	*Cluster 2A*	Sun–Earth system
9 August 2000	*Cluster 2B*	Sun–Earth system
17 October 2002	*Integral*	Gamma rays

Collaborative missions with significant European involvement

22 October 1977	*ISEE 1,2*	Sun–Earth Explorer
26 January 1978	*IUE*	International Ultraviolet Explorer
12 August 1978	*ISEE 3*	Sun–Earth Explorer
16 August 1984	*AMPTE*	Magnetospheric particles
15 October 1997	*Cassini*	Saturn and Titan

5.16 EXPERIMENTAL EUROPEAN COMMUNICATIONS SATELLITES: *OTS*

These were the European scientific missions, the direct descendants of the ESRO science programme. In 1971 ESRO had, amid division, agreed to broaden its satellite programme into applications. Communications satellites were high up on the agenda.

The first tests of satellites for communication purposes were made by the United States as early as 1958 with an orbiting *Atlas* rocket that retransmitted a message from President Eisenhower. In 1965 the United States launched *Intelsat 1*, popularly known as *Early Bird*, beginning a global communications revolution. Demand for 24-hr satellites for television relay, telephone lines and direct broadcasting began to accelerate. Eutelsat was formed as a European public corporation based in Paris to manage European satellite communications on behalf of national telecommunications services. It was formed as an intergovernmental body, 48 countries eventually becoming members. It leased out transponders at between €2.8m and €3.8m a year for television and telephone lines to national and commercial telephone and television companies (Eutelsat was privatized as a French corporation in July 2001, the worldwide Intelsat soon thereafter).

ESA made it a priority to develop both technology-proving and operational communications satellites, both of which would pave the way for wide-scale commercial and private communications services. At an early stage ESA devised *OTS* (Orbital Test Satellite) to make pre-operational tests of a European satellite in 24-hr orbit. Its specification was for a satellite with six antennae able to handle 6,000 telephone circuits and two television channels, test frequencies in the 11 GHz to 14 GHz band (here, atmospheric interference was low) and demonstrate the reliability of its systems. *OTS* weighed 865 kg (444 kg once it had used its motor and arrived on station), measured 2.39 m high and 2.13 m long and was shaped like a hexagonal box with 9-m-wide solar wings on either side.

The television channels used spot beams that could be swivelled to point to the broadcasting footprint, where they could be received by satellite dishes of only 13 m diameter, compared with the then standard of 30 m. With large solar panels over 10 m long it was designed to operate up to five years. Goonhilly Downs in Cornwall was the designated tracking station. *OTS* carried three beams: one to focus on Western Europe (the spot beam), a second to cover a wider range, taking in eastern and central Europe as well (Eurobeam B) and a third to take in north Africa and the Middle East (Eurobeam A). The quality for the user would be the same, but more channels could be handled on the spot beam than the eurobeam. *OTS* was built by Hawker Siddeley, AEG Telefunken and Matra and managed by ESTEC in Noordwijk, the Netherlands. An important aspect of *OTS* was that it should last and that its components should prove durable, resistant and reliable. Because communications satellites are expensive, their operators expect them to last five, six or seven years, not the much shorter periods acceptable for some scientific satellites [103].

Things went badly wrong for Europe's first communications satellite *OTS 1*. Originally to fly in May 1977 its launch was delayed by an accident when one of the *Delta* launch vehicle's nine solid rocket boosters fell off during assembly. Initially, the launch from Pad 17 at Cape Canaveral appeared to go perfectly on 13 September 1977. Then, at 54 seconds a glow began to appear on the nose of the *Delta*'s solid rocket motors. The flame spread and then the side booster was seen to fall off the launch vehicle, rupturing the *Delta*'s main fuel tank. The whole rocket then exploded in an ominous preview of what happened to the shuttle less than 10 years later. Debris spilled into the sea 16 km off the coast. *OTS* was not insured, though the launching was, so ESA got €20m back toward the cost of another attempt. It was the second *Delta* failure that year and NASA salvaged the wreckage to find out what went wrong.

Fortunately, an identical back-up had been built, but when it was brought to the United States for a second launch one of the solid rocket boosters fell on top of the *Delta* tank in the hangar, damaging it and causing it to be dismantled for repair. Was the project jinxed?

Eventually, the back-up *OTS 2* was launched 11 May 1978. Its take-off was also problematical, being delayed by spoof signals in coaxial cables at the launch site and then ground equipment being struck by lightning. In the end all went well. The orbit was circularized on 13 May, and it drifted to its assigned station where it arrived on 3 June at 10°E over Gabon, providing communications for Europe, Iceland, the Canaries, Madeira, the Azores and the Middle East. It was one of the first communications satellites to test out the use of broad wavebands. *OTS 2* had four ground stations: Fucino near Rome; Bercenay near Troyes in France; Goonhilly Downs; and Uisingen near Frankfurt. Fifty European research institutes, some with ground terminals as small as 3 m, participated in the experimental programme.

A year later its operators declared the mission a success. ESA declared the mission successfully over and handed it over to Eutelsat for commercial use. *OTS 2* exceeded expectations in its ability to transmit television and up to 7,200 phone calls at a time and tested out the use of small ground terminals. Eutelsat handed *OTS* back to ESA in 1984, whereupon ESA began a series of testing manoeuvres to see how the satellite would respond when spun out of control and to test prolonged hibernation. *OTS* passed all these tests with flying colours, and in 1988 was reactivated to celebrate 10 years of orbital operations.

5.17 *OTS* FOR THE MARINE: *MAROTS, MARECS*

OTS was also the basis for a series of maritime communications satellite originally called *MAROTS* (MARitime Orbital Test Satellite). The aim was to adapt the *OTS* experimental satellite to service maritime, ship-to-shore services. *MAROTS*, it will be remembered, was one of the core programmes of ESA in 1973–5 on British insistence.

Inmarsat (International Maritime Satellite Agency) ordered communications satellites based on *OTS* to improve the quality of ship-to-shore-based communications. *MAROTS* was later renamed *Marecs*. Inmarsat had been set up by the Intergovernmental Maritime Consultative Organization over 1975–9, start-up funding coming from the United States, the Soviet Union, Britain, Norway and Japan. Headquartered in London it now has 48 member countries. The concept of Inmarsat was an international intergovernmental organization to provide standard maritime safety and communications equipment on an economic, commercial basis. From the 1960s more and more ships began to equip themselves with small-diameter parabolic antennae inside protective, white radomes. The dishes were designed to receive and send all kinds of messages from emergency calls to routine telephone, fax and telex messages.

Marecs was a specifically European, ESA-linked aspect of the Inmarsat system, with ESA developing *Marecs A* in 1981. The first *Marecs* was launched on the fourth *Ariane* test flight on 20 December 1981. *Marecs* was a prism 2.5 m high and 2 m across, 563 kg when on station, providing ship-to-shore communications for Inmarsat. At that time about 1,000 ships were kitted with satellite-based communications systems (each costing about €70,000), but Inmarsat was confident that the market would over time reach up to 70,000 ships and yachts.

Marecs B was lost on the *Ariane V5* failure of 10 September 1982. A replacement satellite *Marecs B2* was then launched in November 1984 and took up position over the Pacific Ocean. It carried up to 50 channels for emergency and routine ship-to-shore messages and was placed over the Pacific where demands for service had become especially strong. By 1990 eight satellites were working for Inmarsat with over 13,000 users: tankers, freighters, ocean liners, drilling rigs, ice-breakers, tugs, yachts and survey ships. *Marecs* paved the way for the *Inmarsat 2* and *3* series, which have since provided maritime communications to ships worldwide, the ships needing to spend €30,000 for the 1-m antenna in a plastic dome. The *Inmarsat 3* programme of four satellites was ordered in 1994.

5.18   OPERATIONAL SERIES: *ECS*

OTS paved the way for an operational series of communications satellites. Not only did *OTS* carry out the technological tests required of it, but it gave European companies the experience in building and operating satellites that they subsequently put to commercial use [104]. *ECS* (European Communications Satellite) was approved in March 1978 with an initial funding of €50m. The aim was to establish a fully operational European telephone, telex and broadcasting system to follow *OTS*. Five satellites were planned and operated for the Eutelsat network. *ECS* had one broad beam, extending from Norway to north Africa and from the Azores to Ankara; with spot beams for Western Europe, Eastern Europe and 'Spot Atlantic' for the Azores and Madeira [105].

From 1983 the first of five *ECS*s was launched, built by British Aerospace. *ECS* satellites were about 1,175 kg in weight, carrying telephone and television for the European Broadcasting Union in 5 beams using 10 transponders. *Ariane 1* (V6) launched *ECS 1* on 16 June 1983. *ECS 1* was located at 10ºE over Gabon, west Africa. The aim of the *ECS* series was to provide data, computer and teleconferencing links to businesses in Europe, Africa and the Middle East. A secondary payload was *Oscar 10*, built for amateur radio enthusiasts. *ECS* had 12 transponders, each able to handle either a television channel or 1,800 phone circuits and five beams, of which three could be moved as spot beams. The solar arrays spanned 13.8 m and generated 1,200 W of electrical power.

ECS 2 was launched by *Ariane 3* in August 1984 and commissioned that October. In a strange accident *Oscar* was hit by the third stage after its deployment and destroyed. Once developed by ESA, *ECS* satellites were made available to Eutelsat for commercial development and renamed *Eutelsat F1, F2, F3* and so on. Typical costs for leases were between €2.8m and €3.8m a year.

With the success of *ECS*, and the continued expansion in demand for telephone and television series, the *Eutelsat 2* series was commissioned at a cost of €1.094bn. The *Eutelsat 2* series offered 16 transponders, 39 TV channels, longer lifetimes, more back-up systems and power to reach Turkey and the Urals. One satellite *2F5* was lost in the *Ariane* failure of 25 January 1995. A *Eutelsat 3* series was announced in 1998. In March 1994 Eutelsat announced the *Hot Bird* series for direct broadcast satellites to dishes only 45 cm across. A further spin-off of *OTS* was the Anglo-French *Satcom* comsat series.

ECS/Eutelsat communications satellites

16 June 1983	*ECS 1*	*Ariane 1*	Kourou
4 August 1984	*ECS 2*	*Ariane 3*	Kourou
12 September 1985	*ECS 3*	*Ariane 3*	Kourou (fail)
16 September 1987	*ECS 4*	*Ariane 3*	Kourou
21 July 1988	*ECS 5*	*Ariane 3*	Kourou
30 August 1990	*Eutelsat 2F1*	*Ariane 44LP*	Kourou
15 January 1991	*Eutelsat 2F2*	*Ariane 44L*	Kourou
7 December 1991	*Eutelsat 2F3*	*Atlas 2*	Cape Canaveral
9 July 1992	*Eutelsat 2F4*	*Ariane 44L*	Kourou
24 January 1994	*Eutelsat 2F5*	*Ariane 44L*	Kourou (fail)

5.19   OLYMPUS

OTS had paved the way for *ECS* and *MAROTS*. ESA now faced the challenge of bringing orbital communications through to the next stage of development: direct broadcasting to homes and business.

The next experimental communications satellite had the dull title of *L-Sat*, receiving the more appealing name *Olympus* once it reached orbit. Built by British Aerospace with Selenia Spazio, Aeritalia, Spar Canada and Fokker, it weighed the then unheard-of weight for a communications satellite of 2,300 kg. *L-Sat* was approved by ESA in 1979, with an initial investment of €388m (the eventual cost of the project was closer to €700m). The principal funders were Britain 39% and Italy 32%.

If it worked it could pave the way for much larger comsats of up to 3.5 tonnes with ever more sophisticated communications capacity for the commercial market. *L-Sat* was designed to carry high-powered steerable beams, able to send down signals to small, home dish receivers only 90 cm in diameter. It was also intended to test such new techniques as videoconferencing and redirecting signals from low-orbiting Earth satellites back down to ground control. Its communications capacities were the most powerful of their day. In scale *L-Sat* broke new ground. It was 5.5 m high, 2.9 m diameter, and its solar panels were 25.6 m from tip to tip, half the length of a football field, generating 3.6 kW. *L-Sat* was ground-tested in Canada, which played an important role in the project (Canada had become an associate member of ESA). To get *L-Sat* from its final tests in Canada to Kourou, ESA chartered the large Belfast freighter aircraft. The support equipment had to be sent in a separate Boeing 747 freighter.

L-Sat was launched and named *Olympus* on 11 July 1989 and reached its destination of 19°W several days later. *Olympus* was the first ESA satellite to use a nitrogen tetroxide/monomethyl hydrazine engine to get to geosynchronous orbit, solid-fuel motors having been used previously. It was checked out by ESOC in Darmstadt before being handed over to the ESA ground station in Fucino, Italy that October. At first all went perfectly [106]. *Olympus* found itself in serious trouble when, after two years, on 29 May 1991 ground controllers lost control. A beacon broke down and the satellite could not be recontacted. It appeared to have lost its lock on the Sun and begun drifting and spinning. There was a real danger that its thermal protection systems would break down and the batteries would drain irreparably dry. As the satellite seemed less and less prepared to respond to commands, pessimism grew. A board of investigation was formed [107].

Not prepared to contemplate the loss of such a valuable (€650m) satellite, the Director General of ESA Jean Marie Luton formed a 50-strong recovery team of engineers during the first week of June. They asked for and got the cooperation of NASA's tracking stations. After several worrisome weeks they recovered control, and by August 1991 had an operational satellite once more. Ground control in Fucino managed to turn the solar panels towards the Sun to begin to regenerate electrical energy, recharge the batteries and thaw out the spacecraft's frozen fuels and pipes. Then, using thruster fuel they managed to get the errant satellite back on station.

Two more years of experiments were conducted (e.g., in satellite-based video-conferencing and in high-definition television broadcasting). *Olympus* was used to help establish BBC world service television. Suddenly, two years after its recovery, on 11 August 1993 *Olympus* lost attitude and began spinning. Although it was partly recovered the satellite was almost entirely out of fuel at this stage, so the final fuel remaining was used to lower *Olympus* to a graveyard orbit.

5.20   *ARTEMIS*

Europe's most recent experimental communications satellite is *Artemis* (Advanced Relay TEchnology MISsion), approved by ESA in 1990. This is a project both to test the feasibility of communicating between satellites in orbit using optical systems and lasers and to develop mobile communications with trucks and cars. *Artemis* was designed to attack the problem of receiving data from satellites when they are out of range of ground stations and ensure that information is not lost. It is an expensive project, costing €800m. Italy was the main contributing country, with Alenia the main contractor supported by 66 companies in 14 countries.

Artemis was big, 4.5 m tall, 24.7 m long, with a mass of 3.1 tonnes. It was to be positioned at 36,000 km at 21.5°E, a good vantage point for communication with the satellites below. The wingspan of its solar panels was 25 m. Signals from *Artemis* were designed to be picked up by dishes as small as 20 cm. Artemis was a step forward, equivalent to that of *Olympus* in its own time [108]. Its objectives were to test out from 24-hr orbit:

- interorbital, laser-based communications relays, retransmitting signals from *SPOT 4*, *Envisat* and a number of Japanese satellites (e.g., *OICETS*);
- develop mobile communications;
- pave the way for a European navigation satellite system;
- demonstrate ion-based station-keeping (UK-10 thrusters).

Figure 5.11. *Artemis* beams *SPOT 4*.

The designers believed that laser-based interorbital communication would be faster, carry more data, use lighter equipment than traditional means, need less power and would be less vulnerable to interference. This was the future for space-based communications. The actual system for transmitting data by laser had first been tested out on the *SPOT 4* Earth resources satellite in spring 1998. For interorbit communications *Artemis* was planned to communicate with the same *SPOT 4*, the European environmental satellite *Envisat* and *OICETS*. If successful, compatible equipment would be added to other satellites later (e.g., the Japanese *Kibo* module of the International Space Station). For the first time for an ESA satellite, *Artemis* used electric propulsion for station-keeping. *Artemis* was equipped with two ion thrusters with 40 kg of xenon fuel, the system drawing 600 W from the solar array. With conventional station-keeping, 400 kg of fuel would have been required.

The launch was originally to have been on *Ariane*, but when costs ran ahead of projections Japan offered a free launch on its *H-II* in exchange for participation in the experiment and a link to its own, lower orbiting, *OICETS* experimental communications satellite. This offer was accepted, but then the Japanese launcher programme entered a period of severe difficulty and the launch was transferred back to *Ariane*. All the containers of *Artemis* shipped out to Japan now had to travel halfway around the world again, this time to South America.

Launch duly took place on 12 July 2001. Then things went wrong. The upper stage of *Ariane 5* failed to fire properly, stranding *Artemis* in an orbit of 590 by 17,487 km. Ground control in Fucino, Italy, went ahead with deploying the solar panels and started a plan to save the mission (*Artemis*'s companion, a Japanese comsat with a single-burning solid-fuel motor, had to be written off completely). The liquid propulsion system was brought into play sooner than expected. A burn on 18 July raised the apogee to 19,164 km and a second one to 31,000 km, ever closer to the magic 36,000 km, though still a long way from a circular orbit. The committee of inquiry attributed the failure to combustion instabilities in the *Aestus* upper stage, which reduced thrust, depleted propellant and caused an early shutdown. By the end of the year *Artemis* had lengthened its orbital period to 19 hours, getting ever closer to the 24 hr intended, using its ion propulsion system. When firing, the engine can raise the orbit by about 1 km an hour. One of the advantages of the ion propulsion system is that it can provide low thrust for long periods. Using this system, it was estimated that it would take up to early 2003 for the *Artemis* spacecraft to reach its 24-hr, 36,000-km orbit.

On 21 November *Artemis* used for the first time its optical transmission system to communicate data with *SPOT 4*. *Artemis* locked on to the French satellite, which began transmitting its pictures at 50 MB a second for forwarding on to the *SPOT* ground station in Toulouse, France. Whether *Artemis* will indeed pave the way for a revolution in space-based communications remains to be seen.

European experimental communications satellites

14 September 1977	OTS 1	Delta	Cape Canaveral (fail)
12 May 1978	OTS 2	Delta	Cape Canaveral
11 July 1989	Olympus	Ariane 3	Kourou
11 July 2001	Artemis	Ariane 510	Kourou

5.21   EUROPEAN WEATHER SATELLITES

The first weather satellite was launched in 1960, the American *Tiros*, and ever since then national and international meteorological organizations came to rely more and more on weather satellites for images and data. Weather satellites came in two forms: polar orbiting satellites at low altitudes (generally around 1,000 km) and satellites orbiting at geostationary orbit of 36,000 km, scanning the globe every half hour or so.

In the mid-1970s Europe, Japan, the USSR and the United States joined in a global programme to improve weather observations from 24-hr orbit. Their idea was that if each country placed a satellite in evenly spaced geosynchronous orbits, then between them they would be able to provide coverage of the whole Earth.

The European contribution was *Meteosat*. The satellite was a drum shape, 2.1 m across and 3.2 m high, weighing 700 kg. The purpose of the mission was to collect and retransmit imaging data for cloud cover, sea temperature, wind, water vapour and radiation. The camera would take a fresh picture every 25 minutes in the visible, infrared and water vapour channels, sending data down to Darmstadt in Germany and to local weather stations. One of the main construction companies was Marconi in Portsmouth [109].

Because *Ariane* was not yet ready, an American *Delta* launcher was used. *Meteosat*'s launch was delayed four times because of spoof radio signals getting into the telemetry system of the *Delta* rocket. These were eventually traced to a ship off the Florida coast using the same frequency as the destruct system for the *Delta* – a worrying thought. All went well and the *Delta* placed *Meteosat* into 24-hr orbit over the eastern Atlantic on 23 November 1977. From its vantage point of 0°, *Meteosat* was able to look down on all of Europe, Africa, the Middle East and the eastern tip of South America. After sending 40,000 images over two years *Meteosat 1* lost its radiometer, the electrics overloaded and it never achieved its normal performance again. Its data interrogation and collection function continued to operate until the satellite drifted off station in 1986.

Meteosat 2 followed in June 1981, but suffered an electrical failure and was replaced by a back-up satellite, which became *Meteosat 3*. This was launched by *Ariane 4* in June 1988. It worked so well that it took over *Meteosat 2*'s former role effortlessly. By the time its fuel had begun to run out in December 1991, it had sent back no less than 284,000 photographs and it was put into a graveyard orbit. At one stage it was moved over Colombia in South America to assist the American weather service.

After this pioneering phase the European meteorological service Eumetsat was set up to oversee an operational system, though ESA continued to have responsibility for construction and operation. Eumetsat became the controlling body in January 1987. There were 17 European participating states eventually. These levels were based broadly-speaking on gross national product and each country had a representative at the Eumetsat council.

Contribution to Eumetsat (%)

Austria	2.23	Italy	15.46
Britain	14.09	The Netherlands	4.02
Belgium	2.7	Norway	1.47
Denmark	1.76	Portugal	0.86
Finland	1.84	Spain	6.96
France	16.78	Sweden	3.2
Germany	22.29	Switzerland	3.33
Greece	0.96	Turkey	1.5
Ireland	0.54		

Meteosat 1, *2* and *3* paved the way for the three MOP (Meteosat Operational Programme) satellites ordered May 1984 (in some listings, they continue to follow the *Meteosat* designator). *MOP 1* was blasted into orbit by *Ariane V29* on 6 March 1989. Launch was delayed several times: by a strike, then by cables being blown off the *Ariane* by high winds and then by the need to await spare parts from France. All went well in the event and *MOP 1* arrived on station over 0°, the Greenwich

meridian. *MOP 1* scanned the Earth every 25 minutes day and night, sending down pictures to over 1,000 registered users in Europe and Africa. *MOP 2*, which flew on *V42* on 2 March 1991, was designed to provide images every 30 minutes in the three wavebands to a detail of 2.5 km. These were sent down to the 30-m dishes at Odenwald, 40 km from Darmstadt, which acted both as mission control and distributor of *MOP* images. *MOP* satellites were drum-shaped and about 700 kg in weight. Their main feature was a 40-cm telescope, whose task it was to image the whole Earth's weather every 25 minutes and transmit the picture to the ground. Its infrared sensors had to be cooled to a temperature of 4°K. Its photographs appeared on many national, European, television station weather forecasts every evening. Signals were sent down from *MOP* to ground control in Darmstadt, Germany and to Fucino, Italy.

Two new generations of meteorological satellites were approved at the ESA council meeting held in Granada, Spain, in 1992. These were three satellites called *MSG* (Meteosat Second Generation) and a further three called *Metop* (Meteorological Operational Polar Satellite) later. *Metop* would not only provide regular weather data but carry out longer term climate studies as well. *Metop* would weigh 4.5 tonnes and carry a high-resolution radiometer, infrared sounder, microwave sounding unit, humidity sounder, inferometer, ozone monitor, radiation budget scanner and wind scatterometer. The contract for the *Metop* series went to Matra Marconi in late 1999 for €791m to build three satellites, the first to fly in 2003, based on the *SPOT 5* and *Envisat* bus. The launcher will be *Soyuz ST*. The contract between Eumetsat and Starsem was signed in 2000.

While awaiting *Metop*, *MSG* was made ready. The first made a triumphant night-time entry to orbit on 28 August 2002 on *Ariane 5*. *MSG 1* was a radical improvement on the first generation, transmitting pictures twice as frequently, every 15 minutes, with 12 spectral banks instead of 7. It was expected that the satellite would lead to big improvements in the forecasting of snow, fog and thunderstorms. The two-tonne satellite, over twice the size of its predecessors, carried an instrument to measure Earth's radiation budget, a vital tool for measuring climate change. Successors are planned for 2005 and 2009.

European weather satellites

23 November 1977	*Meteosat 1*	*Delta 2*	Cape Canaveral
19 June 1981	*Meteosat 2*	*Ariane 1*	Kourou
15 June 1988	*Meteosat 3*	*Ariane 44LP*	Kourou
4 March 1989	*MOP 1/Meteosat 4*	*Ariane 44LP*	Kourou
2 March 1991	*MOP 2/Meteosat 5*	*Ariane 44LP*	Kourou
20 November 1993	*MOP 3/Meteosat 6*	*Ariane 44LP*	Kourou
2 September 1997	*MOP 4/Meteosat 7*	*Ariane 44LP*	Kourou
28 August 2002	*MSG 1*	*Ariane 5*	Kourou

European weather forecast series

1977–1984	*MOP/Meteosat*	2002–2005	*MSG/Metop*

Figure 5.12. *Meteosat Second Generation (MSG).*

5.22 EARTH OBSERVATION

An early and unexpected benefit of space exploration was the finding that orbital altitude provided an ideal point from which to observe human and natural events taking place on the Earth's surface. It proved possible to look down through the atmosphere with unusual clarity. Scanning and enchancing instruments, many developed to look through clouds, provided a wealth of data that could be used for agriculture, fishing, forestry, environmental protection, mining, hydrology, cartography, geology and oceanography. These discoveries opened the way for what have been called Earth observation or Earth resources space application programmes. The first dedicated Earth observation was launched by the United States in 1972 *Landsat 1*. ESA began preparatory studies in 1979.

5.23   *ERS*: RADAR-BASED EARTH RESOURCES OBSERVATORY

ESA then proceeded with the development of *ERS* (Earth Resources Satellite). This project was approved in October 1981, with Germany as the main funder (24%) and the prime contract going to Dornier. In May ESA had secured sufficient commitments for funding for the mission to go ahead the following year. The aim was that Europe should have its own independent ability to monitor the environment from space and make a substantial step forward in our knowledge of seawater and freshwater, their interaction with the atmosphere and land surfaces. The crucial instrument identified for the mission was a radar to map the ground through darkness, clouds and rain [110]. Hitherto, Earth resources satellites had used visual imaging techniques only. Radar would be a radical advance, able to scan both the surface of the seas and the landmass with extraordinary accuracy and do so through cloud cover and darkness. This came at a price, for space radars posed special problems, requiring high levels of energy and returning large volumes of data.

ERS was big. The spacecraft was 11.8 m high and had a wingspan of 11.7 m, with two solar wings providing 2.2 kW from 22,000 cells. *ERS 1* weighed 2.3 tonnes and was made by Dornier. It had four main instruments: the synthetic aperture radar, active microwave instrument, along-track scanning radiometer and radio altimeter.

ERS 1 was launched on 16 July 1991 on *Ariane V44*, entering an orbit of 785 km. The ground station in Alaska failed to acquire the signal, which created much worry, but Perth picked up *ERS* and all was well. The first few months were spent in getting the instruments on the spacecraft properly calibrated. A well-known part of the Netherlands, Flevoland, was used to calibrate the imaging system. The radar was able to measure wind speed by measuring the ripples on the ocean surface. By measuring very small variations in the height of the sea, it could estimate the depth of the ocean floor beneath. The radar altimeter was checked out against lasers in Venice, Graz and Zimmerwald in order to get an accuracy of 5 cm. The along-track scanning radiometer (ATSR) was able to measure the sea temperature to within 0.3°C.

Within days of entering orbit *ERS 1* sent back its first picture – the northerly island of Spitzbergen – and not long afterwards the burning oilfields in Kuwait. *ERS 1* operated for nine years until 16 March 2001 when its attitude control system failed. By then it had made 45,000 orbits and sent back more than 1.5 million radar pictures. *ERS 1* proved to be enormously valuable, especially so in detecting and identifying tankers illegally dumping or spilling oil at sea. It mapped

Figure 5.13. Europe from *ERS 2*.

changing farm use in Europe, spotted ice floes in the Arctic, provided routine weather-forecasting and watched floods in the Netherlands. Its radar was so sensitive it could pick up the minuscule rise and fall in the Earth's crust around the volcano of Vesuvius. Several years later its photographs were still impressing analysts in such ways as tracking the El Niño current and mapping the topography of the northern ice cap. A global sea state map was assembled every three days. Most of the world's seas were coloured blue (calm), but there were some patches of red (waves of 5 m or more) and some parts of the Southern Ocean coloured white (waves more than 12 m) – probably worth avoiding [111].

The *ERS 2* programme was approved in 1991 at a cost of €371m. It was built by 1,000 engineers from 14 nations in 50 companies. Weighing 2.5 tonnes it took over where *ERS 1* left off, although *ERS 1* and *2* operated in tandem at one stage. By operating with *ERS 1* the two satellites between them were able to produce highly accurate digitized elevation maps of considerable accuracy. Within days of its night-time 20 April 1995 launch, pictures were flooding in of Italy's land and relief. *ERS 2* orbited at 774–793 km, 98.5°, repeating the same ground track every 501 orbits or every 35 days.

ERS 2 carried a global ozone monitoring experiment. Its ATSR had the ability to measure sea temperatures to within an accuracy of 0.2°C. Like its predecessor it carried a synthetic aperture radar and a scatterometer to measure wind speed and direction. *ERS 2* played an important role in measuring the degree to which the ice cap on Antarctica was rising or falling. The accuracy of *ERS 2* was 0.5 cm, enough to determine change even over a relatively short period. The outcome: one part of Antarctica was losing ice at 12 cm a year (the Thwaites glacier basin), but the continent as a whole was almost unchanged over the 1990s. In January 2002 *ERS 2* was able to notice a sudden 30% thinning of the ozone level over Europe, with the concomitant dangers of exposure to ultraviolet rays. Ozone maps from *ERS 2* identified 'ozone holes' over different parts of Europe, often associated with warm air moving up from Africa.

5.24 *ENVISAT*: EUROPE'S BIGGEST-EVER SATELLITE

Following the success of *ERS 1*, in 1992 ESA approved the next phase of Earth resources satellites, *Envisat*. The project also had convoluted roots in the polar platform project of the late 1980s that was to accompany the manned *Columbus* space station module [112]. It was one of the few projects of the period that Britain was prepared to lead, several important companies contributing such as Logica, British Aerospace, Matra Marconi and Rutherford Appleton Laboratory. It had originally been intended that the polar platform would be serviced by astronauts in orbit – but with the closure of the shuttle base in Vandenberg, California and with the end of the European spaceplane project *Hermes* (see Chapter 7), this ceased to be realistic.

The project was allocated €1,174m, making it the most-expensive-ever European satellite (some give the final cost as €2bn). The main funders were France 24%, Britain 19%, Germany 18% and Italy 11%. By the time it had been finally defined the polar platform had become a global environmental monitoring project from polar orbit. Even without links to the manned space station programme it was still a viable, large and ambitious project. It had two modules: a service module and a payload module. The service module was based on the *SPOT 4* bus, being responsible for power, attitude, propulsion, control and communications, with a single 7.5-kW solar array attached. The payload module provided 2.4 tonnes of experimental equipment and advanced synthetic aperture radar. The idea of the designers was that the polar platform design could be adapted for a number of different satellites, *Envisat* being the first exemplar [113].

Figure 5.14. *Envisat.*

The final *Envisat* configuration was twice the size of *ERS 1*. It was a huge satellite, 10.5 m high with 10 main instruments, the size of a small space station. Preparing it for launch required moving 400 tonnes of equipment to French Guiana, and once there it took over almost all of the new S5 processing hall. *Envisat* was equipped with instruments for the observation of land, ocean, atmosphere and ice. It was intended that in its 100-min, 800-km polar orbit it would overfly each part of the Earth's surface every 35 days.

To encourage student interest in *Envisat* and its broader environmental mission, ESA ran a competition for a symbol for the mission. The winner's symbol would be painted on the side of the ascending *Ariane* and would win a trip to French Guiana to see the take-off. The lucky winner from 13,000 competitors was 12-year-old Anke Hartmanns from Oldenburg, Germany, who designed a sunflower, each petal illustrating a different part of our fragile environment. In French Guiana *Envisat* became the first satellite to use the new €80m payload processing facility.

Envisat was next in line for an *Ariane 5* launch after the discouraging stranding in the wrong orbit of the experimental communications satellite *Artemis*. A lot was riding on *Envisat* and its huge batch of experiments [114]. Several were a continuation of experiments begun by *ERS 2*. Indeed, to ensure complementarity with *ERS 2*, *Envisat* was set to trail *ERS 2* in orbit 30 minutes later. Data from *Envisat* was to be sent down in three ways: real time, in recorded dumps over tracking stations and via the *Artemis* satellite.

Envisat's **experiments**

MERIS	(Medium Resolution Imaging Spectrometer), measuring solar radiation from the Earth's surface and clouds in 1,150-km strips in the visible and near-infrared
MIPAS	(Mitchelson Interferometric Passive Atmosphere Sounder), to monitor chemical changes in the atmosphere
ASAR	(Advanced Synthetic Aperture Radar), designed to radar-map the Earth's surface and oceans regardless of darkness and clouds in 400-km-wide strips
GOMOS	(Global Ozone Monitoring by Occultation of the Stars), to monitor ozone depletion, aerosols and trace gases between 20 km and 100 km
RA-2	(Radar Altimeter 2), to precisely measure the height of waves, the ocean floor, ice and polar sheets and the height of the land
MWR	(Micro Wave Radiometer), to measure atmospheric humidity
LRR	(Laser Retro Reflector), to gauge the satellite's precise distance above the Earth
SCIAMACHY	(Scanning Imaging of Absorption Spectrometer for Atmospheric Cartography), to investigate pollution, haze, dust storms, burning, industrial plumes and the influence of volcanoes on the atmosphere
AATSR	(Advanced Along Track Scanning Radiometer), to measure the temperature of the sea and the moisture of vegetation
DORIS	(Doppler Orbitography and Radio Positioning Integrated by Satellite), to measure *Envisat's* location to within 5 cm

The launch of *Envisat* for *Ariane* flight *V511* was delayed as ESA tried to debug the *Aestus* upper stage after it left *Artemis* stranded on *V510*. The investigatory committee attributed the failure to combustion instabilities, but was not able to be more precise. Over the autumn the *Aestus* upper stage was tested and fired 200 times and the culprit was suspected to be some water getting into the propellant. Accordingly, for *V511 Aestus* was drained and dried out and new procedures were introduced to guard against contamination or humidity. Would this be enough?

The precautions must have worked for *Envisat* was launched with extreme precision on 28 February 2002. Although *Envisat* had 319 kg of fuel available to adjust its orbit, it only required 10 kg to get it into the precise position required. The first instruments were turned on in March, though it would take until the end of the year to calibrate them all. The first images were returned to the Kiruna station in Sweden. In its first operations it observed the break-up of the Larsen B ice shelf in Antarctica and shoals of plankton off west Africa, the pictures being of excellent resolution.

Envisat will be followed in 2004 by a small satellite called *Cryosat*. This €70m project is to measure with extreme accuracy the thickness of ice sheets in the polar oceans with a double radar scanner and laser reflector from 720-km orbit. The launcher is expected to be *Rockot* and Astrium is the main contractor.

5.25 BIOLOGY AND MATERIALS-PROCESSING: *FOTON*

Space biology and materials-processing have been a relatively low-priority area for ESA's scientific programme. Europe's biological and materials-processing space experiments have flown on the Russian *Foton* and *Bion* programmes. With the International Space Station, though, this area of activity will expand.

From the 1960s the Soviet Union began to fly missions for biological experiments and then materials behaviour. These were part of the *Cosmos* programme and received *Cosmos* numerical designations. In both cases the missions used cabins derived from the *Zenit* and *Vostok* programmes. From the 1980s the USSR opened these missions to international cooperation, but also gave them programme designations in their own right: *Bion* and *Foton*, respectively.

ESA developed two sets of biological equipment to fly on the Russian *Foton* series of satellites, called Biopan and Biobox. The first set of European equipment was called Biopan. Based on a Soviet design from the 1970s Biopan was a 38-cm × 21-cm container that could be mounted on the exterior of the cabin and exposed to space conditions, holding experiments to the weight of 5 kg. The lid of Biopan can be opened in space (it closes in time for re-entry). Biopan is kept on the exterior during re-entry, but facing away from the ablative side of the cabin.

The European variant was built by Kayser Threde of Munich and first tested on *Foton 8* in 1992. The first operational Biopan mission was *Foton 9*, a 17-day mission in June 1994. Here, six experiments were carried to test the effects of radiation on molecules, organisms and DNA [115]. Biobox is much larger, 25 kg, holding three experiment containers and a centrifuge. By contrast, Biobox was carried inside the cabin. Its first operational flight was *Foton 9*, which carried experiments into the effects of microgravity on bones and cells. The results from the second Biobox were unfortunately lost when the recovery helicopter dropped the *Foton 10* cabin from a height after its return to Earth. Most of these experiments were reflown on the *Foton 11* mission (9 October 1997) as part of the arrangement if something went wrong with *Foton 10*. *Foton 11*'s Biopan and Biobox carried beetles, flies, cancerous cells and the CNES Ibis experiments for organic crystals (Crocodile).

Foton 11 also carried German materials experiments devised by Kayser Threde in a small cabin called *Mirka* (standing for microgravity re-entry capsule). *Mirka* was a spherical, 150-kg, 1-m-diameter cabin carrying three experiments: on heatshield instrumentation, heat flow, pyrometers and a new type of ablative material. The *Mirka* capsule was carried on the front of the cabin, separated after retrofire and had its own ablative material, parachutes and beacons. The main cabin carried crystal growth experiments, human cell biology experiments to learn more about cancer, flies (to test for ageing) and beetles (to see how their biological clocks were affected by zero gravity). Both the main *Foton 11* cabin and *Mirka* touched down safely on the brown steppe grass near Omsk on 23 October 1997 [116].

Foton 12 flew in September 1999, carrying 11 experiments from Russia, Germany, China, Belgium, the Netherlands, France, Britain, Italy and Sweden. The European package took up 240 kg of experimental space. The French team was given a real treat – being permitted to watch the launch from a protected shelter only 800 m from the rocket itself. *Soyuz*, as usual, performed as advertised, taking off only 1/100th of a second late. *Foton 12* came back a day early, because the French Ibis experiment shut down prematurely during one of the plant experiments. *Foton 12* landed in the green–brown grasslands 133 km north-west of Orenburg 15 days later, a mere 2,000 m from the welcoming team. It included experiments on cells, fluid physics, mapping, micro-organisms, particles, radiation and meteorites. The last was probably the most unusual. Various types of rocks were exposed to re-entry to test why some types of meteorites survived entry at great speed into the Earth's atmosphere and why others did not. Within minutes rescue teams were alongside unloading the experiments [117].

The 13th *Foton* was launched on 15 October 2002. This was an improved model, able to carry 650 kg of experiments. The satellite was named *Foton M-1* to mark the modifications. *Foton M* was loaded with 44 biological and materials-processing experiments from Europe, the United States, Indonesia and Japan, covering everything from fluid dynamics to growing crystals in microgravity on what was planned to be a 15-day mission.

No sooner had *Foton M-1* cleared the pad than observers noted smoke streaming from one of the strap-on rocket engines. At 29 seconds the engine failed, the computer shut down the whole rocket and the *Soyuz U* launcher fell back into the forest. There was a huge explosion, which killed one of the rocket troops in his cabin and injured many others. The fires took some time to put out. An investigation was ordered. It was the first *Soyuz U* failure in 10 years.

A related experimental system to that flown on *Foton* was called Biorack and it was flown on the space shuttle. Biorack was a container that provided incubators and a glovebox system that enabled a range of biological experiments to be carried out. On STS-81, a mission to the *Mir* space station in January 1997, Biorack flew for nine days to test the effects of space travel and microgravity on plants, fungi and cells.

5.26   EUROPEAN MILITARY MISSIONS

Both the United States and the Soviet Union were quick to recognize the value of spaceflight for military purposes. By the late 1950s both countries were designing military photoreconnaissance satellites: *Discoverer* (*Corona*) and *Zenit* series, respectively.

European military satellites, called *Helios*, were first launched in 1995. This was

Figure 5.15. *Helios* military design.

not an ESA programme (ESA's brief is civil and includes several prominently neutral countries), but a collaborative programme involving three countries. As usual, France took the lead. Dating back to the days of General de Gaulle, France had always taken an independent line within the North Atlantic Treaty Organization and specifically cited the need for an independent military surveillance capacity separate from the United States. There were rows between the two countries when France published photographs of Russian facilities that the Americans felt should have been classified.

Helios was based on the *SPOT 4* platform, with a resolution of 1 m. France launched the *Helios 1A* photoreconnaissance satellite on 7 July 1995. *Helios 1A* was a 2.5-tonne satellite and settled into a 680–682-km orbit. *Helios* was a French programme managed by CNES, with collaboration from other European countries: Italy covered 14% of the cost of the project and Spain took a minor 7% share. Germany at one stage considered joining, but changed its mind. The launching made France and its associates the only other countries with a military surveillance capability after the United States, the Soviet Union and China.

Helios 1A was controlled from the Centre de Maintien de Poste in Toulouse and encrypted photographs were downloaded to Colmar (France), Lecce (Italy) and Maspalomas (Spain). It has been reported that downloading was later sent to a purpose-built centre run by the Western European Union (a bridge organization between the EU and NATO) at the converted American Air Force base in Torrejón, Spain [118]. *Helios 1* was in orbit in time for the Kosovo crisis and French Defence Minister Alain Richard says that its images made a big difference to the French role in responding to the crisis.

Helios 1B was launched on 3 December 1999 carrying an SSTL (Surrey Satellite Technology Ltd) minisatellite for signal intelligence called *Clementine*. It was 2.55 tonnes in weight, with solar panels able to provide power of 2.2 kW. The main change from *Helios 1A* was the use of a solid-state memory able to store 9 GB more data than *1A*. *Helios 1B* was controlled from an operational control centre in Toulouse, supported by the *Helios* centre in Creil, Paris. Each day, military command centres in Madrid, Paris and Rome would send in their data requests and these would then be sorted and uplinked to the satellite. They would be received at image reception centres in Colmar, Maspalomas and Lecce. The French estimate that they saved at least €200m by basing *Helios* on the *SPOT* design.

The only problem with *Helios* is that it provided only cloud-free, day-time observations, limiting data during cloudy winters. Two *Helios 2* satellites are under consideration for 2003–7 and these would carry infrared images to take night-time photographs. The first *Helios 2* will be launched in 2004 and will be based on the *SPOT 5* Earth resources satellite already launched May 2002. France also approved in 2001 a four-satellite constellation of electronic intelligence satellites. These are for 120-kg microsatellites orbiting 10 km apart 680 km above the Earth. Appropriately called *Essaim* ('swarm'), the use of a formation would give them the equivalent of 10 minutes loiter time over a target, the ability to triangulate a signal precisely and the equivalent listening power of a large satellite. Launch is set for 2004 [119].

France also flew a small military test payload to check out the possibilities of conducting electronic intelligence from orbit. *Cerise* (Caractérisation de l'Environnement Radioélectrique par un instrument Spatiale Embarqué) was a 50 kg microsatellite flown piggyback on an Ariane launch. The mission came to a temporary halt on 21 July 1996 when *Cerise* collided with the upper stage of an *Ariane* rocket, making it the first victim of a space collision, but ground controllers later regained control.

The *Helios* systems are likely to be followed by European systems for military communications. In 1997 France, Germany and Britain reached agreement for a unified military space communications system Trimilsatcom, but the following year Britain pulled out and the system had to be abandoned. Britain went ahead with its own *Skynet 5*, leaving France to pursue its *Syracuse* programme, Italy *Sicral* (Chapter 3) and the other countries to make arrangements with them. Some cooperative effort reappeared when in 2000 France and Italy agreed to develop a €1bn military observation system called Cosmos Pleiades, merging the planned two-satellite French system and the four-satellite Italian one. Each satellite would weigh 1,000 kg and have a resolution in the order of 1 m.

5.27   ASSESSMENT AND CONCLUSIONS

Whereas ESA inherited a doubtful rocket project from ELDO (the *L3S*), ESA inherited a lively, healthy and largely successful satellite programme from ESRO. As a result ESA made an early and confident start to its satellite portfolio, with the orbiting of *COS B* (1975), *GEOS* (1977–8) and *Exosat* (1983). Successful cooperative programmes were run with the Americans: *AMPTE* and *ISEE*. ESA's first project in its own right was *Giotto*. It was courageous of ESA to decide on a deep space mission at such an early stage and doubly so to target such a difficult object as a comet. In sending *Giotto* into the heart of comet Halley, fortune favoured the brave, the mission making headlines, seizing the imagination of scientists, engineers and astronomers and stirring public interest in Europe and further afield. *Giotto* paved the way for a wave of successful scientific European-led missions: *Hipparcos*, *ISO*, *Ulysses*, *SOHO* and *Cluster*. Europe was an active participant in the American-led *Hubble* Space Telescope and the *Cassini* probe to Saturn. After a shaky start the *Hubble* mission profoundly reshaped both popular and scientific appreciation of the universe, becoming one of the great telescopes of all time.

Several features of these missions were important. First, ESA developed a framework of long-term planning (*Horizon*), which provided a sense of purpose, stability and development. Second, the missions had a high success rate and produced a substantial scientific return. Such failures as there were took place on the launch vehicles rather than the spacecraft themselves (*GEOS 1*, *Hipparcos*, *Cluster*). Europe demonstrated its robustness by being able to recover the American–European *SOHO* and to reconstruct the *Cluster* mission on the *Soyuz Fregat* at a time when all seemed to be lost. Third, these missions attempted to complement rather than rival the American and Russian missions of the same period. Europe identified niche missions that the two space superpowers had not undertaken or did not plan. *Ulysses* is a good example of a complementary mission that pushed back the frontiers of knowledge in new ways. The end of the period saw the orbiting of two large observatories at the cutting edge of X-ray and gamma-ray astronomy, *XMM* and *Integral*, the former taking up where the American *Chandra* observatory left off.

The question of space applications had been one of the big problem issues in the terminal days of ESRO. Under ESA's new system of optional programmes it was now able to press ahead with applications programmes in the areas of communications (*OTS*, *MAROTS/Marecs*, *ECS*, *Olympus*, *Artemis*), Earth observation (*ERS*, *Envisat*) and weather-monitoring and forecasting (*Meteosat*, *MSG*). Each communications satellite represented a logical progression in the mastery of modern communications, from television retransmission and direct broadcasting to fast, high-volume communications links. *ERS* set a standard for Earth resources observational work, while *Envisat* reflected the growing preoccupation with the fragility of Earth's environment. Materials-processing experiments were carried out in cooperation with Russia (*Foton*), while European military cooperation satellites were developed multilaterally (not within the ESA framework).

Taken as a whole these scientific and applications programmes represented a substantial and successful investment, putting Europe in the front line of space exploration. The main problems encountered were with launchers rather than the satellites or their equipment. There were few problems among the international teams of scientists or engineers, nor politically within ESA itself – another measure of the success of the 1975 formula. These programmes laid a solid basis for ever-more-ambitious science programmes in the 21st century (Chapter 7).

6

European manned spaceflight

Europe's first manned space project was *Spacelab*, one of the core projects that founded ESA (Chapter 4). In this context the first European manned spaceflight was *Spacelab 1*, taken aloft by the American space shuttle in December 1983. Although *Spacelab* was seen as Europe's introduction to manned spaceflight, in fact some Europeans had been flying in space some time before. Before we look at the European Space Agency's (ESA) involvement in manned spaceflight we review the national programmes, which began first.

6.1 EUROPE'S INTRODUCTION TO MANNED SPACEFLIGHT

In 1976 the Soviet Union invited the countries of the Soviet bloc in eastern and central Europe to join in flights to the manned Soviet space station *Salyut*. Prospective cosmonauts were sent to Moscow for training from autumn that year and the first began to fly into space from spring 1978. Typically, guest cosmonauts from the socialist countries flew on week-long missions to the *Salyut* orbital station (*Salyut 6* from 1978–81; *Salyut 7* from 1982–5) and the *Mir* space station thereafter (1986– 2001). First was Czech Vladimir Remek who arrived on the *Salyut 6* orbital station in March 1978, to be followed by guests from Poland, the German Democratic Republic (GDR), Hungary, Bulgaria and Romania [120]. The Bulgarians flew twice because the first mission was aborted. In September 1978 Sigmund Jähn became the first German in space when he flew into orbit on *Soyuz 29* with Valeri Bykovsky on a week-long visit to the *Salyut 6* space station. His space capsule was later exhibited in Dresden, though its subsequent neglect and removal from view was

Figure 6.1. Sigmund Jähn.

First Europeans to *Salyut*, *Mir* (up to 1990)

2 March 1978	*Soyuz 28*	Vladimir Remek	Czechoslovakia
27 June 1978	*Soyuz 30*	Miroslav Hermaciewski	Poland
26 August 1978	*Soyuz 31*	Sigmund Jähn	GDR
10 April 1979	*Soyuz 33*	Georgi Ivanov	Bulgaria
26 May 1980	*Soyuz 36*	Bartalan Farkas	Hungary
14 May 1981	*Soyuz 40*	Dmitru Prunariu	Romania
24 June 1982	*Soyuz T-6*	Jean-Loup Chrétien	France
7 June 1988	*Soyuz TM-5*	Alexander Alexandrov	Bulgaria
26 November 1988	*Soyuz TM-7*	Jean-Loup Chrétien	France/Argatz

France was the first Western country to make an arrangement to fly to a Russian space station. Although the mission was seen later as a French response to a Soviet invitation, it was in fact the French who first asked the Soviet Union about the possibility. CNES (Centre National d'Etudes Spatiales) first raised the question of a French manned flight to a Soviet orbiting station at the annual meeting of the Soviet–French cooperation commission in Kiev in 1974. The Russians initially said no, but an invitation was subsequently issued by President Leonid Brezhnev to President Valéry Giscard d'Estaing in April 1979. Granted that the United States was about to fly the first Western Europeans on board *Spacelab*, *Salyut* gave the Russians the opportunity to fly Western Europeans ahead of their traditional rivals. The agreement was signed on 20 October 1979. This was the beginning of Western European involvement in manned or piloted spaceflight. As usual, France led the way.

6.2 THE FRENCH MISSIONS

Following the Soviet–French agreement candidates for the mission were invited by advertisement within the aerospace industry and the military in September 1979. The requirements were for pilots and engineers who were less than 82 kg in weight and less than 195 cm in height. A total of 173 people applied (26 were women) and this was screened to 72. This group was then sent to the Brétigny flight centre for medical tests. The worst of these tests was being tossed around in a rotating chair. Only 25 got through, who were then put through the centrifuge. Only 7 passed and of these 5 were sent to Moscow for final training. These were:

French selection for *Salyut 6* mission

Test pilot	Major Patrick Baudry
Test pilot	Air Force Lt. Col. Jean-Loup Chrétien
Test pilot	Patrouille de France leader Jean-Pierre Job
Air France pilot	Gérard Juin
University of Marseilles lecturer and pilot	Françoise Varnier
Pilot	Hélène Lacour

Figure 6.2. Jean-Loup Chrétien.

Figure 6.3. *Salyut 7.*

It is believed the French favoured one of their women candidates for the mission, but the Russians were insistent on a military pilot. So, in the end Jean-Loup Chrétien was selected with Patrick Baudry as his back-up. Both had over 3,000 hours of flying time, Jean-Loup Chrétien being head of the Mirage F-1 fighter test programme. They arrived in Moscow for training in July 1980. When they got to Moscow they had to learn Russian and familiarize themselves with Russian space systems. CNES supervised French experiments for the mission, which totalled 500 kg in weight and cost €6m. These experiments were loaded into *Salyut 7* ready for launch in April 1982. The main problem during training was a political one: there was pressure from the scientific community in France to call off the flight as a protest against the Soviet invasion of Afghanistan. The new socialist government of President Mitterand rode out the storm, but chose not to publicize the flight instead.

In June 1982 Jean-Loup Chrétien duly made a week-long visit to *Salyut 7* on the *Soyuz T-6* spacecraft. He took off in darkness, floodlights bathing the rocket for its night-time launch. With Jean-Loup Chrétien on board, there were five cosmonauts on the orbital station and they spent a week carrying out 37 medical, biological, materials-processing, astronomical and Earth observation experiments. When resident cosmonaut Anatoli Berezovoi's diary was published later, he remembered the Frenchman's visit most for breaking the tedium of Russian space food with French delicacies of crab, cheese, hare and lobster followed by strawberries.

The mission was regarded as a success. What happened to his fellow astronauts? Jean-Pierre Job went back to the air force and became a general. Gérard Juin retired from the air force and now lives on a small island in Brittany. Sadly, Françoise Varnier died when her aerobatics plane crashed in 1995.

In October 1985 Soviet Leader Mikhail Gorbachev offered French President François Mitterand a long-duration mission. This was signed on 7 March 1986, just two weeks after the new space station *Mir* had entered orbit. The long-duration mission was given a code name Argatz.

For this mission France had available a new group of astronauts recruited in 1985 for a series of forthcoming European, American and Russian missions. This time they were divided into an engineers/pilots group and a scientists group. In advance an astronaut training centre had been set up in Toulouse and opened in October 1983. For the new selection 715 people applied, of whom 140 were called for interview. These were then reduced to 19 engineers/pilots and 20 scientists. The latter group was varied, including astronomers, medical doctors, physicists, teachers and an agronomist. Those selected were:

French astronaut selection, 1985

Engineers/Pilots	
CNES engineer	Jean-François Clervoy
Brétigny chief test pilot	Jean-Pierre Hagneré
Test pilot	Michel Tognini
Scientists	
Rheumatologist	Claudine-André Deshays (later Hagneré)
Nuclear physicist	Jean-Jacques Favier
Medical doctor	Frédéric Patat
Vet	Michel Viso

From this group four were chosen for the second Soviet mission: Chrétien, Tognini, Hagneré and Clervoy. When Clervoy was disqualified due to a minor medical problem, a replacement had to be found. CNES went back to its files from the 1985 selection and chose Antoine Couette, a French naval flyer who piloted Crusaders and Super Etendards. He was not chosen (Chrétien got the flight) and on his return became a DC-10 airline pilot based in Lyons. Of the 1985 group all but Patat, Viso and Couette reached orbit. Patat left due to medical problems and became a professor in the University of Tours. Viso had hoped to fly on the third shuttle life sciences mission, but it never took place and he eventually went to the CNES microgravity division.

President Mitterand flew to Baykonur to attend the launch of *Soyuz TM-7* with Jean-Loup Chrétien on board on 26 November 1988. Two days later Jean-Loup Chrétien was on board *Mir* for almost a month of experiments. On 9 December, with Alexander Volkov, Jean-Loup Chrétien became the first non-American, non-Russian to make a space walk. His task was to erect a carbon-plastic truss structure called ERA made by the French space agency CNES. Samples were deployed on the outside of the station and the spacewalk lasted six hours. Jean-Loup Chrétien returned to Earth on 21 December 1988. It was a historic homecoming, not so much for him, but his two companions Musa Manarov and Vladimir Titov who were coming back after a year in space, a record 366 days. The mission cost the French €18m.

6.3 SERIES OF FRENCH MISSIONS TO *MIR*

The two flights of Chrétien were one-off, pioneering French missions to Soviet space stations. Possibilities now opened up for a series of regular flights to the *Mir* space station. By this stage the time of the free or low-cost mission was over and the increasingly cash-strapped Russians were beginning to see guest missions as a means of maintaining their orbital space station *Mir*. The cost of flying Jean-Loup Chrétien on the Argatz mission was probably only about 10% of the real cost to the Russians if Western accounting had been used. Although the flights now cost CNES money, they were no longer based on goodwill, but on contractual obligations, making them routine and giving the French the right to carry their own experiments.

A preliminary agreement to a new mission was reached during the Mitterand visit to Baykonur and confirmed when, in July 1989, Mikhail Gorbachev visited France. A 10-year agreement was signed that December. The USSR agreed to fly Michel Tognini in 1991–2 for a fee of €10.9m; and signed an agreement in April 1990 for a second mission. A total of four missions were envisaged in the new framework.

Michel Tognini's mission code-named Antares went ahead on 27 July 1992 when, with Anatoli Soloviev and Sergei Avdeev, he flew from the summertime steppes on *Soyuz TM-15*. The launch was carried live on French TV. They boarded *Mir* two days later and Tognini spent 12 days carrying out 10 experiments on the space station, principally focused on medicine (muscles, fitness, the heart system, radiation).

The second (Altair) mission for his colleague Jean-Pierre Hagneré had been scheduled for two years later, but was brought forward when a seat vacancy arose a year earlier. Jean-Pierre Hagneré flew to *Mir* with Vasili Tsibliev and Alexander Serebrov on *Soyuz TM-17*. Six medical and two technological experiments were carried out, some using equipment brought up during the Antares mission. The cost was about €12m.

These two successful missions led the French Space Agency CNES to confirm the two further missions already pencilled in (Cassiopeia and Pegasus). For the first one, Claudine-André Deshays was scheduled to fly to *Mir* with Gennadiy Manakov and Pavel Vinogradov in August 1996. Claudine-André Deshays was to become one of France's best known astronauts. She had won her place in the French space corps during the 1985 competition. Among a total of 700 applicants she was the only woman chosen from the 7 finally selected. She had resolved to become an astronaut the night she watched Neil Armstrong walk on the moon. She got her baccalaureate at 15 and the first of several university degrees at 24 (medicine, followed by biology, sports medicine, aeronautical and space medicine, and

Figure 6.4. The great space station *Mir*.

rheumatology). Following her visit to *Mir*, she married veteran *Mir* cosmonaut Jean-Pierre Hagneré (and took his surname). Originally, he instructed her, but she fast overtook him. She combined ambition (always top of her class), determination (essential to handle cosmonaut training), a love for contemporary art and Mozart with personal charm (she was well suited to the French media).

A week before lift-off, Manakov failed his final medical (doctors diagnosed a possible heart irregularity) and he was taken off. Claudine-André Deshays flew with the back-up Russians Valeri Korzun and Alexander Kaleri instead. Their spacecraft *Soyuz TM-24* reached the orbital complex on 19 August 1996 for 14 days of medical and biological experiments. Already on board were Yuri Onufrienko, Yuri Usachov

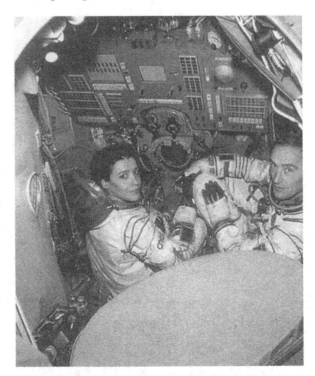

Figure 6.5. Claudine and Jean-Pierre Hagneré.

and American spacewoman Shannon Lucid. Claudine-André Deshays brought an extensive range of experiments to *Mir*, in life and physical sciences. These were designed to measure blood pressure and cardiac frequency, the egg-laying of sala-manders, the behaviour of fluids in microgravity and vibration on orbital structures, to name a few. Claudine-André Deshays returned to Earth after a 16-day mission with Yuri Onufrienko and Yuri Usachov, whose mission had lasted 195 days. The recovery squad comprised two rescue vehicles, 11 planes and 18 helicopters near Akmola in Kazakhstan. The only anxious moment of the flight, she recalled after-wards, was when the parachute came out and the cabin swayed back and forth for 90 seconds before stabilizing itself. She made a number of presentations of her mission afterwards and drew huge press attention.

For the second mission in this part of the series Pegasus, Leopold Eyrharts was ready to fly to *Mir* in August 1997 on *Soyuz TM-26*. He was a fighter pilot, subse-quently test pilot from Biarritz in France. His mission Pegasus was designed to continue the experiments undertaken by Claudine-André Deshays on Cassiopeia. In the event the mission coincided with a period of extreme difficulty on *Mir*, the summer starting with a fire, a near collision and then a real collision between *Mir* and a *Progress* freighter, depressurization of the station and an extended period of emergency repairs. *Soyuz TM-26* had to carry up emergency equipment for repairs in effect taking up his seat, and Pavel Vinogradov and Anatoli Soloviev left Earth without him. Leopold Eyrharts got his chance six months later when he flew on 1st February 1998 with Talgat Musabayev and Nikolai Budarin on *Soyuz TM-27*. He spent nearly three weeks on *Mir* before returning to a snowy landing on 19 February 1998. The cost of Cassiopeia and Pegasus amounted to about €12m each.

By the time of his return, funding for the *Mir* station had begun to run out as Russia increasingly invested its diminishing resources in the International Space Station, whose hardware was now nearly completed. In mid-1998 CNES agreed to pay €18m to the Russian Space Agency (RKA), to include a French cosmonaut on what would be the final decommissioning crew. This crew was announced as Viktor Afanasayev and veteran Jean-Pierre Hagneré, the mission code-named Perseus.

Hagneré duly went into orbit on *Soyuz TM-29* with a Russian commander Afanasayev and Slovak cosmonaut Ivan Bella on 20 February 1999. This was Jean-Pierre Hagneré's second visit to the station and he extended the programme of experiments undertaken by Cassiopeia and Pegasus. Despite it being a decommissioning mission, Viktor Afanasayev, Jean-Pierre Hagneré and a cosmonaut from the previous mission, Sergei Avdeev, spent a busy time launching small satellites, carrying out space walks and observing that summer's spectacular solar eclipse. Hagneré made a 6-hr, 19-min spacewalk with Afanasayev in April to collect samples from outside the station that were used to measure the Leonid meteor shower. The experiments were called Physiolab, Cognilab, Castor, Alice, Comet, Exobiology, Genesis and Spica (for France), WSG and Titus for Germany, and BSMD and Mirsupio for ESA. They covered a broad range, from salamanders to the cardiovascular system, fluids, meteors, the human spinal column, metal alloys and bone density.

Originally, Jean-Pierre Hagneré's mission was to last to 1 June 1999, but it was extended by two months. Russia had been hoping to find the resources to keep *Mir* further occupied. These came to nothing – for the time being. Jean-Pierre Hagneré spent 188 days on *Mir* in the end, and his mission marked almost 10 years of continued manned occupancy of *Mir*. Their spacecraft plunged back to Earth on 27 August, with, as Afanasayev said, 'grief in our hearts' to leave the space station unoccupied. In the event a final crew occupied *Mir* for the summer of 2000, but resources finally ran out soon thereafter and the great space station *Mir* finally deorbited in a blaze of flames over the Pacific in March 2001.

6.4   FRANCE FIRST TO THE INTERNATIONAL SPACE STATION (*ISS*)

It was only appropriate that France, first to *Salyut 7*, first to *Mir*, should also be first to the *ISS*. In October 2000 France's Research Minister Roger-Gérard Schwartzenberg negotiated a week-long mission to the *ISS*, costing CNES about €15m. CNES selected *Mir* veteran Claudine-André Deshays. After her first mission she had stayed some time in Russia, returning there in 1999 for further training and

Figure 6.6. Andromeda crew, 2001.

to take her *Soyuz* flight commander exam, so no-one could have been better qualified. This made her eligible to be the first foreigner to lead a Russian space mission – though whether this will ever happen is another matter. On this mission, she flew as the flight engineer (#2 position) rather than the more humble research engineer (#3 position). *Soyuz TM-33* flew to the *ISS* on 21 October 2001, the launch transmitted live on the CNES website by videostream from Baykonur and was given the code name of Andromède (Andromeda). Prime Minister Lionel Jospin was in the TsUP control centre when *Soyuz TM-33* docked with the space station two days later. Claudine-André Deshays companions were Viktor Afanasayev, mission commander and Konstantin Kozeev, flight engineer and they spent a week with the station's resident crew of Frank Culbertson, Vladimir Dezhurov and Mikhail Tyurin. She carried out a range of experiments in the areas of space medicine, crystals, salamanders, frogs and computers, as well as an educational programme for schoolchildren. Experimentation weighing 75 kg was sent up in advance on *Progress*, a further 15 kg with her on *Soyuz TM-33*. The cost to France for the mission was about €15.2m.

French missions to *Mir*, and *ISS* (1992–2002)

27 July 1992	*Soyuz TM-15*	Antares	Michel Tognini
1 July 1993	*Soyuz TM-17*	Altair	Jean-Pierre Hagneré
17 August 1996	*Soyuz TM-24*	Cassiopeia	Claudine-André Deshays
1 February 1998	*Soyuz TM-27*	Pegasus	Leopold Eyrharts
20 February 1999	*Soyuz TM-29*	Perseus	Jean-Pierre Hagneré
28 October 2001	*Soyuz TM-33*	Andromède	Claudine Hagneré

The summer after her return from the *ISS*, Lionel Jospin's socialist government in France collapsed to be replaced by a republican government under Jean-Pierre Raffarin. The new Prime Minister appointed Claudine Hagneré, out of the blue, as one of 10 women in his new government formed in June 2002. She became the new minister for research, a choice widely and warmly applauded.

In the course of these missions France managed to build up a considerable manned spaceflight experience on Soviet and Russian space stations, 292 days on *Mir* alone. French cosmonauts flew once to *Salyut 7*, six times to *Mir* and once to the *ISS*. All this had come at the comparatively inexpensive price of just over €110m. But France was not the only country to seize such opportunities. The other countries were Germany, Austria and Britain and later, ESA itself.

6.5 GERMAN *MIR* MISSIONS

In 1990 Germany accepted an invitation issued by President Gorbachev to fly a cosmonaut to *Mir*, the cost involved being about €15m. Germany already had five astronauts in training for the American–German *Spacelab D2* mission: the Russians said they were quite happy to accept the level of proficiency that they had attained without the need for additional, general, space flight training (they would still need to learn about *Soyuz* and *Mir* and learn Russian). *Mir* offered much less science than *Spacelab*, for the visiting German could bring up only 100 kg of experimental hardware, but it would give Germany a grandstand view on how to live on a space station.

Two German astronauts were selected for training for the mission in October 1990. They were Reinhold Ewald from Möenchengladbach, a student of experimental physics and subsequently a test pilot, and Klaus-Dietrich Flade. Reinhold Ewald wrote a thesis on the spectroscopy of interstellar matter, studied medicine in his spare time and became a research assistant at the University of Cologne's 3-m radio telescope. The two sent to Moscow complemented one another: a scientist and a test pilot [122].

The mission duly took place in March 1992 and was the first piloted space flight by Russia, as distinct from the Soviet Union. Klaus-Dietrich Flade was selected for the flight, together with Alexander Viktorenko and Alexander Kaleri, and he had prepared experiments in the areas of medicine, biology, engineering and Earth observations. As a sprinkling of snow lay on the ground, his rocket took off from Baykonur decorated down its side with the bars of the flag of Germany and for the

Figure 6.7. Boarding *Soyuz*.

first time ever the red, white and blue flag of Russia, an event marking the passing of
the Soviet Union. While on *Mir* he carried out 13 medical and 1 materials science
experiments. When Flade returned to Earth a week later, he came back with the
previous crew of Alexander Volkov and Sergei Krikalev, the latter still with his party
membership card in his pocket. He had been launched 312 days earlier by a country
that no longer existed.

Agreement was reached between Germany and Russia for a second mission in 1996, though it slipped into early 1997. Unlike the French who scheduled a series of missions, the Germans took each one at a time. Reinhold Ewald and Hans Schlegel were sent to Moscow for training in 1995. Hans Schlegel, from Uberlingen, had studied physics at the Rhine Westphalia technical high school. The purpose of the mission was to repeat and extend the experiments carried out by Klaus-Dietrich Flade in 1992. Ewald had been back-up to Flade and was familiar with the procedures, and he was selected. Twenty-seven experiments were prepared for his mission.

Reinhold Ewald arrived at Baikonour cosmodrome for his mission at the end of February 1997 on a Tupolev 134-A-2 with 'Yuri Gagarin Cosmonaut Training Centre' painted on the side. Three days later he watched the *Soyuz TM-25* roll down to the launch pad on the railway system, the *Mir 97* logo displayed on its side, in a temperature of −15°C, something that seemed to bother no-one. On 10 March the crew took the bus to the pad. In a dramatic contrast to Cape Canaveral, anyone was free to walk around the fuelled rocket, from VIPs to press to launch officials and their children. It was bitterly cold and ice had formed on parts of the rocket. By the time they lifted off it had got dark [123].

Ewald flew to *Mir* with Vasili Tsibliev and Alexander Lazutkin. Little could he have realized that he had joined the most troubled, difficult and dangerous period of the entire *Mir* operation – though he was to escape the worst of it. When he arrived on *Mir* on 12 February 1997 he greeted the resident crew of Alexander Kaleri, Valeri Korzun and Jerry Linenger. The plan was for Reinhold Ewald to carry out experiments for the German space agency for three weeks. Midway through, on 24 February, one of the Russians went to release an oxygen canister, a standard procedure to renew the air in the station. Instead, the canister exploded and burned like a blowtorch for 90 seconds. As in a submarine, a fire is one of the greatest dangers in a space station – one cannot simply open the window to let out the smoke. The fire took place in the *Kvant* module, which meant that one of the return craft was inaccessible. The six cosmonauts – four Russians, a German and an American – donned masks and tried to fight the fire with extinguishers as dense smoke quickly filled the station. Eventually the canister burned itself out and the environmental control system cleared the air. All this happened over Africa out of reach of ground control, who were duly alarmed on the next pass to hear that the crew had been firefighting. Ewald survived the scare to return on 2 March. For his colleagues Vasili Tsibliev and Alexander Lazutkin, this was the beginning of a nightmare summer in which there was a collision, the loss of the *Spektr* module, repeated computer crashes and moves by the nervous Americans to pull out of their joint missions, which brought precious funding.

German missions to *Mir*			
17 March 1992	*Soyuz TM-14*	*Mir 92*	Klaus-Dietrich Flade
10 February 1997	*Soyuz TM-25*	*Mir 97*	Reinhold Ewald

6.6 AUSTRIAN MISSION TO *MIR*

The high point of Austria's space history was the €20m *AustroMir* mission, agreed between Austria and the Soviet Union in 1988. Two Austrians trained for the mission: Clemens Lothaller, a 27-year-old Viennese doctor, and Franz Viehböck, a 29-year-old electrical engineer from Peichtoldsdorf. Viehböck flew to the *Mir* space station on *Soyuz TM-13* with Alexander Volkov and Toktar Aubakirov and stayed on board *Mir* from 2 to 10 October 1991. Viehböck was seen off by Chancellor Vranitski who travelled to Baikonour for the occasion. Viehböck could be seen waving a small Austrian flag as he floated into *Mir* to the amplified taped music of Strauss's waltz *The Blue Danube*. At a personal level it was an emotional time, for his daughter was born on his launch date and the first pictures of her were relayed up to him by television five days later [124].

This project was coordinated by Joanneum Research in the city of Graz, which organized 15 experiments to be carried out on the *Mir* space station, of which 11 were medical, 3 technological and 1 for remote-sensing. The experiments weighed 150 kg and were sent up in advance by the *Progress M-9* freighter. They involved the study of blood, motor systems, the heart, eyes, body fluids, videoconferencing and observations of Austria itself. The mission attracted commercial sponsorship from 54 companies, such as Austrian Airlines and Kodak. The mission took place against a background of some uncertainty, for the coup had taken place in Moscow only a few months earlier. *Soyuz TM-13* turned out to be the last Soviet-period manned space mission. From the point of view of the Austrians, it all seems to have gone very smoothly, the scheduled experiments were carried out and Viehböck was in good condition when he landed near a chilly, cloudy Arkalyk (Kazakhstan) a week later.

6.7 *JUNO*

Ironically, at a low point in the development of British space history (Chapter 3), Britain at last managed to get its first astronaut into space [125]. This was done courtesy of the Soviet government, the UK government maintaining a haughty distance from the project and the official broadcasting company, the BBC, largely ignoring the event.

When a British parliamentary delegation visited Moscow in May 1986, Soviet cosmonaut Georgi Beregovoi made an offer on behalf of the Soviet government for a Briton to fly on a Soviet spaceship. Little happened for the next few years until, in June 1989, a commercial project was agreed between the Soviet space promotion agency Glavcosmos and the London-based Moscow Narodny Bank, the only Russian bank outside the USSR, to fly a British astronaut. At this stage the USSR had begun to commercialize *Mir*. Gone were the days of goodwill and Western countries like France flying to Soviet space stations for free. The first seat to *Mir* was sold to the Japanese, the Tokyo Broadcasting Corporation paying over €10m to fly its chief foreign news editor to *Mir*. Commercial imperatives created a

Figure 6.8. Helen Sharman's spacesuit.

number of uncertainties, and over the next three years the British mission went through lengthy phases of being on and then off.

The mission was to be paid for by sponsorship, which would cover the cost of the mission, set at UK£16m. Heinz Wolff was appointed science director for the project and he quickly devised a set of 26 experiments into medicine and biological development. A company, Antequera, was set up to manage the mission and it made a call for astronauts that autumn. French-style, the project was called *Juno*. Some 3,000 people duly applied, of whom 150 were selected for medical tests. This enabled the selection to be cut to 25 by October 1989, 16 by the following month and then a last group of 8. The two finalists were selected on 25 November 1989 – Helen Sharman and Maj. Timothy Mace – and left for Moscow for 18 months' training a fortnight later.

Not long after they were there, the financial end of the project unravelled. The funding package for the project collapsed on 27 March 1990. The problem was that *Juno* had failed to attract the financial sponsorship expected. In November 1990, with the launch only six months away, the mission directors announced that they had saved the flight with a revised financial deal. In reality the Soviet government made up the shortfall on the basis that the British astronaut carry out the experiments designated by NPO Energiya on the Soviet side (in the event most of the British experiments were deemed to be sufficiently interesting to be worth flying anyway). The *Juno* project was lucky: when deals for other, prospective space station visitors fell financially short several years later, there were no bail-outs.

The crew selections were made in February 1992, with Helen Sharman as the lucky winner, selected to fly with Viktor Afanasayev and Sergei Krikalev. Her mission duly lifted off on 18 May 1991. All went smoothly and there were telecasts from space showing Helen Sharman carrying out experiments and speaking to a school science class in southern England. British media coverage was subdued and the mission received much less coverage than earlier, but governmental plans to fly Britons to the space shuttle never materialized. They contrasted sharply with French media coverage of their *Soyuz* missions, the key points of which were televised live. Helen Sharman co-wrote a book about her experiences afterwards, but otherwise stayed shy of media interest on her return.

6.8   THE *EUROMIR* MISSIONS

Until 1994 all the European visiting missions to *Salyut* and *Mir* were organized on a national basis, with bilateral agreements between Russia and the country concerned. In summer 1993 ESA reached agreement with Russia for two ESA missions to the *Mir* space station. Both were arranged at relatively short notice: the Russians needed cash badly. Trained European astronauts were already available and required only instruction in Russian and Russian space systems. These were called the *Euromir* missions. For ESA they provided valuable opportunities for long-duration experience in advance of its forthcoming participation in the *ISS*. The cost to ESA is believed to have been up to €100m.

Four astronauts were assigned to the *Euromir* programme: Pedro Duque of Spain, Christer Fuglesang of Sweden and Thomas Reiter and Ulf Merbold from Germany. They arrived in Star Town (Moscow) for training in August 1993. The preparations for the mission involved the astronauts in 160 hours of language instruction, 175 hours of spacecraft familiarization, 315 hours of technical instruction, 325 hours of biomedical training and 530 hours of mission-specific training. On top of that there was water survival training in case they came down in the sea or in a lake (like *Soyuz 23*).

The first mission was to last 30 days, the second 135 and include a space walk. The purpose was to build up experience on long-duration missions that could be put to good use on the *ISS*. ESA quickly put together a set of 30 experiments for the first flight, concentrating on the cardiovascular system, the neuro-sensory system, muscle system and materials science. It was planned to bring back 100 bags of frozen blood and urine samples. In some cases ESA took advantage of equipment already known to be on *Mir* (e.g., from the 1991 Austrian flight).

Figure 6.9. *Euromir.*

ESA had longer to prepare for the second duration flight, which focused on changes in bone density in orbit and the reaction of biological samples to radiation. Experiment time was 100 hours on the first mission and 400 hours on the second. In advance of the first mission 150 kg of 29 experiments were ferried up by the *Progress M-24* freighter craft and then a further 350 kg for the second mission. Each astronaut was allowed to bring up 10 kg with himself. This indeed was one of the contrasts with the space shuttle and *Spacelab*. With the Americans, one could bring up a much greater range and heavier weight of experiments and bring them back – but the missions were much shorter.

As *Soyuz TM-20* was wheeled out for the first *Euromir* mission, the ESA logo could be seen stamped on its side. Clanking along the railway to the pad the rocket passed the growing rusting debris of an ever-less well-maintained cosmodrome. The flight began at night-time on 4 October 1994 a historic date at Baikonour cosmodrome. On board *Soyuz TM-20* were Germany's Ulf Merbold, mission commander Alexander Viktorenko and flight engineer Elena Kondakova. Ulf Merbold joined the very small club of people to fly on both American and Russian spacecraft. The mission began a month of experiments on *Mir*. Normally *Mir* had a crew of either two or three cosmonauts, more during the week-long handovers, but this was the longest continuous time six people had lived there. One of the long-duration *Mir* crew was Valeri Poliakov, setting the longest-ever spaceflight of a year and a quarter. Merbold was able to report on the progress of his experiments through 20-min daily video broadcasts to the main ESA space facilities in Paris, Cologne and the Netherlands. The mission was not entirely trouble-free, for *Mir* experienced some power and battery problems, but there were no emergencies [126]. Toward the end of his

Figure 6.10. Ulf Merbold.

mission, 3 November 1994 became a red-letter day for ESA, for on that day Merbold's fellow astronaut Jean-François Clervoy blasted off on the space shuttle, meaning that two European astronauts were in orbit at the same time.

On 4 November 1994 Ulf Merbold returned with the previous *Mir* crew, Talgat Musabayev and Yuri Malenchenko. They had a rough landing with strong winds blowing them 9 km from the designated landing point, which was 79 km north-east of Arkalyk, and the cabin then bounced on impact. He brought back with him 16 kg of samples from his medical experiments, including some saliva, blood and urine (the rest had to wait for later returning missions). Merbold's original crewmates returned the following March after 169 days, bringing back Poliakov after his historic 438 days. Soon after his return Ulf Merbold was appointed Head of the European Astronaut Centre in Cologne. Later, he was able to contrast the shuttle experience with the *Mir* experience. The pace on *Mir* was more relaxed, he said and there was always time on its lengthy missions to do an experiment the next day. But *Mir* required a lot of maintenance, was underpowered, communications with the ground were intermittent and the space available for experiments was more limited. *Mir* had also accumulated a huge amount of items, equipment and hardware over the years, which sometimes took time to find, while the shuttle was much more orderly [127].

The second *Euromir* mission, *Euromir 95*, lifted off in daylight on 3 September 1995. The ESA crew-member was Thomas Reiter and he accompanied mission commander Yuri Gidzenko (later the commander of the first mission to the *ISS*) and flight engineer Sergei Avdeev (later to hold the record for days spent flying in space – 742 days). Forty-one experiments were scheduled for *Euromir 95*—18 in life sciences, 5 in astrophysics, 8 in materials science and 10 for technology. They included a biokit to measure human changes in reaction to weightlessness and

Figure 6.11. Thomas Reiter.

another machine to measure bone density. The three *Soyuz TM-22* cosmonauts docked two days later.

Highlight of the early part of the mission was a 5-hr space walk by Thomas Reiter on 20 October, the first ESA space walk. In advance, Thomas Reiter and his back-up Christer Fuglesang had spent much time practising in the 12-m deep water tank in Star City. Thomas Reiter exited *Mir* with his colleague Sergei Avdeev. By coincidence the ESA council meeting was taking place that day in Toulouse and he made a broadcast to their meeting. The main function of the EVA was to install cassettes on an exposure facility on the outside of the station and to bring in cartridges from a Swiss experiment. They were at some distance from the hatch and the job had to be done slowly and carefully. The samples were designed to trap dust and tiny particles to assess the level of hazard to the *ISS*. During the mission the Earth passed through the tail of a comet and it was hoped to trap some of its debris. The experiments were covered over during shuttle and *Progress* arrivals at the station to save them from being contaminated.

When the mission took off ESA directors in Baykonur hailed the mission as a quantum leap forward in Europe's experience of long-duration flight. More so than they had imagined. Originally, Thomas Reiter was to have returned on 16 January 1996, but not long after he reached *Mir* it was clear that the rocket for the next crew switch-over would not be ready in time. Accordingly, his return was delayed by about 44 days, though there was no extra charge to ESA for the extension. Reiter soon settled down to the routine of experiments on board the orbital station [128]. For the first time a daily record of his mission was posted on the Internet by mission control. Reiter was actually in charge of *Mir* on 8 December 1996 when his two Russian colleagues left the station for a space walk to refit a docking cone. Reiter became the first European to see in the new year in orbit and posted a Christmas message on the ESA Internet website to tell of the work done on the orbital station so far. One of the most spectacular sights of the mission was to see an undocked *Progress* freighter burn up high in the atmosphere. First, he saw flames, then explosions as the residual fuel was ignited and then the rest of the debris burning up in shreds.

A big bonus for Reiter was a second space walk, this time to recover the samples put out the previous October. The space walk took place during the extended part of the mission. Thomas Reiter and Yuri Gidzenko left through the *Mir* hatch on 9 February 1996. Again, they made their way carefully along the complex to retrieve the samples, which ESA hoped would provide valuable data on microscopic debris in Earth orbit and also trap some cometary samples. The spacewalk lasted 3 hours 6 minutes. Reiter eventually returned to Earth on 29 February 1996. The temperature in Arkalyk was −18°C, so they were quickly wrapped in warm clothes and put into the recovery helicopters. Reiter had been in orbit 179 days, making him the most travelled non-American, non-Russian ever. Little wonder that ESA was delighted with its progress in manned flight. The results of the mission were made known at a series of presentations later that year. There was substantial medical data; for example, the level of bone decalcification was between 5% and 10%. The external samples picked up a huge amount of dust – *Mir* seemed to hit a dust cloud every 10 hours or so and at one stage met 5,000 tiny particles in 10 minutes. Following his landing Reiter returned to Moscow to undertake a 600-hr course to qualify as a *Soyuz TM* commander, receiving his certificate in 1997.

6.9   NEW EUROPEAN MISSIONS

There had been some discussion of a mission called *Mir 98*, a third *Euromir* flight, but this was never firmed up. In October 2000 the *ISS* was occupied with a resident crew for the first time. Soon after, spring 2001 saw the signing of an agreement between ESA and RKA. These missions were called taxi missions, for they brought the emergency escape vehicle or lifeboat up to the *ISS* – a *Soyuz* spacecraft. A central aspect of the station was that a *Soyuz* had to be attached to the station at all times, to facilitate an emergency return to Earth should this be necessary. Because the fuel and electronics on board had a guaranteed life of only six months, these lifeboats had to be changed every six months. Each change-out on what became known as the taxi missions took a week, which gave ESA the opportunity to carry out a week's experiments on the *ISS*. This provided an initial five missions by European astronauts on week-long missions to the *ISS*. The costs were divided between ESA and the national space agency concerned. The first mission went to Italian Roberto Vittori, with later missions assigned to Frank de Winne of Belgium, André Kuipers of the Netherlands and Pedro Duque of Spain. Vittori's flight marked the mission of the second space tourist Mark Shuttleworth. In the French style of giving names to missions, Vittori's was called Marco Polo.

European missions to *Mir*, *ISS*, 1990–2002

18 May 1991	Helen Sharman	*Soyuz TM-12* to *Mir*	Britain
2 October 1991	Franz Viehböck	*Soyuz TM-13* to *Mir*	Austria
16 March 1992	Klaus-Dietrich Flade	*Soyuz TM-14* to *Mir*	Germany
27 July 1992	Michel Tognini	*Soyuz TM-15* to *Mir*	France
1 July 1993	Jean-Pierre Hagneré	*Soyuz TM-17* to *Mir*	France
4 October 1994	Ulf Merbold	*Soyuz TM-20* to *Mir*	ESA
3 September 1995	Thomas Reiter	*Soyuz TM-22* to *Mir*	ESA
17 August 1996	Claudine-André Deshays	*Soyuz TM-24* to *Mir*	France
10 February 1997	Reinhold Ewald	*Soyuz TM-25* to *Mir*	Germany
1 February 1998	Leopold Eyrharts	*Soyuz TM-27* to *Mir*	France
20 February 1999	Jean-Pierre Hagneré	*Soyuz TM-29* to *Mir*	France
28 October 2001	Claudine Hagneré	*Soyuz TM-33* to *ISS*	France
25 April 2002	Roberto Vittori	*Soyuz TM-34* to *ISS*	Italy/ESA
30 October 2002	Frank De Winne	*Soyuz TMA-1* to *ISS*	Belgium/ESA

6.10 *SPACELAB*: EUROPE'S OWN MANNED SPACEFLIGHT PROGRAMME

These were the national and ESA flights to Soviet and Russian space stations. *Spacelab* was Europe's own original contribution to manned spaceflight and was the beginning of Europe's own indigenous role in manned spaceflight.

Spacelab has its origins in what was called at the time the post-*Apollo* programme. Having achieved the first manned landing on the Moon, NASA (National Aeronautics and Space Administration) was anxious to put together a credible, publicly and politically supported programme to capitalize on the achievements of *Apollo*, to use and develop the *Apollo* hardware and to take the next quantum step forward in space exploration. The post-*Apollo* programme came not a moment too soon, for political support for space exploration had quickly waned as the United States became involved in pressing economic issues and attempting to extricate itself from the war in South-East Asia. NASA did prevail upon President Nixon to support the shuttle programme, which was approved over 1970–2. Resources were not there, however, to support a space station programme as well, which crudely meant that the shuttle had nowhere to shuttle to. The expensive development period for the shuttle funded only the shuttle itself, not any of the ancilliary programmes that should have been developed at the same time.

Here, NASA saw a funding gap which it hoped its European partners could fill, to their mutual benefit. NASA representatives, including its administrator, visited Europe to explain American thinking on the post-*Apollo* programme, the shuttle and a possible European role in it. Initially, the Americans appeared very open as to which way Europe could contribute to the shuttle. For example, Europe could build a major part of the orbiter, or a module, or related functions.

In responding to the invitation to post-*Apollo*, Europe's main interest was the space tug. This would be an upper stage, probably cryogenic, to fire payloads from the shuttle to higher or geostationary orbits. Other versions might be reusable and ferry satellites or cargoes from one orbit to another. Europe took the view that the tug project would offer the most work over the longest time period, be technologically challenging, of most benefit to Europe's own rocket development and, while expensive, just within reach. Tug feasibility studies were carried out accordingly over 1972, even Britain contributing [129]. ELDO (European Launch Development Organization) commissioned two studies on space tugs and the view was taken that the technology developed could be useful for the later European launchers. The space tug held out the prospect of a production line of 20 to 30 units over a decade valued at €500m [130].

At the same time as European space tug studies progressed, the Americans began to float proposals for Research and Applications Modules (RAMs) to operate in the payload bay of the shuttle, or 'sortie cans' to be ejected from the shuttle as a substitute for a space station. These were an important aspect of the shuttle. The shuttle's crew space, while generous by comparison with the *Apollo* cabin, was insufficient for serious scientific research in orbit. If there was no money for a space station, a laboratory in the back of the shuttle could provide an attractive interim solution.

The Americans then let it be known that they were not keen on Europe doing the space tug. Officially, the space tug was 'too difficult'. There may have been many other reasons, such as its possible use for the US air force to lift military cargoes, the desire to keep Europe out of building rockets, lack of confidence in ELDO as a manager and the presence of American contractors anxious to build a tug. In 1972 the Americans vetoed European participation in the space tug, saying that *the* appropriate form of cooperation for Europe was the RAM.

6.11   DIFFICULT NEGOTIATIONS

The space tug was never actually built as envisaged. A derivative, high-energy upper stage was developed by the Americans and was almost ready to fly from the shuttle in 1986, but was cancelled. The post-*Apollo* negotiations came at the worst possible time for ELDO, then riven by dissent about the continuation of the *Europa* programme. Germany was the country keenest to take part in a post-*Apollo* programme. France was intensely suspicious of the Americans and wished to invest all Europe's efforts in an independent launcher [131]. Europe found it enormously difficult to reach a decision on participation in the post-*Apollo* programme, procrastinating in the face of repeatedly extended American deadlines [132]. Europe had hoped and expected to get some of the subcontracting work for the shuttle orbiter. But, as a European decision was delayed, the better positioned and faster acting American contractors sourced this work at home.

The RAMs or sortie cans eventually became known as *Spacelab*. This was a project that went through many evolutions. Although it did eventually provide time

Figure 6.12. *Spacelab.*

in orbit for European astronauts in a European-built laboratory, the ultimate outcome fell far short of the heady expectations of the mid-1970s when the project was conceived. Some Europeans are even bitter about the experience, feeling that the Americans managed to get a high level of European financial investment into the shuttle, for which the Europeans got little in return.

The Americans estimated that getting an American contractor to build *Spacelab* would have cost $250m [133]. This was money that NASA simply did not have, so it was natural for the Americans to press Europe into making good the biggest single gap in the shuttle programme. Europe agreed to the *Spacelab* proposition, largely because it was the only option open by 1973. *Spacelab* would be a laboratory designed to provide a shirtsleeve environment for four astronauts up to a month at a time. Later on, versions of *Spacelab* would be free-flyers, to be visited by orbiting shuttles. At the time NASA was hoping to make almost weekly flights of the shuttle, so up to 50 *Spacelab* missions were projected (indeed, as late as 1984 NASA was still manifesting 41 *Spacelab* missions). In 1978 NASA was planning 468 shuttle flights between 1979 and 1992, of which 51 would be *Spacelab* [134].

The *Spacelab* programme was agreed in a memorandum of understanding between NASA and ESA on 14 August 1973. The cost of the project was now budgeted at US$456m. But what about the terms?

● Europe would deliver one flight-ready *Spacelab* free of charge to NASA along with two engineering models, three sets of ground support equipment and spares.

● Of the European costs, 52% would be paid for by Germany, 10% by France, 6.3% from Britain and the rest by the others.

● Europe would receive a half share of the payload for the first *Spacelab* mission. NASA would launch the mission free of charge to Europe.

- At least one European would fly on the first *Spacelab* mission. European astronaut assignments for *Spacelab* missions would be negotiated on an *ad hoc* basis [135].
- After the first mission *Spacelab* would revert to NASA.
- NASA would later buy at least one complete *Spacelab* (this was duly ordered in 1980, NASA paying $189m for a complete module, five pallets, data-handling systems and integration facilities).
- Europe could fly future experiments and astronauts on *Spacelab*, but would have to pay the shuttle launch costs.

There was some discussion of a European-owned and/or a German-owned *Spacelab*, but this idea did not progress [136].

Europe's financial commitment was €308m, but in the event building *Spacelab* cost Europe a lot more than expected, €750m [137]. Despite this, *Spacelab* offered Europe the opportunity to get its own laboratory into orbit, design its own experiments and have them operated by a squad of its own astronauts. All this could be achieved without having to build, or worry about, the development of a launcher system. The agreement was hailed in Europe as a lifeline to the most important space programme then being planned, one permitting Europeans to fly into space as partner astronauts [138]. Although Europe had been vetoed out of the tug project, the Americans gave some assurances that Europe would not have problems getting access to Americans launchers (though it is doubtful if the French were convinced). For the Americans the shuttle now had a useful cargo, something valuable to carry and the programme appeared to be more scientific.

6.12 BUILDING *SPACELAB*

Spacelab was designed to be operated in three versions:

- fully pressurized laboratory, filling most of the length of the cargo bay;
- smaller laboratory module, with pallets filling the rest of the payload bay;
- pallet-only version, with no pressurized space, operated from the main cabin.

Different elements and combinations of elements could be flown on different missions. Whenever the shuttle returned to Earth, *Spacelab* would be lifted out and then kitted out in a different combination for the next mission.

One of the first things that Europe now had to settle was: Who would build the *Spacelab*? The initial €200m contract involved 50 companies in 10 countries. It was obvious that the contract would go to a German company, as Germany was the paymaster for the project. VFW-Fokker Erno in Bremen became the prime contractor, and there were important roles for Aeritalia (pressurized shell) and British Aerospace (pallets). Programme management was given to the European Space Technology Centre (ESTEC) in Noordwijk, the Netherlands.

Quite apart from their *Spacelab* role, the pallets were also used to hold routine equipment and experiments for ordinary shuttle missions. British Aerospace in

Stevenage supplied 10 pallets to NASA for this purpose, the first being used as early as the second shuttle mission in 1981.

The *Spacelab* pressurized laboratory actually consisted of two segments: a core segment 2.7 m long with life-support equipment, data-processing systems and workshop space, and a 4.6-m-long tunnel connecting to the main part of the shuttle orbiter. When the fully pressurized laboratory was flown, this extended 5.4 m down the payload bay and weighed 11 tonnes.

6.13   EUROPE'S FIRST ASTRONAUTS

In 1974 Europe set about recruiting its first astronauts. At that time the first shuttle flight was set for 1979, with *Spacelab* as the seventh mission and it was assumed that four Europeans would fly with three Americans on this mission. In January 1977 NASA set 15 July 1980 as the date for the first *Spacelab* mission.

These would not be career astronauts. They were not being invited to fly the shuttle, rather to operate the experiments on the *Spacelab* in the payload bay. NASA called them payload specialists, and in addition to their generic astronaut training they would devote considerable time to familiarizing themselves with the experiments to be flown on their mission. *Spacelab 1*, for example, was to carry 40 experiments, evenly divided between NASA and ESA, into stratospheric and upper atmospheric physics, materials-processing, biology, medicine, astronomy, solar physics, Earth observations, technology applications and space plasma physics. NASA's manifest had 462 civilian shuttle flights and 133 military missions between 1979 and 1990. In summer 1977 NASA announced that two scientists would be onboard the first *Spacelab*, which would fly for a week. *Spacelab 2* was set for an 11-day mission in March 1981.

Each country went through its own selection procedure, working from common rules. For example, in France 401 people applied for selection and their numbers were reduced in stages to 133, then 45, then 20 and finally 5. These were:

First French astronaut selection [139]

Engineer	A. Chantal Levasseur Regourd
ONERA engineer	Jean-Jacques Dordain
French navy submariner	Philippe de Guillebon
Engineer	Laurent Stieltjes
MATRA engineer	Jacques Susplugas

In the event none were selected for the ESA squad though A. Chantal Levasseur Regourd reached the shortlist. Regourd became a lecturer at the University of Paris. Jean-Jacques Dordain became Director of Launchers for ESA. Philippe de Guillebon sailed around the world and became a commander of deep-sea submarines, Laurent Stieltjes became a research manager in Marseilles and Jacques Susplugas headed Alcatel in Monaco.

The selection went to ESA for consideration in summer 1977. Twelve countries contributed 53 prospective candidates to ESA's first astronaut corps. For example, Britain sent 5, France 5, Ireland 2. They were put through a further round of medical and psychological tests, as well as a test for proficiency in English, which would be the mission language. This number was whittled down to 12 in the autumn. In September they reported for further testing and training to be done both in Europe and the United States. ESA then selected them down to a final 5, of whom 1 would fly on the first Spacelab with 2 as back-ups. Of these, 2 were German, 1 Italian, 1 Belgian, a French woman (the only woman) and a Briton. In the very end, on 22 December 1978 ESA selected the following group:

The first ESA astronaut selection, 1977

Franco Malerba	Aged 31, Italy, a computer specialist from Milan
Ulf Merbold	Aged 36, Germany, originally from the GDR, which he fled at the age of 19. He worked at the Max Planck Institute for Metal Research in Stuttgart and an expert in solid state and low temperature physics
Wubbo Ockels	Aged 31, a physics expert from Groningen, the Netherlands
Claude Nicollier	Aged 33, from Vevey, a researcher, Swissair pilot, Hawker Hunter air force pilot from Switzerland. He had studied supergiant stars at Geneva University and then worked in ESTEC

They at once went to Houston to begin training. At the same time NASA announced the names of six astronauts who would be finalists in the competition to be the American specialists for Spacelab.

By this time it was clear that only one European would have a seat on the first Spacelab. NASA adopted a procedure whereby it would select one European and one American from a group of two Americans and three Europeans. Ulf Merbold was the lucky European chosen. The Americans now had a big astronaut corps of their own, but when the four Europeans were chosen NASA was down to only 30 astronauts, either Apollo veterans or Apollo period hopefuls who had never flown. With a new programme of annual recruiting in 1978 NASA had over 100 astronauts by 1983, all clamouring for a place.

6.14   PREPARING FOR SPACELAB 1

The programme entered a period of considerable difficulty in the late 1970s. Development proved more difficult than anticipated, and there were serious problems in integrating Spacelab systems with the American shuttle, which was itself now running about three years late. In May 1978 the first two Spacelabs were delayed to December 1980 and April 1981, respectively. Europe's costs had now risen to €800m and additional financial support had to be sought from the participating countries. The first flight-ready module was not delivered until February 1982.

ESA made a call for experiments to fly on the first and subsequent Spacelab missions and indicated an interest in experiments concerned with material sciences and processing, life sciences, mission to Earth and deep space observations. Each applicant or his or her country was expected to carry whatever costs were involved. For experiments on the first *Spacelab* flight, 2,000 applications were filed to NASA, with 222 candidates from 16 countries. For *Spacelab 1* each of the two partners had 1,390 kg of experimentation weight available. Twenty-four ESA experiments were selected and 13 from NASA.

Even as this happened, the shuttle programme encountered more and more difficulty. There were problems with the external tank, then the main engine and in March 1979 a problem arose with its heat-resistant tiles and 31,000 had to be replaced. These delays had put the shuttle programme 19% over budget by 1979. The very tight budgets of the Carter presidency essentially forced NASA to resolve these difficulties within existing budgets – and that could only be done by stretching the timescale. Several times NASA came to within six months of prospective launch dates – only to put them back by a further three months.

The shuttle eventually flew on 12 April 1981, by coincidence the 20th anniversary of Yuri Gagarin's first flight into space. By that autumn, when NASA published its manifest of shuttle flights, the projected 50 to 60 missions a year were down to about 20 a year. Not only did the shuttle require a lot of work to get ready for its first flight, it needed a lot of work for any flight. The turnaround period had lengthened from original estimates of 10 days to nearer 60.

The first *Spacelab* had now slipped back to the ninth mission, the fourth operational flight. Priority had been given to getting tracking and data relay system satellites into orbit first. The costs of *Spacelab* had grown considerably by the time the first mission reached orbit, by which time ESA had spent well over €1bn.

6.15 *SPACELAB 1*: HAZARDOUS SUCCESS

The first *Spacelab* was eventually flown on the ninth shuttle mission STS-9 from 28 November to 8 December 1983. This was a habitable module with an instrument pallet carrying an atmospheric physics experiment, electron generator, far ultraviolet experiment and mirowave sensor. The launch was delayed when it was discovered that the solid rocket motor on the previous mission STS-8 had almost burned through. NASA studied the problem, which took on a renewed, retrospective significance many years later, an advance warning of what happened to *Challenger*.

This delayed STS-9 until the end of November when the northern hemisphere was more cloudy and there was an old Moon in the sky. Clouds and thunderstorms cleared just in time for *Columbia* to head into a clear blue sky, arcing over the Atlantic to the coast of France. *Spacelab 1* was the first American mission to fly non-astronaut scientists (payload specialists) and was designed to be a comprehensive test of the *Spacelab* system. The launch was watched by the two foremost science fiction film-makers of the day, George Lucas (*Star Wars*) and Steven Spielberg (*Close Encounters Of The Third Kind*), both of whom grudgingly admitted that the real thing was better than TV or even the movies.

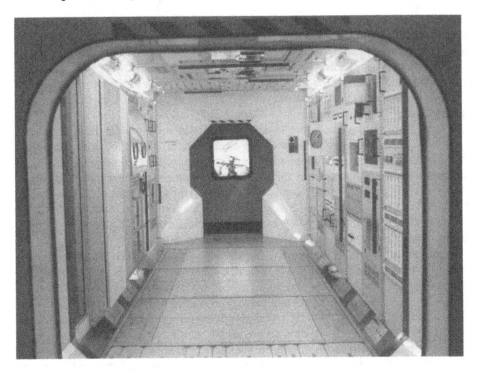

Figure 6.13. Inside *Spacelab*.

This was a 10-day mission, then the longest of the shuttle programme [140]. The orbital path was 248 km, circular, at 57°, travelling far more to the north and south than previous shuttle missions. The mission commander was the very first shuttle commander John Young, then on his sixth mission into space. No person could have been better qualified to take charge.

Spacelab 1 carried 72 experiments into atmospheric physics, Earth observations, space plasma physics, solar physics and astronomy. The laboratory weighed 15.7 tonnes, including 2.8 tonnes of experiments. There were many television relays from *Spacelab* showing several astronauts floating and flying around the inside of the laboratory, with the wall crammed with instruments and controls. Shuttle astronauts, used to the more confined pilot's cabin and lower deck, loved the extra space. They talked by teleconference to President Reagan in the White House and Chancellor Kohl at a European summit in Athens. Viewers heard about the many experiments carried out on board, like the effects of zero gravity, the role of the ear in determining balance, X-rays in the upper atmosphere, ultraviolet starlight, a furnace producing alloys at over 2,000°C, the sampling of blood and the growing of fungi. Data were transmitted back to Earth – a total of 3 trillion items altogether. Shuttle astronauts communicated with amateur radio hams from the laboratory and there was a hook-up with ESA in Cologne. The astronauts divided into two teams, the blue team and the red team, so that the full laboratory was used around the clock. Awake astronauts had to move quietly around the lower deck, because their companions were sleeping there at the same time.

The mission was not without its problems, some of them potentially quite serious [141]. The astronauts suffered space sickness for the first two days. It took six astronauts to wrestle open the hatch into the *Spacelab* tunnel. The metric camera jammed and then *Spacelab*'s high-rate data recorder jammed, but both were fixed. A power surge blacked out TV transmissions. The telescope was pointed at the wrong stars. The pallet-mounted remote acquisition unit functioned intermittently. The tunnel to the *Spacelab* made loud bangs as it expanded and contracted with the extremes of orbital heat and cold.

The return to Earth was the most dangerous phase. Because of the low use of consumables, NASA had extended the mission to a 10th day. Then things began to go wrong. When John Young fired the thrusters to get *Columbia* into the right orientation for re-entry, the thrusters blasted like a howitzer and the #1 computer crashed. When he tried the manoeuvre again, the second computer crashed out. Now there was no computer control over the ship. Young, a person not easily shaken, later admitted his extreme anxiety at the situation. As NASA postponed re-entry so as to consider the situation, the guidance platform went out. Ground control was afraid that all the five computers might go down and then not even John Young would be able to get them back alive. Computer #3 was taken off-line to be ready in reserve. Ground control coaxed #2 back into erratic life, but #1 remained dead for the duration. As they came through re-entry, one of the space shuttle's main engines caught fire, damaging the auxiliary power units that powered the control surfaces, before it was extinguished automatically. They eventually made it down, mission control glossing over the difficulties when they told Young at wheel stop that they had already drunk all the beer and there was none left.

The next shuttle was delayed by a month to sort out these problems. But *Spacelab* itself had come through with flying colours. The experiments were regarded as successful and the amount of data returned was enormous. The artificial aurorae produced by the electron generator, the observations of the telescope and the measurements of the atmosphere were considered especially valuable. The furnace cooked metal alloys at 3,800°C. They found deuterium in the upper atmosphere. Many blood samples were taken. The value of *Spacelab* as a working space laboratory could never have been more evident, and its success may have influenced President Reagan to announce the American space station *Freedom* only a month later.

6.16 THE EARLY *SPACELABS*

Due to a variety of factors the *Spacelab* series got out of sequence, with *Spacelab 2* flying after *Spacelab 3*. The second *Spacelab* – *Spacelab 3* – was flown on STS-51B *Challenger* on 29 April to 6 May 1985. Here, 15 experiments were selected, 13 American, 1 Indian and 1 French. Experiments were conducted into life sciences, materials-processing, fluid mechanics, astronomy and atmospheric physics. This eight-day mission flew at 57°, 352 km. The lab carried 2 squirrel monkeys and 24 rats as well as other animals. The mission was best remembered for its smell. Animal waste began drifting toward the cabin, the astronauts had to don face masks and try and vacuum it up, but were never entirely successful. 'Not a lot of fun', commented commander Robert Omermyer to mission control.

Spacelab 2 was designated a pallets-only astrophysics mission. Because of the British experiments on *Spacelab 2*, two British trainee astronauts were considered possible candidates to fly: Bruce Patchett, 30 and Keith Strong, 26. When the *Spacelab 2* crew was chosen in December 1978, six payload and mission specialists were chosen, but no Europeans. If the Europeans thought there was anything odd about only Americans flying with their equipment, they kept their comments to themselves. Twelve experiments were selected, 10 American and 2 British. *Spacelab 2* carried four telescopes.

Spacelab 2 instruments

Coronal helium abundance experiment	Cosmic ray nuclei experiment
Solar optical universal polarimeter	Superfluid helium experiment
Vehicle charging and electrical experiment	Plasma diagnosis package
X-ray telescope	Solar ultraviolet spectral incidence monitoring
Infrared telescope	High-resolution telescope and spectrograph

Spacelab 2 had more than its fair share of launch problems. On its first launch attempt, on 12 July 1985, a coolant valve failed to close on the #2 main engine. The three main engines were all running from $T-6.6$ seconds and the closedown command was given at $T-3$ seconds. Once it was eventually launched on 29 July 1985, the shuttle lost one of its three main engines 6 minutes into the flight. This was the first ever abort-to-orbit situation. A brave ground controller correctly read the failure as a sensor problem commanding the engine to switch off, not a real engine failure. She ordered the shuttle to ignore future sensor indications of failure ('inhibit the limits,' she called) and the shuttle burned the other engines for longer to get into orbit, one much lower than planned at 315 km, some 67 km less than hoped for. The pallet carried 13 experiments in solar and infrared astronomy, life sciences, the atmosphere, plasma physics and technology. Despite the low orbit, the mission was extended one day and most mission objectives were carried out. The German pointing system for the telescope broke down initially, but suddenly jolted back into life, enabling the crew to take in thousands of spectacular images of sunspots, solar explosions and magnetic fields. Good results were obtained from the solar optical universal polarimeter, which spotted gas jets shooting out of the Sun, and the plasma depletion experiment, which found that the orbiter's manoeuvring system depleted plasma in the ionosphere for an hour after a firing. More images were obtained of the Sun than during the *Skylab* mission in 1973–4.

However, although it was the third *Spacelab*, only one European had flown. The next *Spacelab* flew on a dedicated German mission. This was called *Spacelab D* (D

Figure 6.14. Reinhard Furrer on *Spacelab*.

for Deutschland). The *Spacelab* was kitted out to fly 75 German experiments
tackling such problems as materials-processing, crystal growth, cell functions,
fluids and human adaptation to weightlessness.

Spacelab D orbited from 30 October to 6 November 1985. Three European
specialists were carried: Ernst Messerschmid, Reinhard Furrer, both from
Germany and Wubbo Ockels of the Netherlands. This was the longest shuttle
mission to date, with the largest crew (eight). The shuttle flew in an orbit of 322–
333 km, 57°. Significant results were obtained from experiments into unicellular
organisms, bacteria, blood cells and insects. A 68-kg relay satellite called *GLOMR*
was launched by the crew. For the first time, part of the mission was controlled from
outside the United States, with mission control for the *Spacelab* being the Operations
Control Centre at Oberpfaffenhofen, near Munich. Germany paid NASA a fee of
$64m for the mission. Sadly, Reinhard Furrer died in September 1995 when the
Messerschmitt 108 he was flying crashed at the end of a classic air show in Berlin.
Ernst Messerschmid later became a professor in the University of Stuttgart.

Germany followed this up with a second dedicated *Spacelab*. For the *Spacelab
D-2* mission, a €500m project, Germany held a competition in 1986 and 1,400 people
applied, of whom 20% were women. The following candidates were selected in 1988:

German astronaut selection, 1988	
Heike Walpot	Gerhard Thiele
Ulrich Walter	Hans Wilhelm Schlegel
Renate Brummer	

Figure 6.15. Ernst Messerschmid.

From these, Ulrich Walter and Hans Wilhelm Schlegel were chosen. Germany paid NASA $150m for the mission. Ninety experiments were loaded on board, a much greater capacity than the *Mir* mission and for a day longer than previous *Spacelabs*, concentrating on advanced fluid mechanics, Earth observations and the Galactic plane. *Spacelab* was operated in two 12-hr shifts, resulting in one German working on each shift, one as a member of the red team and the other the blue team. The STS-55 mission got off to a bad start. The shuttle counted down and lit its main engines on 22 March 1993. The space shuttle main engines roared into life, but the next thing viewers knew was that there was a pad abort. The engines were turned off at $T - 2.6$ seconds, just before the solid engines were due to light. The fire protection system sprayed the shuttle's tail and the crew was evacuated. *Columbia* eventually took off a month later on 26 April 1993, making an entirely successful mission, which was extended by one day, returning after 10 days to Edwards Air Force Base in California. As before, *Spacelab* was controlled from the Space Operations Control Centre in Oberpfaffenhofen.

  With the end of *D-2*, the squad was stood down. Heike Walpot should have become the first German woman in space. She was a champion swimmer, a member of the German Olympic team and a qualified doctor and anaesthetist. After *D-2* she worked on the *Hermes* programme and then became a pilot for Lufthansa [142]. A third mission *Spacelab D-3* was postulated in 1989 for *Columbia* on STS-75, but this did not materialize [143].

Figure 6.16. Hans Schlegel.

6.17 THE LATER *SPACELABS*

Spacelab missions were suspended for five years following the loss of the *Challenger* space shuttle. The Space and Earth Sciences Committee of the NASA Advisory Council concluded a report in September 1986, which had a gloomy prognosis for the *Spacelab* programme. By this stage only four *Spacelabs* had flown (*1, 3, 2, D-1*). The more cautious rate of future shuttle launchings meant that the number of *Spacelab* missions would be much reduced, it concluded, even once the backlog of other launches had been cleared. The *Spacelab* system would be seriously under-utilized with the concomitant difficulty in maintaining scientific teams.

Despite all this the programme began to look up in the 1990s. Eighteen *Spacelab* missions were flown, Europeans flying on several of them. The scientific returns were substantial.

Astro 1 was the first *Spacelab* mission after the *Challenger* disaster. Originally listed for May 1986 it eventually reached orbit four years later. *Astro 1* was a pallet-only *Spacelab* carrying 7.8 tonnes of telescopes, making it one of the most intense astronomy missions ever put into orbit. It carried an ultraviolet telescope, imaging telescope and broadband X-ray telescope. The *Astro 1* mission was a difficult one. The computer controlling the three ultraviolet telescopes failed, meaning that they could not be controlled. One hundred out of 250 planned observations had to be cancelled. Due to deteriorating weather at Edwards Air Force Base the mission then had to be cut short by a day. There was a fine photographic epitaph to the mission: a picture of the observatory swivelling in the payload bay of the shuttle against the night-time background of the stars, dominated by the constellation Orion. The mission was so successful that a second *Astro* mission, which had been thought doubtful, was approved immediately afterward. *Astro 2* was launched 2 March 1995 as a reflight of the successful *Astro 1* pallets, except that the broadband X-ray telescope was no longer needed. The use of the extended duration version of the orbiter meant that the flight STS-67 could be extended to two weeks.

The *SLS* (*Space Life Sciences*) *Spacelab*, carried 29 rats and 2,478 jellyfish on *Columbia* on STS-40 on 5 June 1991. This was a nine-day mission by seven astronauts to study the effects of weightlessness on the human body and on other life forms. They performed experiments to test the effects of zero gravity on the heart, lungs, kidneys, hormones, bones, muscles and blood vessels.

IML 1 (International Microgravity Laboratory) carried 55 life sciences and materials-processing experiments. The specialists on *IML 1* were Ulf Merbold, the *Spacelab 1* veteran, and Canadian Roberta Dunbar. *IML 1* carried a menagerie of frog eggs, hamster kidney cells, fruit flies, worms, stick insects, mould and bacteria. The mission was hailed as an outstanding success, with hundreds of biological samples brought back for analysis [144]. *IML 2* had 80 experiments from 13 countries into heat, protein crystallization, electrophoresis, aquatic animals and particles.

ATLAS (Atmospheric Laboratory for Applications and Science) was the first of what was intended to be 10 missions devoted to the study of the atmosphere, the Sun, solar plasma and astronomy. On board *ATLAS 1*, which lifted off on 24 March 1991 on flight STS-45, was the first Belgian astronaut Dirk Frimout from Poperinge [145]. Dirk Frimout was 16 years old the day *Sputnik 1* was launched, and he decided that he had to follow. He studied applied physics in Ghent and then in Colorado in the United States, later working on sounding rockets. NASA heralded the mission as 'the most intense assessment of the Earth's atmosphere ever mounted.' *ATLAS* carried 11 instruments on a pallet, most designed to look at the changing chemical composition of the atmosphere (e.g., the level of ozone depletion) and there were also some solar physics and astronomical experiments. They used an electron gun to make artificial aurorae. Over a thousand billion bits of data were collected during the mission, a new record. Data were relayed back and analysed in real time, one of the first times this was done, giving experimenters the opportunity to modify experiments as they went along. After his mission Dirk Frimout wrote a book *In Search of the Blue Planet*.

ATLAS 2 was brought aloft by *Discovery* at night on 8 April 1993 on mission STS-56. Three of seven instruments were developed by European institutes. One measured the effect of the Sun on the Earth's atmosphere, while another tried to detect ozone and chlorine monoxide in the atmosphere. The mission concluded successfully after nine days. Jean-François Clervoy of ESA got to fly on *ATLAS 3* on a mission in November 1994 to study the Earth's atmosphere and ozone layer. STS-66 carried two important Belgian and French experiments – SOLCON (solar constant) and SOLSPEC (solar spectrum) – designed to measure changes in the Sun's energy and radiation. They were controlled from the Belgian Space Remote Operations Centre in Brussels, in a telescience experiment to test control over experiments from locations other than the normal mission control. Jean-François Clervoy deployed and retrieved the German *CRISTA-SPAS* free-flyer from the shuttle. The German free-flyer made an independent flight of eight days to measure the atmosphere. Because it lacked the capacity for manoeuvring, the shuttle actually moved ahead of *SPAS* (Shuttle PAllet Satellite) and then turned round for retrieval. Clervoy was engaged in a range of other experiments, such as a heat pipe experiment and the test of a chair designed to assist long-duration returning astronauts through re-entry.

USML (United States Microgravity Laboratory) carried 31 experiments in materials, fluid physics, combustion and biotechnology. *Columbia* was launched for the mission on 25 June 1992 and stayed aloft over 13 days, the longest ever shuttle mission at that time. Intended as the first of three missions, the concept of *USML* was to offer microgravity and space physics experimental opportunities for science and industry in advance of the space station. *USML 1* was flown on STS-50 with a full *Spacelab* pressurized module and with a new extended duration facility that extended shuttle flights beyond the normal single week to between 13 and 16 days. As usual, the *Spacelab* end of the mission was controlled by a special mission control facility in the Marshall spaceflight centre in Huntsville, Alabama.

MSL 1 (Materials Sciences Laboratory) flew the *Spacelab* module on STS-83 on 4 April 1997. Carrying 19 major experiments the mission had barely got under way when Fuel Cell #2, one of three, went down. The shuttle must have a minimum of two fuel cells if it is to have enough fuel to manoeuvre during re-entry. This left no margin if there were to be a further failure, so they had to return early, coming back on 8 April. Only 20% of the experiments had been carried out. NASA took the unusual step of reflying the whole mission with the same crew. The payload was left in the shuttle cargo bay and the shuttle readied for what became mission STS-94 on 1 July 1997. This time *Columbia* flew the full 16-day profile, carrying out a range of materials-processing and related experiments. It was also a vindication of the *Spacelab* system as a method of undertaking science projects in orbit.

On *LMSL 1* (Life and Microgravity Sciences Laboratory), which was mission STS-78, the European astronaut was Jean-Jacques Favier, one of the CNES selection of 1985 and formerly a scientist with the nuclear facility in Grenoble, France. He was a doctor in engineering, metallurgy and physics, born 1949 in Kehl, a small town in Germany close to the French border. He worked for many years in the French nuclear energy commission and had been involved in space experiments from *Salyut 6* onward. STS-78 was flown by *Columbia* on 20 June 1996, with 41 experiments devised by American, French and European scientists. This was the longest-ever space shuttle mission, almost 17 days [146]. ESA provided five of the research facilities and over half the 41 experiments flown on the mission, with an agreement that their utilization be also open to NASA. The mission made extensive use of telescience, with experiments being relayed by video teleconference direct to ground facilities in Naples, Brussels, Aachen, Toulouse and Turin. This was something that ESA intended to develop further on the *ISS*. The experiments ranged from physiology (13 experiments) to microgravity and materials science, one of the last being called the 'bubble, particle and drop' unit. The plant and animal experiments involved rats, fish, pine and fir seedlings.

The final *Spacelab* was *Neurolab*, the first manned space mission to focus exclusively on the human nervous system. Launched on 17 April 1998 it carried seven astronauts on a 16-day mission along with 7 medical doctors. The mission STS-90 carried an ESA-developed rotating chair to test visual and vestibular systems, with 7 of the 26 experiments devised by scientists from France, Germany and Italy. Other experiments involved snails, rats, crickets, mice and even an oyster toadfish.

This marked the end of the *Spacelab* programme. The cost for Europe had been about four times more than expected. The joint programme was signed off at this stage. Altogether, *Spacelab* flew 22 times. In addition, the pallets-only model was used 10 times. On 16 April 1999 the NASA administrator returned the second *Spacelab* to Europe, formally handing it over to ESA Director General Antonio Radota at Bremen airport, near where it was originally built. This version was used for the last *Spacelab* mission – *Neurolab* – and the German *D-1* and *D-2* missions. The original *Spacelab*, used on the first mission, was given to the Air & Space Museum in Washington, DC.

Spacelab missions

Number	Date	STS	Designation	European astronauts
1	28 November 1983	STS-9	*Spacelab 1*	Ulf Merbold
2	29 April 1985	STS-51B	*Spacelab 3*	
3	29 July 1985	STS-51F	*Spacelab 2*	
4	30 October 1985	STS-61A	*Spacelab D-1*	Ernst Messerschmid, Wubbo Ockels, Reinhard Furrer
5	2 December 1990	STS-35	*Astro 1*	
6	5 June 1991	STS-40	*SLS 1*	
7	22 January 1992	STS-42	*IML 1*	Ulf Merbold
8	24 March 1992	STS-45	*Atlas 1*	Dirk Frimout
9	25 June 1992	STS-50	*USML-1*	
10	12 September 1992	STS-47	*Spacelab J*	
11	8 April 1993	STS-56	*Atlas 2*	
12	26 April 1993	STS-55	*Spacelab D-2*	Hans Schlegel, Ulrich Walter
13	18 October 1993	STS-52	*USMP 1*	
14	2 March 1994	STS-62	*USMP 2*	
15	8 July 1994	STS-65	*IML 2*	
16	3 November 1994	STS-66	*Atlas 3*	Jean-François Clervoy
17	2 March 1995	STS-67	*Astro 2*	
18	27 June 1995	STS-71	*SLM 2 Mir*	
19	20 October 1995	STS-73	*USML 2*	
20	20 June 1996	STS-78	*LMS*	Jean-Jacques Favier
21	4 April 1997	STS-83	*MSL 1*	
	1 July 1997	STS-87	*MSL 1* reflight	
22	17 April 1998	STS-90	*Neurolab*	

6.18 EUROPEANS ON THE SHUTTLE: BILATERAL MISSIONS

Just as ESA participation in European missions on *Mir* was paralleled by bilateral national missions to *Mir*, so too did Europeans fly on bilateral missions on the space shuttle. As time went on and shuttle missions grew more frequent again, NASA found it easier to earmark seats for qualified European astronauts. The European astronauts were required to undertake a certain amount of training in Houston, quite apart from whatever specific tasks were required for the mission in question.

Jean-Loup Chrétien's back-up Patrick Baudry became the first Frenchman to fly the shuttle, participating in the *Discovery* mission launched 17 June 1985. Called Mission 51G this launched three satellites, and Baudry carried out research into physiology and life sciences. This mission arose as a result of a personal invitation by President Ronald Reagan in 1984 to President Mitterand when he was visiting Washington, DC. The fact that the French already had a trained, qualified astronaut (albeit for a Soviet mission) meant that this could be organized quite quickly.

Others followed. French astronaut Jean-François Clervoy flew on the STS-84 visit to *Mir* in May 1997, while the ubiquitous Jean-Loup Chrétien flew on the shuttle in 1997 on the STS-86 September 1997 visit to the *Mir* space station. His colleague Michel Tognini flew on *Columbia* in July 1999 on STS-93 on the mission to deploy the *Chandra* space observatory. Tognini was involved in the *Chandra* deployment (an ultraviolet telescope to study the stars) and experiments on beetles and aphids. Mission commander was Eileen Collins, the first woman commander of a multi-crew flight.

Two Europeans flew on the *Hubble* repair mission of December 1999 – Jean-François Clervoy and Switzerland's Claude Nicollier, the latter becoming the first European to make a space walk from the space shuttle. Nicollier spacewalked with British-American *Mir* veteran Michael Foale to install new computers, thermal blanketing and fine guidance sensors in *Hubble*.

Pedro Duque was the first Spaniard to go into space and flew on the STS-95 mission in October 1998. This was a shuttle mission flying a diverse range of astronomy and biology experiments, but was publicly known as the John Glenn flight, heralding the return to orbit of American's greatest space hero at the tender age of 77. Duque had been born in March 1963, just over a year after John Glenn's flight and was the youngest crew member on the mission. He was a graduate of the Spanish higher technical school of aeronautical engineers and a member of the precise orbit determination group of the European control centre in Darmstadt. On STS-95 Duque was responsible for the payload bay doors, the *Spacehab* mini-module, the shuttle's computers and a battery of ESA experiments carried on the mission.

In March 2000 German Gerhard Thiele flew as ESA astronaut on the Shuttle Radar Topography Mission (SRTM), STS-99. This mission was dedicated to the ecological mapping of the Earth's surface, creating a highly accurate 3D map of the surface from 58°N to 58°S using German and other instruments [147]. A 60-m mast was extended from the shuttle to make the 16-m detail of the land surface possible. Gerhard Thiele was born in 1953 in Heidenheim an der Brenz (Germany), obtaining

Figure 6.17. Shuttle radar mission.

physics degrees from the universities of Munich and Heidelberg. He trained for space-walking and was ready to go out, if for some reason the radar mast failed to deploy and the job had to be done manually. The radar weighed 14.5 tonnes, took 17 minutes to unfurl and filled the cargo bay. It was able to provide 300 times greater detail than anything provided before, providing data at the rate of 270 MB a second. By the time the mission was over they had covered 99.98% of the landmass between 60°N and 56°S. The crew came back with so much data that they needed 20,000 CDs just to store it. The mission was one of the shuttle's greatest applications successes.

Europeans flying on other shuttle missions

17 June 1985	STS-51G	Patrick Baudry
31 July 1992	STS-46/*TSS 1*	Franco Malerba, Claude Nicollier
2 December 1993	STS-61/*Hubble* repair	Claude Nicollier
22 February 1996	STS-75/*TSS 2*	Maurizio Cheli, Umberto Guidoni, Claude Nicollier
15 May 1997	STS-84/*Mir* visit	Jean-François Clervoy
25 Sepember 1997	STS-86/*Mir* visit	Jean-Loup Chrétien
29 October 1998	STS-95/John Glenn	Pedro Duque
23 July 1999	STS-93/*Chandra*	Michel Tognini
19 December 1999	STS-103/*Hubble* repair	Jean-François Clervoy, Michael Foale, Claude Nicollier
11 February 2000	STS-99/SRTM	Gerhard Thiele

6.19   EUROPEAN ASTRONAUTS

ESA had made its first selection in 1977 – although disappointingly few Europeans flew in space as a result. During the 1980s ESA had drawn on national space squads to supply astronauts for the *Spacelab*, shuttle and *Mir* missions. ESA decided on a second, European – level selection and this took place in 1992. The criteria were:

- height of 1.53 m to 1.9 m;
- age from 27 to 37;
- good health;
- three years' postgraduate science experience.

As an example of a small country, Ireland advertised in spring 1992, offering a salary of €90,000 for those selected. There were 325 completed applications, with the selection being cut down to 250, then 46, then 10, then the final 4. They were sent for training to Cologne – an anaesthetist, military pilot, airline pilot and surgeon. Britain sent three candidates: cosmonaut Helen Sharman, her back-up Timothy Mace and another Juno candidate, Gordon Brooks. However, given the lack of British support for manned flight, it was hard to see much chance that any of the three would be selected.

There were several thousand applicants from throughout the 13 member states. Each country was asked to reduce its selection to five and make nominations by 30 April 1991. In the end there were 59 finalists from 13 countries. The surviving 59 were then scrutinized at European level. From these the following six were selected, no more than one from each country being allowed.

ESA's second astronaut selection, 1992

Maurizio Cheli (Italy)	Christer Fuglesang (Sweden)
Jean-François Clervoy (France)	Marianne Merchez (Belgium)
Pedro Duque (Spain)	Thomas Reiter (Germany)

Maurizio Cheli was a liutenant colonel in the Italian air force. Fuglesang, born in Stockholm in 1957, worked on anti-proton collisions at CERN (European Centre for Nuclear Research). Thomas Reiter, who came from Frankfurt, was a test-pilot from England's Boscombe Down pilot school and had more than 1,500 hours on 15 different combat aircraft to his credit. Belgian Marianne Merchez later married Italian astronaut Maurizio Cheli and changed her surname accordingly, so a listing of astronauts called 'M. Cheli' could refer to either. Clervoy was born in 1958 in Longueville (France), a polytechnic engineer from the armaments corps, avionics expert; selected by CNES in 1985, he had trained in Star City and in his spare time was a parachutist.

They were first sent to the European Astronaut Training Centre in Cologne for basic training before being sent on to the manned spaceflight centre in Houston, Texas for training as payload specialists. The Astronaut Centre in Cologne even had a number of mock-up spacecraft. From 1984 ESA had used aircraft for weightless training – flying astronauts in parabolas in specially padded cabins to give them a limited experience of zero gravity. They used an American KC-135, then a French Caravelle and then a Russian Ilyushin 76, before in 1997 acquiring a French Airbus 300 that used to operate out of Bordeaux [148]. The largest water tank facility in Europe was built at the training centre to simulate conditions for space walks by getting astronauts used to working outside the *ISS*. Near to the Astronaut Training Centre is the DLR Aeronautical Medicine Institute. This has carried out experiments on how space crews survive long-duration missions. In 1990 six people were isolated as if in a space station for a month (the experiment was done in the Norwegian Underwater Technology Centre). In 1992 ESA ran a 60-day isolation test for four volunteers.

6.20   EUROPE'S ASTRONAUT SQUAD

On 26 March 1998 ESA decided to integrate the European astronaut teams and the national astronaut teams. In effect, this meant a merger of the different teams at France's CNES, Italy's Space Agency ASI, Germany's DLR and the rest. They would all be based at the European Astronaut Training Centre in Cologne. The idea was that by 2000 there would be a team of 16 astronauts, all receiving common training and available to fly to the *ISS* either on the shuttle or *Soyuz*. Every two years thereafter there would be a fresh recruitment to replace those who had left or retired. There would be an 'appropriate' representation of each of the member states in the corps. At that stage there were five members of the European Astronaut Corps: Jean-François Clervoy, Pedro Duque, Claude Nicollier, Thomas Reiter and Crister Fuglesang. They were then joined by five who transferred from their national teams: Hans Schlegel and Gerhard Thiele from Germany, Umberto Guidoni from Italy and, from France, Leopold Eyrharts and Jean-Pierre Hagneré. As the group built up they had their first official photograph taken in Cologne in September 1998, in blue spaceflight overalls against a backdrop of a dark sky and a cloud-covered Europe at their feet. There were no female members. An anomaly was that when the French nominations were made, they were limited to Clervoy, Eyrharts and Jean-Pierre Hagneré. The obvious missing person was Claudine-André Deshays, now married to Jean-Pierre Hagneré. She had taken the opportunity of living in Russia to train as a *Soyuz* commander in her own right, qualifying in summer 1999, which was more than acceptable for Russian missions, even if she did not qualify as a European astronaut. She did join the ESA group on 1 November 1999 with Michel Tognini.

The national team-members were joined by a small number of European-level recruits. The first of the new ESA selections were Paolo Nespoli and Roberto Vittori from Italy and André Kuipers from the Netherlands. André Kuipers was a medical doctor, born 1958 in Amsterdam, who was an expert in the vestibular system and had studied accidents caused by disorientation for the Dutch air force. He became an expert in space sickness, balance and blood flow and had been involved in

Figure 6.18. Europe's astronaut squad.

experiments from *Spacelab D-2* and *EuroMir*. He was the first astronaut selected from the Netherlands. Next to be selected was the Belgian Frank de Winne, who would be the second Belgian in space after Dirk Frimout. Frank de Winne was a military man from Ghent with a background in the Belgian air force where he flew Mirages and F-16s, picking up many distinctions and awards en route to becoming a test-pilot and squadron commander.

European astronaut corps, 1999 (third selection), with later finalists

Jean-François Clervoy	Jean-Pierre Hagneré
Pedro Duque	Paolo Nespoli
Claude Nicollier	Roberto Vittori
Thomas Reiter	André Kuipers
Crister Fuglesang	Frank de Winne
Hans Schlegel	Michel Tognini
Gerhard Thiele	Claudine Hagneré
Umberto Guidoni	Reinhold Ewald
Leopold Eyrharts	

Spacelab D-1 veteran Ernst Messerschmid became Head of the European Astronaut Training Centre in 2000. Within ESA itself the Head of the astronaut division was Jean-Pierre Hagneré.

In May 2000 the European Astronaut Training Centre in Cologne celebrated its 10th anniversary [149]. For the occasion all the 16 active members of the squad were gathered from their distant locations. By this stage 27 members of ESA countries had flown in space on 38 different occasions, 34 if we include the former Soviet bloc states. When the Centre had been sent up, there were only three astronauts on the roll (Merbold, Nicollier and Ockels). At the time of the tenth anniversary Cologne was relatively quiet. The *Spacelab* programme had come to an end, *Mir* was in its final stages of operation, the *ISS* was not yet operational and *Columbus* had yet to fly. However, the picture was about to change. An indication of what was to come could be seen on German national space day in Cologne, on 24 September 2000, when the German ESA astronauts made a presentation of the work of the Centre. The crowds were so thick that the building had to be closed to prevent a crush.

How many flights the European astronauts get remains to be seen. Hopes of large numbers of Europeans flying in space were disappointed with *Spacelab*. When the European Astronaut Corps was formed in 1989, its first director anticipated that Europe would need 38 astronauts at the turn of the century: yet when that century turned only one was in active training for a space mission (Guidoni). However, if European hopes for its *Columbus* laboratory are realized, that will change.

6.21   *EURECA*

There was a subtext to the *Spacelab* programme, the *Eureca* space platform. In the early 1980s ESA began to give consideration to programmes that could bridge *Spacelab*, and a prospective manned American space station in which Europe could play a part. A retrievable carrier was felt to be one of the most useful ideas to emerge from the *Spacelab* follow-on programme. A retrievable free-flying platform in orbit could offer long-duration experiments in microgravity (six months being suggested) without crew interference or movement [150].

In May 1982 ESA approved in principle project *Eureca* (EUropean REtrievable CArrier). This was a 4.5-tonne, free-flying *Spacelab* pallet with its own solar arrays, remote control systems, attitude and thermal control systems and solar wings for electricity, to be deployed in orbit by the space shuttle and then retrieved six months later. Once in orbit it would manoeuvre from the shuttle's altitude of 400 km up to its own operating altitude of 500 km. *Eureca* would be tracked from ESOC (European Space Operations Centre – mission control) in Darmstadt throughout its mission, but the experiments would be assisted by the Microgravity User Support Centre in Cologne. This comprised the DLR institutes for space simulation, aerospace medicine and materials research. The manufacturer was DASA (Deutsche Aerospace) in Bremen, and it would be returned there for refurbishment after each mission. Eight member states joined the project that had an initial investment of €156m. The construction contract was signed in June 1985. It was intended that *Eureca* make five flights over 10 years.

Eureca arrived at Cape Canaveral in late 1991 in an environmentally controlled container in a Lufthansa 747 from the Erno workshop in Bremen. The satellite was accompanied by 20 technicians responsible for the final assembly of the spacecraft and its fuelling with 600 kg of hydrazine. *Eureca* was a functional-looking, box-shaped platform with solar panels. On its first flight it carried 26 individual experiments located at five facilities on the platform. Most were in microgravity research, followed by space science and space technology.

Eureca made its first flight in August 1992 when it was put into orbit by shuttle STS-46, the first of the Italian tethered satellite missions. Its carrier *Atlantis* was launched into orbit on 31 July. Responsibility for the deployment of *Eureca* fell to Swiss ESA astronaut Claude Nicollier. On the first day of the mission he lifted *Eureca* out of the payload bay. Holding it there, he commanded the unfurling of its 20-m solar wings. Deployment was delayed because of interruptions to the data from the free-flyer but eventually it took place on 2 August. *Eureca* could be clearly seen moving away over the Cape Canaveral Kennedy Space Centre – its landmass and Canaveral point standing out, with the sea covered in cloud. There were further problems when *Eureca* fired its engine to ascend to its operational 507-km orbit. The spacecraft pitched over 6 minutes into the 24-min firing and the burn was aborted [151]. It transpired that this anomaly was due to computer programming rather than a technical fault by the engine or attitude control system. The commands were reconfigured, and eventually *Eureca* manoeuvred into a 508-km circular orbit a week later and began operations on 18 August under ESOC control in Darmstadt. Early on, a link was set up with the *Olympus* experimental communications satellite.

Once in orbit *Eureca*'s mission went smoothly and three-quarters of the planned experiments had been carried out by February the following year. Around 800 commands were being sent up to *Eureca* each day and the data system was sending back about 35 MB a day. The carrier has its own guidance, navigation, power supply and propulsion system, the latter to bring it up to high orbit of 500 km and then back down to a suitable level for shuttle recovery at 315 km.

Eureca was recovered by the space shuttle *Endeavour* on STS-57, launched 21 June 1993. The *Endeavour* astronauts snared *Eureca* three days later as it flew over north-east Australia, and within three hours had gently manoeuvred the carrier back into the payload bay, retracted its solar wings and reconnected its electrical umbilical. The carrier weighed 4.2 tonnes, including a payload of 1.5 tonnes. *Eureca* was back on the ground by 1 July 1993.

Overall, though, the *Eureca* project proved much more expensive than anticipated and the nine subsequent flights never happened. In October 1995 ESA abandoned the project and sold *Eureca* back to the factory that made it, DASA, for a nominal €1! It can be seen in the hangar there in Bremen.

6.22 GERMAN FREE-FLYERS

Eureca was Europe's flagship free-flyer, but Germany developed a smaller and arguably more successful free-flyer called *SPAS* (Shuttle PAllet Satellite). This was a precursor free-flyer, weighing only 1.5 tonnes including a 900-kg payload, measuring 2 m long by 5 m high. *SPAS* was first flown on STS-7 in August 1983, on Mission 41B in 1984 and again on STS-39 in 1991. It carried telescopes and other experiments, but unlike Eureca it was limited to the duration of a shuttle mission and had to be brought back by the same shuttle that placed it in orbit. This *SPAS* was later bought by the Missile Defence Organization for €20m for star wars tests, but in the event never flew again.

Subsequently, Germany developed the *ORFEUS* and *CRISTA* versions of the *SPAS* free-flyers. The German space agency DARA invested an initial €115m and made an agreement with NASA whereby the free-flyers would have a free launch in exchange for carrying American experiments and access to data.

ORFEUS (Orbiting and Retrievable Far and Extreme Ultraviolet Spectrometer) was built by MBB (Messerschmidt Bölkow Blohm). *ORFEUS* was released by STS-51 in September 1993 for six days before retrieval and took images of hot galactic objects. The mission achieved photographic fame, for an IMAX camera was mounted on *ORFEUS* in order to film *Discovery* moving against the background of Earth for the film *Destiny in Space*. *ORFEUS* observed more than 120 objects, including novae and intergalactic gas clouds. The telescope became badly contaminated during landing, but it was subsequently cleaned up and found to be undamaged. It flew again on STS-80 (*Columbia*) on 19 November 1996 on a 16-day mission. It was released on the second day of the mission for 14 days of free flight and made 422 observations of the life cycle of stars through its 1-m ultraviolet telescope. After release the shuttle then flew about 80 km ahead to avoid any possible collision. It was reberthed on the 15th day of the mission.

CRISTA (CRyogenic Infrared Spectrometer Telescopes for the Atmosphere) was a different type of telescope flown on free-flyer missions. It measured trace gases by scanning the atmosphere. Kayser Threde in Munich was the prime contractor. The telescope was 4 m long with an aperture of 1,000 mm. *CRISTA* was carried on STS-66 in November 1994 for eight days and again on STS-85 (*Discovery*) in August 1997. The procedure was for *CRISTA* to be deployed early in the mission, the shuttle to establish a safe distance from the free-flyer – about 70 km – and then the shuttle to retrieve it the day before re-entry 10 days later. On STS-85 *CRISTA* made observations of radiation and the Earth's atmosphere. Astronaut Jan Davis lifted the platform free on the second day for 200 hours of independent observations. During free flight the platform communicated to ground control using the shuttle as a relay station. The platform also included instruments for measuring ultraviolet radiation in the atmosphere and the presence there of hydroxyl. Mission commander Curtis Brown brought *Discovery* back to the free-flyer on the 10th day for a smooth retrieval.

Germany also developed a free-flyer for use with the *Mir* space station. This had a different purpose: it was a small free-flyer designed to fly around orbital space complexes to inspect them for problems, wear, tear and damage. Appropriately called *Inspektor* and made by DASA in Bremen, it was brought to *Mir* on the *Progress M-36* freighter. It was let free on 17 December 1997 as *Progress* left *Mir*. Unhappily, *Inspektor* lost its solar sensor and found itself unable to manoeuvre to the extent that it began to pose a minor danger to the station itself. *Mir* took an evasive manoeuvre and *Inspektor* had to be abandoned.

Eureca **and the German free-flyers**

18 June 1983	*SPAS 1A*	STS-7	*Challenger*
3 February 1984	*SPAS 1B*	STS-41B	*Challenger*
28 April 1991	*SPAS 2*	STS-39	*Discovery*
31 July 1992	*Eureca*	STS-46	*Atlantis*
12 September 1993	*ORFEUS 1*	STS-51	*Discovery*
3 November 1994	*CRISTA 1*	STS-66	*Atlantis*
19 November 1996	*ORFEUS 2*	STS-80	*Columbia*
7 August 1997	*CRISTA 2*	STS-85	*Discovery*
17 December 1997	*Inspektor*	*Progress M-36*	*Mir*

6.23   ITALIAN TETHERS

Italy came to be a major contributor to the *ISS*. Before this, Italy made a spectacular entry to manned spaceflight through the use of tethers on the American space shuttle. The space tether is an intriguingly unconventional idea [152]. A tether towed either above or below a space station has the possibility of generating electrical energy. Inspired by the work of Giuseppe 'Bepi' Colombo, the ASI commissioned Aeritalia to devise a plan for the deployment of space tethers with probes at the end under the space shuttle. The cost was put at €150m. The tethered satellite system was built by Alenia, working with Laben, Piaggio and Martin Marietta.

The plan was to unwind a 518-kg, ball-shaped, white aluminium globe 1.6 m in diameter from a girder in the shuttle payload bay to a full 100 km below the shuttle's orbital path where it could troll through and measure the upper atmosphere, at an altitude where it was otherwise unreachable. Second, the tether could also serve as a means of generating up to 250 V of electricity through a copper coil in the middle (indeed, such a system was later planned, but never flown, on the *Mir* space station). Ultimately, the idea was to fly a 96-km tether operationally, generating 15,000 V.

Two tests were made of the tethered satellite system. The purpose was to see if the system in principle worked and, if it did, the amount of electricity generated. The experiment was very much an adventure, for nothing like this had been done before. Indeed, there were some concerns that the experiment might even endanger the shuttle, cause all kinds of unpredictable motions or even wrap itself around the cargo bay [153].

On the first mission (STS-46 in August 1992), the modest aim was to extend the tether to 20 km. However, the winding mechanism fouled up at 256 m. Despite strenuous efforts the astronauts were unable to unreel it any further. On board was Italy's first astronaut Franco Malerba representing ASI (Claude Nicollier of Switzerland, for ESA, was also on board). The spherical, tethered, 1.5-m diameter satellite was placed on a girder boom on the bottom of the payload bay. Unfortunately, the astronauts encountered nothing but problems. First, an umbilical would not disconnect from the tethered system and it took some time to shake it free. Then the first time that they began to reel it out, the satellite began to sway and it had to be reeled in. The second time they played it out, the satellite jammed at 179 m down the cable. Again, they rewound and began to play it out again. This time the tether jammed at 256 m. The cable was 20 km long, so they had made little progress so far. At one stage they seemed to have difficulty in getting the satellite back in and a space walk was considered. They settled for recovering the satellite, but gave up further attempts to extend its cable. Several frustrating days of the eight-day mission had been spent on the project. On his return Malerba left the astronaut corps, entered politics and was elected *Forza Italia* MEP to the European Parliament in 1994.

The tethered satellite experiment was flown again on STS-75 on 22 February 1996. For the first time NASA flew three European astronauts together on the space shuttle: Claude Nicollier again (ESA) with Maurizio Cheli (ESA) and Umberto Guidoni (ASI). Guidoni had been back-up to Franco Malerba during the first tether mission. Other European equipment was carried on the mission, such as the French Mephisto solidification research facility and the Dutch/British glovebox handling facility.

This time the experiment appeared to go well at first and electrical current of 3,500 volts was generated. Then suddenly the tether snapped at 19 km, the cord snaking and twisting away into the distance. The break took place just above the release point on the shuttle and was filmed by the astronauts as it happened. The ball-shaped satellite continued to transmit using its batteries, but these began to run down and eventually died. The astronauts later relocated the satellite, but by then it was in a much lower orbit and close to burning up. A NASA/ASI investigation subsequently attributed the break in the tether to an electrical arc that burned through a weakened part of the tether. The tether was required to be light, strong and electrically conductive, but either a foreign body got into the system or there was some form of defect. There was no funding for further experiments, and on this disappointing note the project was abandoned.

Italian tether missions

31 July 1992	*TSS 1*	STS-46	*Atlantis*	Franco Malerba (ASI) and Claude Nicollier (ESA)
22 February 1996	*TSS 2*	STS-75	*Columbia*	Umberto Guidoni (ASI), Claude Nicollier (ESA) and Maurizio Cheli (ESA)

6.24 *HERMES*: ORIGINS

Spacelab was Europe's introduction to manned spaceflight on its own account. As may be seen, the *Spacelab* involvement led, directly or indirectly, to a number of spin-off missions like *Eureca*, *SPAS*, *ORFEUS* and *CRISTA*. With *Hermes* and then the *ISS*, Europe tried to take the next big leap forward in manned spaceflight.

The idea of a European spaceplane was first floated in October 1978 by the French space agency CNES – ironically at the time when the American space

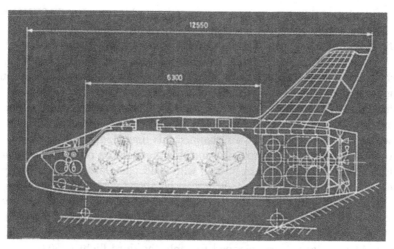

Figure 6.19. *Hermes* design, 1977.

shuttle was going through its greatest period of pre-launch difficulty. Preliminary models had been tested at the Evry test centre in 1976. CNES had done some conceptual studies of a spaceplane with the Dassault aircraft company and considered some small space laboratories (the *Minos* and *Solaris* projects). Originally the spaceplane was called the 'Système Ariane Véhicule Habitable', but it soon acquired the classical name of *Hermes*.

ONERA (Office National d'Etudes et de Recherches Aéronautiques) put a *Hermes* on an *Ariane 4* model in its wind tunnel and was publicly revealed at the next Paris air show. In 1981 the French government formally committed itself to the project. For CNES *Hermes* would complement the *Ariane 5*, the forthcoming American space station and any European module that might be attached to it, giving Europe direct manned access to orbit of its own. *Hermes* was now assigned to the new, larger *Ariane 5* booster flying out of Kourou. *Hermes*, as revealed at this early stage, would weigh 9,500 kg, be 11.55 m long, carry a crew of five or cargo of 1.5 tonnes, and orbit the Earth at 200 km, 30°. Underneath *Hermes* would be a 2.1-tonne solid-fuel rocket, which in the event of a mishap during the launch would blast *Hermes* free of the exploding rocket and parachute it into the sea.

France soon pressed for *Hermes* to be adopted as a European project. A coordination group was established in 1984 to see this process through. The proposal for *Hermes* to be adopted was presented to the January 1985 ESA council meeting in Rome. ESA did not adopt *Hermes* as a programme, authorizing *Ariane 5* and *Columbus* to proceed instead, but asked France to keep it 'informed' about *Hermes*. At that Rome council meeting ESA committed itself, though cautiously, to the notion that Europe should 'gradually' acquire a capability in manned flight. Germany was hesitant, taking the view that the American shuttle would provide sufficient access to space for Europe until at least the end of the century. The British were sceptical, possibly hoping that Europe might adopt the *HOTOL* (HOrizontal Take Off and Landing) proposal instead.

The French made it clear that they would proceed with the idea anyway. France wanted *Hermes* to be adopted as an ESA project, but did not seem publicly perturbed about ESA holding back. The member states had been sufficiently encouraging, and there was the prospect of endorsement later. In the case of *SPOT* (Satellite Pour l'Observation de la Terre), France had proceeded without European support, but several other countries had joined later on a bilateral basis (Belgium and Sweden), so the same might well happen this time [154].

To build *Hermes* was a brave decision, for the histories of both the Soviet and American space programmes are littered with spaceplane projects that had proved too difficult, lost political support or had to be abandoned. The French hoped to break the spell by going for a small spaceplane project. This would not rival the American space shuttle (nor the Soviet *Buran*, which was unknown in the West at the time). Using *Ariane 5*, the weight of the spaceplane would be limited to 17 tonnes in any case. At one stage it was planned for only the third flight (V503) of *Ariane 5* [155]. Only a few expressed concern about the dangers.

6.25   BUILDING *HERMES*

The new specification for *Hermes* suggested that it would be 18 m long, with a wingspan of 11 m, would carry 2.5 tonnes of its own propellant and a crew of two, up to four scientists or a payload of 5 tonnes in a payload bay measuring 35 m^3. It had grown a little since the original specification. *Hermes* would fly up to 400 km high for 90-day missions attached to a space station or 30 days in independent flight. It was designed to refresh space stations with new crews and supplies, help in the assembly of space structures and carry out orbital maintenance and repair. Manoeuvrability in orbit would be ensured by two 20-kN engines at the rear. Launching from French Guiana it could return there or alternately land at the French air force base of Istres near Marseilles, southern France, and for this reason *Hermes* was designed to have a manoeuvring capability of 2,500 km as it glided back. The crew would have ejector seats for use in emergencies. About six missions a year could be expected and up to three *Hermes* spaceplanes would be built.

In 1985 CNES named Aérospatiale as the prime contractor with Dassault responsible for the aerodynamics design. This was a strong team because Aérospatiale had the experience of building *Concorde* and *Airbus*, making it experienced in supersonic flight and fly-by-wire automated controls. Dassault, the second main contractor, had developed delta-winged jets for the French air force and had worked with the American Grumman company to develop the thermal tiles for the space shuttle. Heat protection was a major issue in the *Hermes* design, much as was the case with the American shuttle. Because of *Hermes'* shape and size, temperatures during reentry were expected to be 1,820°C, about 300° higher than the American shuttle. A *Hermes* directorate was set up in Toulouse and 150 staff were assigned there. Other

Figure 6.20. *Hermes* design, early 1980s.

contractors were Matra (electronics), ANY (telemetry), MBB (propulsion), Dornier (life support and fuel cells), ETCA (electrics) and Aeritalia (thermal control). Although France wanted *Hermes* to be a European programme, France also made it clear that it also wished to use the spaceplane for national missions, including docking with the Russian *Mir* space station.

Hermes was eventually brought into the ESA programme at the council meeting in October 1985, with the provision of €58m for a preparatory programme. Of this Britain even contributed 7%. The extra money made quite a difference and enabled extensive wind tunnel work to be undertaken in Stuttgart, Germany and Emmen, Switzerland. The dimensions of *Hermes* had begun to settle now, at length 15.5 m, width 12.2 m and payload 4.5 tonnes. It would now have a 7-m remote arm as well.

Figure 6.21. *Ariane 5* launching *Hermes.*

As a result of the *Hermes* preparatory programme, its weight rose to 21 tonnes, up from 17 tonnes, largely due to new safety requirements. Following the *Challenger* disaster the necessary decision was made that the entire crew cabin should be ejectable so that the pilots could survive a *Challenger*-style catastrophe (this alone added 1.5 tonnes). *Ariane 5* was uprated to take account of the greater weight, with core and two booster weights approved for 190 tonnes and 230 tonnes each, respectively. *Hermes*' payload was now three tonnes into a 500-km orbit.

6.26   DEVELOPMENT PHASE, 1987

In November 1987 France finally won ESA approval to go ahead with a development phase, despite strenuous British objections. ESA allocated €8.5bn until 2000 for the three linked programmes of *Hermes*, *Ariane 5* and *Columbus*.

Financing for *Hermes* (%)

France	42.65	Switzerland	1.98
Germany	26.7	Austria	0.5
Italy	12.47	Denmark	0.45
Belgium	5.86	Norway	0.26
Spain	4.45	Canada	0.45
Netherlands	2.4		

With confidence growing, France now selected a group of astronauts for the *Hermes'* programme, receiving 157 applications. Those selected were:

1989 CNES *Hermes* selection

Test pilot	Leopold Eyrharts
Fighter pilot	Jean-Marc Gasparini
Air force commandant	Philippe Perrin
French naval pilot	Benoît Sylve

Of these, Eyrharts subsequently flew on *Mir* and Perrin on the shuttle (2002). Gasparini left the squad to become a test-pilot for Dassault. Benoît Sylve returned to the navy and came to command a frigate, the *Aconit*. Jean-Loup Chrétien was appointed Head of the Hermes astronaut programme, later went to work for NASA and retired in 2001.

Hermes necessarily involved the building of a ground infrastructure to support the programme. The main elements involved were:

- crew training for *Hermes* astronauts at the Cologne training centre;
- familiarization on mock-ups in Toulouse;
- pilot and simulator training in Brussels airport;
- space-walking facilities in Toulouse;
- engineering support facilities in Turin;
- robot arm development in Noordwijk;
- rescue training facilities in Trondheim;
- flight mission control in Toulouse.

The development phase lasted from 1988 to 1990. The aim was to reach a final configuration, calculate the costs remaining and lay down a timetable for completion. The heavier *Hermes* got, the bigger its wings had to be. The larger they were the more they threatened the stability of the rocket as it went through the thickest layers of the atmosphere. By 1989 *Hermes* had grown even more. Its weight had risen to 24 tonnes and the *Ariane* booster had to be repeatedly upgraded in order to be able to carry it [156].

6.27   REDESIGN, 1990

At this stage designers then made a crucial decision – to make parts of the *Hermes* non-reusable, even though this undermined the principle of the reusability of space shuttles. The rear part of the spacecraft would be a propulsion module, docking module and resources module – all of which would be expendable and dropped before re-entry. Despite these difficulties CNES and ESA presented an upbeat picture of *Hermes*, and publicity material was issued showing *Hermes* being launched from French Guiana on *Ariane 5*, circling the Earth and making a blazing re-entry to land either in Almería in Spain or Toulouse in France. The designers argued that Europe would still have a spaceplane able to fly 12-day missions to the American space station and 30-day missions to the man-tended free-flyer, well worth the cost and effort. At this stage (late 1989) the schedule was:

Hermes schedule, late 1989

1994	Construction of flight model
1996	Atmospheric landing tests on Airbus 310
1997	Unmanned orbital flight
2001	Manned flight

The promoters of *Hermes* were so confident of progress that in autumn 1990 they set up an embryonic new company for the management, development and production of the spaceplane, called Hermespace. The company was made up of Dassault, Aérospatiale, DASA and Alenia under the auspices of CNES. At this stage 3,000 people from 300 different companies were working on the project. Claude Nicollier was mentioned as the most likely pilot for the landing tests [157]. The cost of getting *Hermes* into orbit was now estimated at €4,535m, of which:

- €105m was earmarked for the preparatory programme;
- €530m was earmarked for phase 1: design programme;
- €3,900m was earmarked for Phase 2: development and construction.

The materials to be used for *Hermes* presented some of the greatest challenges [158]. The company charged with resolving these difficulties was Dassault. In 1990 the company had over 300 working on the problem of thermal protection in its test centre in Blagnac, Toulouse. Essentially, the problem was how to bring *Hermes* out of orbit at Mach 29 (30,000 km/hr), with a body temperature of $-100°$C, through re-entry at $1,800°$C down to a landing speed on an airfield at 250 km/hr in 40 minutes. The thermal pressure would be greatest on the nose cap and the winglets. Dassault went to companies involved in ceramics in Spain and elsewhere for different forms of thermal-testing.

At this stage, late 1990, there were some further redesign decisions, all on the conservative side. Instead of the ejectable crew cabin there would be individual ejector seats, a well-proven system. *Ariane* would put the entire *Hermes* into orbit, rather than use *Hermes*' own engines. The spaceplane would now be made of

Figure 6.22. *Hermes*: the final design.

aluminium alloy, not titanium or carbon-based composites that were lighter but more expensive, more difficult to produce and unproven. This is a summary of the new 1990 design:

- landing weight reduced to 15 tonnes;
- payload reduced to 1.5 tonnes;
- landing size reduced to 12.7 m length, wingspan 9 m;
- jettison propulsion system before re-entry;
- loss of payload bay;
- replaced by a rear-facing, expendable resource module;
- crew down to maximum of three: commander, pilot, mission specialist;
- individual ejection seats, based on the system used in the Soviet shuttle *Buran*, rather than an ejectable cabin.

The design was now frozen and the revisions were not well received [159]. The introduction of expendable elements would add to operating costs and would undermine the concept of a reusable system. Doubts began to creep into the project. British Aerospace suggested that an *Apollo*-class capsule might be a cheaper way for Europe to get its astronauts to and from the American space station. ESA tried to respond to these concerns and asked Aérospatiale and British Aerospace to make a feasibility study of an unmanned resupply craft that could bring cargo to the space station, rather like the way the Russians supplied their space stations with *Progress* freighters. This supply craft was initially given the name of *ATV* (*Ariane* Transfer Vehicle, later called the Automated Transfer Vehicle) [160].

6.28 *HERMES* IN SERIOUS TROUBLE: A FURTHER REDESIGN

In spring 1990 Germany had already indicated that, because of financial difficulties, it might have to withdraw from the project. Germany had an early intimation that reunification would impose a severe burden on the national finances and there was no reason why spaceflight should be made an exception to across-the-board restrictions. In June 1990 Germany informed its colleagues that it would have to cut its space budget by between 15% and 20% for 10 years. In November 1990 Germany let it be known that it was entering a period of retrenchment associated with the costs of reunification. The German space agency DARA hinted to ESA that Germany would have to drop its commitment to *Columbus* by 20% and stretch out the programme, and that the *Hermes* programme would also have to slow down.

By summer 1991 the project was in serious trouble as Europe finally discovered, as had the USA and the USSR earlier on, that designing and building a spaceplane was an expensive undertaking. At 30% over budget the June 1991 ESA council meeting was forced to make cuts [161]. The changes were as follows:

- payload down to 1 tonne;
- first flight not until 2000 (unmanned), followed by manned (2001) and operational missions (2005);
- costs up 30%;
- missions limited to one a year;
- to save more costs the spaceplane would be taken from France to French Guiana by boat, not plane;
- delay in building a second model until 1996 at least.

The use of a boat as transporter because an aircraft was too expensive showed just how desperate the situation had become. These changes seemed to put the project below the level of viability, but the French held on tenaciously. The future of *Hermes* went for debate at the November 1991 ministerial meeting of ESA in Munich. This was the most important ministerial council since the controversial Hague meeting on November 1987 that approved *Ariane 5*, *Hermes* and *Columbus*. In advance, the Director General of ESA Jean-Marie Luton put forward a further set of proposals to address the design difficulties of *Hermes* and the increasing costs. They amounted to a second stretching of the programme that year. These were:

- delay *Hermes* till 2002;
- delay the free-flyer till 2003;
- *Hermes* to fly on an upgraded *Ariane 5*;
- review in 1995.

These proposals were, in effect, a compromise to keep the Germans (who wanted out) still on board and convince the French (the main backers) that the programme would still run. These proposals by the ESA secretariat, circulated before the ministers arrived, were called the Darmstadt scenario. However, ESA was aware that the more the programme was stretched the more rather than the less expensive it became and that at a certain stage one had to build and fly. Separate trilateral meetings were held between France, Italy and Germany, who were paying for 80% of the project, to try to resolve these difficulties.

When the ministers arrived at the old Bavarian *Alte Residenz* venue (Munich), they were greeted with a large mock-up of the newly launched *ERS 1* (Earth Resources Satellite). However, the model was not sufficient to bring the ministers to agreement [162]. By this stage the full extent of the *Hermes* problem had become apparent. Of the three programmes agreed at the Hague in 1987 *Hermes*' costs had risen much the most. *Ariane* was up from €3,496m to €3,694m (5.7%), not unreasonable for a programme of its complexity and ambition. *Columbus* had risen from €3,713m to €4,239m, up 14.2%. But in the case of *Hermes* the costs were up from €4,429m to €6,222m, a full 40.5%.

Germany was not prepared to commit to further spending to an additional stretch in the programme. Equally, the meeting was not prepared to cancel the project. A decision was therefore put off until another ministerial meeting the following year (until then ministerial meetings had been irregular, but it was decided they should now be more frequent). Largely to facilitate Germany, the overall ESA budget was cut by 5%, (€120m off the projected €2,427m). The French were hugely disappointed with the meeting and furious that, having conceded the Darmstadt scenario, the Germans did not maintain their commitment to the project. The Germans insisted they were not trying to cancel *Hermes*: they were trying to keep costs under control and make sure that the programme was able to justify itself. The French politely suggested that the Germans were trying to maintain their influence while reducing their funding. The British were the only people who came away happy, for Munich agreed to proceed further with the polar platform. It was clear though that the consensus between France and Germany on *Hermes* was breaking down.

6.29 LAST-DITCH EFFORT

In an effort to keep *Hermes* alive Jean-Marie Luton formed a strategy group under Jean-Jacques Dordain to present options to the next ministerial meeting. Among the options they considered were turning *Hermes* into a crew carrier only (no payload), making *Hermes* the lifeboat for the space station, cancelling the free-flyer version of *Columbus* and buying space shuttle components from Russia. This plan was called X-2000, a stripped down, 20-tonne, unmanned *Hermes* technology demonstrator that would cost 40% less. It could make an automatic flight (possibly suborbital only) in 2000. In another strand to Europe's efforts to cast around for alternatives there was some discussion over 1992–3 about a joint spaceplane project with Russia. The Russians were interested in considering development of their *Mir 2* with the Europeans, but they were cautious about a spaceplane – they probably realized the difficulties too well. The sum of €5m was approved to explore the possibilities of Russian–European cooperation on the programme.

Hermes hit fresh financial trouble in May 1992 when Germany indicated it would have to reduce its overall financial commitment to *Hermes* from €2.5bn to

€750m. The falling axe could not be delayed much longer. The decisions postponed in Munich went to the next ministerial summit held in Granada (Spain) a year later. Here is what was agreed:

- Conclude the development phase of *Hermes* in December 1992. Continue *Hermes* in cooperation with Russia (up to €338m), 'reorient' the programme and review the situation in 1995 (€94m).
- Study an assured crew return vehicle for the American *Freedom* space station (€45m).
- Cut by 5%, but continue the *Columbus* programme, delay a year until 1999, negotiate a more satisfactory relationship about its use on the American space station.
- Abandon the man-tended free-flyer *Columbus*.
- Develop a data relay satellite to be used with *Columbus*.
- Make overall savings of €4bn compared with the Munich proposals.

The polar platform effectively mutated into *Envisat*, a successor to *ERS 1* and *ERS 2*. As such, it commanded strong support throughout the period (Chapter 5).

It is difficult to establish a point at which *Hermes* was actually terminated. It seems to have withered during the 'reorientation' period mandated by the Granada summit. Certainly many people in France interpreted it as abandonment. Presidential candidate Jacques Chirac attacked his government for caving in and promised to restore *Hermes* if he was elected [163]. 'Reorientation' appears to have been a bureaucratic term for abandonment, for too explicit a term would have been humiliating for those concerned. Whatever the nature of the 'Reorientation', *Hermes* was never heard of again. It was effectively mothballed at this point. Despite the larger sums spoken about, only €90m was used to conclude the development, which had a final cost of €567m [164].

If it was any consolation to Europe, the Japanese spaceplane of the same period met an identical fate. Between 1988 and 1998 Japan spent Y43bn (€383m) on a broadly equivalent project called *HOPE* (H-II [rocket] Orbiting Plane) and a number of flanking projects, including the test of demonstrators at Woomera in Australia. *HOPE* was a spaceplane, unmanned at first, to ride the new H-II rocket to rendezvous with the space station, dock with the help of its manipulator system and return to a runway landing. The aspiration was that later versions might eventually be manned.

When *HOPE* proved expensive and late, it was renamed *HOPE-X* and scaled down as a 12-tonne automated space shuttle able to deliver 3 tonnes to the *ISS*. *HOPE-X* was 15 m long, with a 9.7-m wingspan, a height of 5 m and with twin tails. After its mission to the station it would head nose-first into re-entry, protected by its carbon and ceramic tiles, before touching down on a 1,800-m runway on Christmas Island. At this stage *Hermes* was gone, so there was a real prospect that Japan might be the third country to develop a shuttle. In the end *HOPE* shared not only similar dimensions but also a similar fate to *Hermes*. The launch date slipped to 2000, then to 2003, before being suspended indefinitely.

6.30 *SAENGER*

Hermes was not the only European spaceplane project of the 1980s, though it was the one that got furthest. Germany and Britain also put forward spaceplane projects, *Saenger* and *HOTOL*. They were presented as national projects, with the aspiration of adoption as European projects.

Germany's spaceplane was named *Saenger* after the famous Austrian space designer, Eugen Saenger. This was an MBB project whereby a 295-tonne, sleek, black, twin-tailed, winged ramjet with two 300-kN engines would fly to Mach 1 at 10 km on afterburner. Ramjet propulsion would bring the vehicle up to 28 km at Mach 4. Here, there would be a 3,500-km cruise phase, taking the vehicle out over the ocean to 37 km altitude at Mach 6.6 for staging. Then the main stage would release a 91-tonne, small, stubby, liquid-hydrogen-propelled spaceplane with a 1,200-kN cryogenic engine called HORUS, enabling two to six astronauts to fly on into orbit.

Saenger was introduced to ESA at a presentation in June 1986, though it was behind *Hermes* in the queue. In March 1989 the German government allocated financial support for the project. DLR provided €43m, MBB €7.5m, German industry €7.5m, the German state research organization €10m and the government the balance of €110m toward the project. There would be a research phase (1989–93), followed by a technology development period (1994–8) and a development flight in 1999, with a first operational mission in 2006. The Germans claimed that *Saenger* would carry much heavier payloads than *Hermes*, but at only 20% of the cost. Unlike *Hermes* it could fly from and return to any long airfield in Europe itself. ESA found the proposal interesting and saw it not as a competitor to *Hermes*, but as a possible successor. The Germans argued that *Saenger* had the advantage of being a substantial step forward ahead of *Hermes*, but without the complications, costs or unknowns of *HOTOL*. It had much greater efficiency and simplicity than *HOTOL*, not bringing any deadweight engine systems into orbit. The big problem was that of the cost, estimated to be at least twice that of *Hermes*, at a time when the French project was stretching Europe's budget beyond its limits [165].

6.31 *HOTOL*

A rival British project at this stage was *HOTOL*. The idea of *HOTOL* may be traced to an early 1980s British Interplanetary Society (BIS) space transportation symposium where some radically new ideas were put forward by Alan Bond of the UK Atomic Energy Authority and Bob Parkinson of British Aerospace.

The concept was formally announced at the Farnborough air show of 1984, where *HOTOL* caused quite a stir [166]. This was for a 240-tonne, carbon-built spaceplane with canards launched from a trolley and powered by Rolls-Royce RHB 545 hydrogen slush engines [167]. It would be able to put 7–8 tonnes into orbit before returning to Earth, the weight of the landing vehicle being 42 tonnes. *HOTOL* would climb like an aeroplane using a liquid-hydrogen, air-breathing engine for 4–5 minutes. It would rely on hydrogen and air-breathing ramjet oxygen for the lower stage of the ascent. Take-off would be at 400 km/hr, a run of 2,300 m, with ascent at an angle of 24°. Supersonic speed would be achieved in two minutes. About 80% of the lift-off weight would consist of fuel. It would accelerate to reach Mach 5 at 26 km after nine minutes. Once there the liquid-hydrogen-fuelled rocket engine would take over to propel the spacecraft to orbit. There it would deploy a satellite payload, fly a mission up to 50 hours and make a gliding return to Earth to land on a conventional runway.

HOTOL held out the prospect of a substantial reduction in the cost of space travel. Initially the project would fly automatically, and if it were proven then a piloted version could be developed [168]. It would be smaller than the American space shuttle and slightly larger than Concorde. For thermal shielding HOTOL would not have tiles, but 1-m skin panels of carbon, titanium and nickel.

HOTOL was much publicized in the British specialized and popular media and was promoted as Britain's best chance to get back into the space business. In 1985 the British government through the British National Space Centre (BNSC) and British Aerospace provided £1.5m to each for a proof-of-concept study of HOTOL. Alan Bond was the driving force behind HOTOL and he developed the all-important hybrid engines for the system in conjunction with Rolls-Royce. Tests of the propulsion system were carried out in the Rolls-Royce plant at Ansty in Coventry that summer and by the following year satisfactory progress had been reported.

The proof-of-concept studies duly concluded in 1988 with a positive outcome, convincing British Aerospace and Rolls-Royce that the concept was achievable and economically sound [169]. HOTOL had been put through five wind tunnels at simulated speeds from subsonic to the almost orbital. They led to some rebalancing of the various systems on board, such as the canards, intakes, wing thickness and shape, all of these refining drag, mass and propulsive efficiency. The thermal profile for re-entry was predicted. The study indicated those areas that next required work, such as command and controls, computerization and the further development of the propulsion system. HOTOL held out the prospect of costing as little as €6m a launch.

Despite this glowing future, on 25 July 1988 British Industry Minister Kenneth Clark declared HOTOL to be too expensive for Britain to develop alone and he could not contemplate allocating it funds. The Ministry estimated development costs to be in the order of £4.5bn. He invited HOTOL to find a European partner. Indeed, Alan Bond was prepared to take the idea abroad, but this was problematical, given that its engine systems were classified by the same government that would not fund its development. Roy Gibson attacked the decision as being 'consistent in its short-sightedness', adding that 'Britain had now lost its cutting edge.' Former minister Geoffrey Pattie said he was 'appalled'. It was Ken Clark's last decision and four hours later he was made Minister for Health. But the decision stood.

Rolls-Royce pulled out of the project in 1989. Various attempts were made to keep the HOTOL project alive in the 1990s. Its engine design remained classified until 1991. In order to encourage others to develop the system, Alan Bond then refused to patent it. In summer 1991 there was a feasibility study by the Soviet Ministry of Aviation to launch *HOTOL* on the back of the huge Antonov 225 aircraft, the outsize carrier used to transport the Soviet shuttle *Buran*. As ever, attracting the finance appeared to be the core problem.

Following this Alan Bond and his colleagues formed Reaction Engines Ltd. This tried to keep together the expertise that had developed *HOTOL*. An early project for Reaction Engines was *Skylon*. This looked like *HOTOL* at the front, but with delta wings at the mid-point and air-breathing and rocket-propelled Sabre engines affixed to the ends of the wings [170]. *Skylon* was an unmanned spaceplane designed to put 10–12 tonnes into 300-km orbit from Kourou, and was in the designers' words the resolution of the technical compromises that had eroded the payload performance of *HOTOL*. Also unmanned, this 83-m-long spaceplane would be a single-stage-to-orbit spacecraft using dual-mode engines, with a 25-m wingspan and would take off from conventional runways.

Nobody knows at this stage if the *HOTOL* project would have worked. In the late 1990s the Americans invested considerable resources in their own single-stage-to-orbit reuseable spacecraft, the *X-33*, abandoning the project as unfeasible several years later after considerable theoretical and practical study (Chapter 7). The problem is that neither *HOTOL* nor *Saenger* were ever given anything like an equivalent chance.

6.32 ASSESSMENT AND CONCLUSIONS

Although Europe may not popularly be seen as a manned spacefaring entity, in fact Europe's piloted experience in space was quite substantial by the turn of the century. Many Europeans had flown bilateral missions on the Soviet and Russian orbital stations, the leading country being France, followed by Germany, with a solo mission by Austria and a non-governmental one by Britain. France was by far the most extensive user of bilateral missions, flying cosmonauts to *Salyut*, *Mir* and the *ISS* no less than seven times between 1982 and 2001. Europe, through ESA, participated in two lengthy missions on the *Mir* space station – the two *Euromir* missions – and made agreements with Russia for five *Soyuz* taxi missions over 2002–3. In addition, Europeans had flown on bilateral missions with the United States on individual shuttle missions. Here, Italy constructed two adventurous, tethered satellite missions. The Italians were not rewarded with success, but one day engineers will return to the project.

The *Spacelab* experience proved to be a difficult one for Europe. This was Europe's first venture into manned spaceflight. There is no doubt that Europe proved its ability to build a space laboratory module. *Spacelab* was an engineering success, performing faultlessly on its 22 missions and enabling a wide range of experiments and observations to be carried out in orbit during the relatively short, but intensive shuttle missions. Whether Europe got a fair deal from *Spacelab* is another matter. In exchange for delivering a ready-to-orbit space laboratory free of charge to the Americans, Europe got one astronaut in space for 10 days. There seems to have been an underlying assumption that there would be many opportunities thereafter for Europeans to fly on *Spacelab*. In reality, there were very few and Europe had to pay quite high costs for doing so. The Americans clearly got the better side of the deal, though the debt was rebalanced when several Europeans flew on the shuttle in the mid-1990s.

The *Hermes* programme was the outcome of some of the frustrations arising from *Spacelab*, confirming French suspicions about the 'perfidious' Americans. However, it is more than likely that France would have proposed a European space-plane anyway. Despite the difficulty it was probably logical to propose a spaceplane rather than a manned space vehicle of *Apollo* or *Soyuz* vintage, even if that might have been easier. *Hermes* was probably the right project, extending European space engineering by a quantum step forwards, but at the worst possible time. Under the straightened financial circumstances of the early 1990s, getting the project done on time and within budget proved a bridge too far for Europe. Thankfully, even as the *Hermes* opportunity closed, a new one opened in the *ISS*. This is one of the key areas of Europe's space future, as we will see in the next chapter.

7

The future of Europe in space

Even though the *Hermes* project proved abortive, Europe was already moving ahead with participation in the new American space station programme. In looking at Europe's future in space, we look in this chapter at European participation in space station development and in the exciting range of unmanned missions planned for the next 20 years. Finally, there is an examination of Europe's current role, budget and infrastructure in the context of the world's space programmes.

7.1 EUROPE AND THE SPACE STATION: ORIGINS

When the shuttle had started flying, it had nowhere to shuttle to. Construction of an American space station was eventually announced by President Reagan a full 12 years after work had been under way on the space shuttle and three years after the shuttle had first flown. The announcement was no surprise. NASA (National Aeronautics and Space Administration) had been pressing for such a commitment for 10 years. It convened a space station symposium in Washington, DC in July 1983, one attended by ESA (European Space Agency) and representatives of the German and Italian space agencies. By then NASA had a range of groups studying potential space station designs – indeed, up to a thousand people already working on a project that had not yet received any governmental approval.

From the start, NASA indicated its desire that this be an international project – both for its own sake and to avoid duplication and competition [171]. This was reiterated when President Reagan formally announced the venture in his State of the Union message in January 1984 and when he dispatched the administrator of NASA, James Beggs, to Europe soon thereafter to encourage a positive response. In Europe the most straightforward area of cooperation suggested itself: to build on the *Spacelab* experience by developing a space station module of similar dimensions.

Such an idea was explored by Aeritalia and MBB–Erno in 1984 (MBB standing for Messerschmidt Bölkow Blohm).

At the January 1985 ESA council meeting, the agency decided to support the idea of a European module called *Columbus*, to be either attached to the American space station or free-flying in the vicinity. The estimated cost to Europe would be about €2.4bn. As was the case with *Spacelab*, Germany agreed to be the main sponsor, contributing 38% of the costs. The year 1985 saw a memorandum of agreement. The terms of access to the station were: United States 71.4%, ESA 12.8%, Japan 12.8% and Canada 3%. However, these arrangements were complicated by internal access arrangements. In June 1985 the Director General of ESA Reimar Lüst signed an agreement with NASA Administrator James Beggs for parallel detailed studies of Europe's prospective contribution to the space station on the basis of Europe contributing €80m. These were called the Phase B studies.

Britain agreed to be involved in *Columbus*, but on a condition. Although the main product of the programme would be the space station module, the programme was also to cover man-tended modules, free-flyers and a polar platform. Britain's participation was conditional on being awarded leadership of the polar platform. Generally, ESA preferred not to do business this way, but it was overlooked in the welcome for Britain's reinvolvement in the programme.

Figure 7.1. *Columbus* module.

The negotiations with the United States echoed some of the disputes a decade earlier about *Spacelab* and the space tug. Here, the bone of contention was the man-tended free-flyer. Europe wanted *Columbus* to exist in three versions: one attached to the space station, a free-flyer to be detached from the station for microgravity research but to return and an unmanned polar-orbiting version. NASA objected to the man-tended free-flyer on the basis that repeatedly docking it to and undocking it from the space station would destabilize the space station [172]. It may not have escaped NASA's notice that a man-tended free-flyer could be also serviced by *Hermes*, and this could be Europe's route to an independent space station.

NASA released its space station design in summer 1986. It proposed that there be modules for the station from Japan and ESA in addition to four of its own, and four free-flying unmanned platforms, of which two would be provided by NASA, two by ESA. The station soon acquired a name *Freedom* bestowed on it by President Reagan.

Further talks about European access to the space station were held in February 1987 for two days in the US State Department. Europe, Japan and Canada participated, and it seems the talks were less than satisfactory with foreign access to the station likely to be limited, the Americans emphasizing their role in controlling the station and keeping parts of it off-limits to visitors. Despite this, agreement was reached in March 1988 for ESA, Japan and Canada to have a laboratory module and a free-flying, man-tended module. These arrangements were formalized in June 1988 through an intergovernmental agreement, signed in September 1988.

Formal approval of the *Columbus–Ariane 5–Hermes* trio took place at The Hague Council meeting in November 1987, Britain opting out of all three elements. In April 1988 there were indications that Britain might rethink its participation in *Columbus* [173]. Later that month Britain agreed to contribute 5.5% of the *Columbus* programme (£250m over 10 years) and indicated its interest in building the polar platform version. At this stage Dornier and Matra had already bid for the polar platform, making British Aerospace a late entry.

7.2 SPACE STATION IN TROUBLE

In early 1988 the first indications arose that the American end of the project would be delayed. The insufficient level of funding meant that some of the key milestones would slip and that the station could not now be manned until 1995. NASA blamed Congress for trimming the budgets sent over by the White House.

In 1989 NASA began the first of what turned out to be many redesigns. These changes infuriated the Europeans. At a practical level the agreed design had been based on mutual agreements about power levels and the downlink of data. The first redesign cut the power from 75 kW to 38 kW and the downlink from 500 MB/sec to 300 MB/sec: both cuts would affect the performance of the attached *Columbus* module. Second, Europeans were used to drawing up long-term plans and sticking to them. The NASA plan and schedule changed every year as Congress modified it this way and that, making NASA an unreliable partner.

In summer 1989 NASA engineers met at the Langley Research Center to 'rephase' the station, but ESA was told about it only after the process had begun and was debriefed afterwards – but was not invited to participate. Launching the *Columbus* module was delayed until 1997. ESA's Director General Reimar Lüst protested to a Congress committee about the way ESA had been treated. Then in summer 1990 a further assessment found that the redesigned station was overweight, underpowered and would require a huge amount of space-walking to make operational [174]. At the end of the year Congress slashed the station's budget, ordering a redesign so that it could be built incrementally. The danger grew that ESA might pull out of the station, whose design was becoming less and less recognizable to what the Europeans had originally agreed [175]. When the Augustine report was published in late 1990 it suggested a further downscaling of the space station [176]. NASA's redesign was published in spring 1991, proposing a smaller and less expensive station [177]. For ESA this meant reducing the size of the *Columbus* module by 20%, kitting it out before launch (rather than equipping part of it on orbit) and delaying the free-flying version until 2001. Despite these drastic cuts, Congress was less than impressed and the House Appropriations Committee voted in May 1991 to scrap the whole project.

Italy then circumvented all these altercations by announcing a separate, bilateral deal with the United States over the space station *Freedom*. Under an agreement signed a couple of weeks later, 8 December 1991, Italy agreed to build and supply two (later three) small logistics modules for the space station. The idea of the logistics modules, each of which was effectively a small space station module, was that three tonnes of cargo would be put in the module before it was ferried up to the space station in the shuttle's payload bay. It would be swung into place in a docking port, unloaded by the station's crew and filled with experimental results (and probably rubbish) that were Earthbound. After the shuttle visit, the module would be swung back into the payload bay and brought back to Earth. In exchange, the shuttle would bring the Italian astronauts to the space station and they would have access to the experimental facilities on board. Italy would build a technical centre to support the logistics modules in Turin, later called ALTEC (Advanced Logistic & Technological Engineering Centre).

This was a bilateral deal between the Italian Space Agency (ASI) and NASA outside the ESA framework. The logistics modules, to be built by Alenia Spazio, were to be 4.6 m in diameter, 6.4 m long, weigh 4.1 tonnes and carry 10 tonnes of cargo. There would be a docking tunnel at one end and trunnions underneath so as to stow the modules in the shuttle cargo bay. The agreement envisaged 13 logistics module flights during station assembly and a total of 75 by the end of its lifetime. The first module was completed in 1996 and shipped to Cape Canaveral. As construction of the modules got into its stride the Alenia plant in Turin became a hive of activity. A neutral buoyancy tank was built so that divers could test out access into and out of the tank under water. Three modules were built, named after Leonardo da Vinci, sculptor Donato di Niccolo di Betto Bardi and Raffaello Sanzio (*Leonardo,*

Figure 7.2. *Raffaello.*

Donatello, Raffaello, respectively). Each had 16 equipment racks within. The modules had 3.2-mm aluminium metal walls surrounded by heat shielding and then an anti-meteoroid barrier. Once connected to the stations, the modules would draw their electrical power from the station. Indeed, these modules were to prove so successful that, when NASA's space station's finances came under ever-greater strain, the agency considered turning to Italy to build the station's delayed habitation module.

One area where Europe was able to offer some help to the Americans was in the Assured Crew Return Vehicle (*ACRV*). *ACRV* was essentially a means whereby the crew of the space station could evacuate and return to Earth quickly should a serious problem arise (e.g., fire, meteorite strike, depressurization). An evacuation in such circumstances would need to be immediate and there would not be time to await the arrival of a shuttle, which could take weeks. *ACRV* would be a spacecraft designed for a down-only journey, though obviously it would have to be brought up in the first place. In summer 1992 ESA contracted Alenia Spazio and Aérospatiale to examine whether and how Europe could build an *ACRV* for the station. With its tight budgets, there was no way NASA would be given such an opportunity by Congress. Accordingly, the contractors came up with an *ACRV* design looking like a modernized *Apollo* capsule.

ESA followed this with a demonstration test of a prototype manned spacecraft to be used on the international space station (*ISS*). In September 1994 ESA gave the go-ahead for an ARD (Atmospheric Re-entry Demonstration) test. Such a test was intended to take advantage of flight opportunities presented by the *Ariane 5* demonstration programme. Aérospatiale was appointed prime contractor of ARD, an *Apollo*-shaped cabin 2.8 m in diameter, weighing 3 tonnes, designed to be lofted to a great height and then flown through the full re-entry profile. The first test of ARD was carried out from Trapani Balloon Centre, Sicily, in July 1996. The balloon brought ARD to a height of 24 km, whence it was dropped, free-falling 10 km until parachute deployment and recovery from the ocean 10 minutes after splash-down. The second took place on the second *Ariane* test flight in 1997 and was entirely successful (Chapter 4).

7.3 INTERNATIONAL SPACE STATION (*ISS*): ORIGINS

The American space station went through so many redesigns that by the end of 1991 it had managed to spend all its budget without building a single bit of hardware (the first metal was eventually cut in 1992). Congress moved in for the kill, but newly elected President Clinton ordered NASA to come back with a revised, much cheaper design. This redesign coincided with the stalling of the Russian *Mir 2* project, but for the opposite problem: the Russians had a good design, but no funding.

At this stage, the prospect opened up that Europe might abandon its partnership with the collapsing American project and make a deal with the Russians. There were quite serious discussions between ESA and the Russian Space Agency in spring 1993, which could have made *Mir 2* a Russian–European project. Had these discussions been wrapped up quickly, the American station might have been left out to dry and might well have collapsed. The Granada ministerial council meeting had proposed joint studies with the Russians on *Hermes*. In fact, the level of discussion went much further than this [178]. Working groups considered the modernization of the *Soyuz* and *Progress* spacecraft, joint space stations and crewed missions, new technologies, robot arms and space-walking.

No-one knows where this might have led – except that the process was com-pletely upstaged on 1 September 1993 when the United States and the Russian Federation signed an agreement for the merger of their respective space station programmes. The *Mir 2* core module became the core module for the new station (*Zvezda*), the Russians also supplying a functional control block (FGB), later called *Zarya*, and unmanned resupplies (*Progress*). The Americans would scale down the *Freedom* design (the name was gone too) and Russia would in *Soyuz* provide the assured crew return vehicle. Although these negotiations had been going on for some six months or so, no-one had expected so swift, clear and decisive an outcome.

It was called the *ISS* (sometimes also *Alpha*), with a permanent crew alternating between two Russians and one American and vice versa, with alternating comman-ders. Europe and the other participating nations (Japan and Canada) played very little part in this superpower renegotiation. Their contribution was presumed to be the same as had been negotiated for *Freedom*. Although the layout of the station was modified substantially, the dimensions of the modules, docking tunnels and other equipment remained unaffected. The first meeting of the international partners for the project (Russia, the United States, Canada, Europe and Japan) did not take place until November 1993, well after Russia and the United States had agreed the new design.

Europe's negotiation for involvement in the *ISS* was completed in November 1995 at the ESA council meeting in Toulouse. Europe was asked to contribute 10% of the cost of the *ISS* (€2,651m). The agreement between the member states was to

Figure 7.3. International Space Station (*ISS*).

divide this between Germany 41%, France 28%, Italy 19% and less than 3% each for Belgium, Denmark, Norway, Netherlands, Spain and Switzerland. Britain did not participate. Europe's contribution was itemized as *Columbus*, the data management system for the FGB, the European Robotic Arm and the Automated Transfer Vehicle (*ATV*). In a final amendment to the agreement in 1997, NASA agreed to launch Europe's *Columbus* laboratory, in exchange for Europe providing two docking nodes for the *Columbus* and Japanese links to the station. Responsibility for the nodes went to ASI and in turn to Alenia. The legal agreement for the governing of the *ISS* was signed on 29 January 1998, replacing the 1988 agreement. ESA was represented by Antonio Radota, the Director General, alongside NASA's Dan Goldin and the Russian Space Agency's Yuri Koptev.

The *ISS* got under way in November 1998 with the launch of the FGB *Zarya*, which was visited by American shuttles three times over the next 18 months. Construction began in earnest in July 2000 with the launch of the service module *Zvezda*. Here Europe contributed the data management system and, for launch later, the 10-m European Robotic Arm (ERA). The data management system was in effect the computer control system and was supplied in exchange for Russian docking technology that Europe would use on the ATV.

7.4   *COLUMBUS*

The roots of the idea of *Columbus* can be traced to a collaborative project between the German Ministry of Research and Technology and the Italian Ministry of Research and Technology. They funded joint studies between MBB/ERNO and Aeritalia to examine a follow-on to *Spacelab*. They drew up the idea of a *Spacelab*-type, pressurized manned module that could be attached to the forthcoming American space station. They also looked at a version that could be flown as a free-flyer for brief periods and then redocked, a pallets-only-type free-flying platform and a resource module. They also put forward the idea of *Columbus* floating free from the American space station to become a self-standing European space station. They estimated the cost at €1.75bn and Germany volunteered to pay 50% and Italy 25%. The programme was formally adopted by the ESA council meeting of January 1985 as Europe's response to the American invitation and confirmed at The Hague in 1987.

Columbus, or to give it its full title the *Columbus* Orbital Facility (COF), is an orbital module built on the experiences of *Spacelab* and, later, the Italian logistics modules. It is the equivalent of a two-segment *Spacelab*, with 15 experimental racks, able to provide a working environment for three astronauts at a time. *Columbus* is 6.7 m long and 4.46 m in diameter. Its weight will be between 12 tonnes and 18 tonnes, depending on the amount of experiments installed at launch. *Columbus* will have a volume of $10\,m^3$, drawing 20 kW from the *ISS*. The basic manufacturing cost of the module was estimated at €628m. *Columbus* has walls 4.8 mm thick, surrounded by heat-shielding and an anti-meteroid barrier. The two main builders are DASA (Deutsche Aerospace), Germany, the main contractor, and Alenia of Italy. *Columbus* has 10 active equipment racks with electrical connections, designed for use in microgravity experiments, physiology, fluid sciences, biology and materials science. The laboratory has a data management system developed by Matra. There are also external attachment points where experiments can be placed by space-walking astronauts.

The experiments on *Columbus* will be carried on the payload racks of the inside of the module. Each rack can hold 700 kg of equipment. As time goes by, some racks will be removed and replaced and others will be modified. The idea is always to have at least one European on the station at a time, with that person working in the *Columbus* module, for shifts of about 90 days at a time. The line-up of experiments for *Columbus* includes the following:

- materials science laboratory – for research into metal and alloy solidification, semiconductors and the thermo-physical properties of materials;

Figure 7.4. *Columbus* completed.

- fluid science laboratory – to study the dynamic properties of fluids and surface tension;
- physiology laboratory – to study bone demineralization and the operation of heart and lungs in zero gravity;
- up to eight specialized drawers of materials;
- externally fitted experiments for exobiology and stellar observations.

The precise usage of *Columbus* was a sticking point during the early negotiations with the Americans, NASA insisting that the Department of Defense have access to *Columbus* for defence experiments. Regardless of the financial contributions to the space station, there is a second set of arrangements for internal access to laboratory space in which Russia has 100% access to its own laboratory space and the United States 97% access to its laboratory space. The United States has 46% access to both the Japanese and European modules and Canada 3% access to non-Russian space.

7.5   THE NODES AND THE CUPOLA

Two of the space station's nodes are being built in Europe, nodes 2 and 3. Nodes are junctions where modules join up. For example, on *Mir* the five docking ports of the front were called 'the nodes' and there is a similar junction on the *ISS*'s *Zvezda*. The first node on the *ISS* was the American module *Unity*. For the expansion of the space station, two more nodes were required, nodes 2 and 3. These were to have been built by Boeing, but in 1997 responsibility was transferred to Alenia in Italy because of its experience in building pressurized modules.

Node 2 is to be attached to the American *Destiny* laboratory in early 2004 and serve as a junction there with *Columbus, Kibo*, the visiting logistics modules, visiting shuttles and the centrifuge module. Node 3 is to be the port for the American habitation module and the crew rescue vehicle (though both these projects fell substantially behind schedule). Each node will have two bays with eight equipment racks and four radial ports. Each node is 7.19 m long, 4.48 m in diameter and provides power supplies, data and communication links, pressurization and environmental control systems for the attached modules and vehicles. Node 3 will also house the system for extracting water from the station's atmosphere and showers, and additional cabins for the crew.

Europe is also responsible for the cupola that will be attached below node 2. The formal purpose of the cupola is to be an observation point for the space station. Fully pressurized for a shirtsleeve environment, it will have a top window and six trapezoidal side windows, all protected by external shutters. From there, space station crews will be able to use the robotic arm to move equipment along the outside of the window. However, it is widely accepted that it will be the most popular location in the whole space station. Instead of watching the Earth go by through relatively small portholes in *Zvezda* or the other modules, astronauts and cosmonauts may well gather there for hours on end to watch the day-time and night-time Earth roll by below and admire the heavens above.

7.6   EUROPEAN ROBOTIC ARM (ERA)

Europe got responsibility for building one of the robotic arms on the space station. The contract went to the Dutch company Fokker. ERA started life as the *Hermes* robotic arm, but when that project collapsed in 1993 it became a joint project with the leading Russian space company responsible for manned spaceflight, RKK Energiya. ERA will weigh 630 kg, be 11.3 m in length and will be brought up on the shuttle along with the Russian Science Power Platform. It will be fixed to the tower, which in turn will be located on top of the junction between *Zvezda* and *Zarya*. ERA is extremely versatile, for each end can act as the base while the other is the effector. The effector can then moor itself on different parts of the station and the roles can be reversed. The effector has seven joints, thus giving a high level of dexterity. The arm can be controlled either from a panel on the outside

Figure 7.5. Europe's Robotic Arm.

by a space-walking astronaut or from a laptop computer on the inside. ERA will normally be mounted on the Russian service module *Zvezda* and will be used for the assembly of the power system, the movement of solar arrays, inspection of the station's exterior and the replacement of parts. Its lifting ability is in the order of 8 tonnes [179].

7.7   AUTOMATED TRANSFER VEHICLE (*ATV*)

ATV is the bland title for an important part of the European contribution to the *ISS*. When the Soviet Union built the first semi-permanent space station in Earth orbit *Salyut 6* in 1978, one of the first tasks was to provide a regular resupply by automatic spacecraft of fuel, food, water and equipment for the two-person crews. This the USSR achieved by stripping down a *Soyuz* manned spacecraft, taking out the environmental control system and putting in fuel tanks, pumping systems and racks for supplies. Called *Progress* these spacecraft became the basis of resupply missions approximately every two months to *Salyut 6, 7* and later *Mir* and then the *ISS*.

The idea of the *ATV* was provisionally approved at the Granada ESA Council of 1992, the meeting that scuppered the *Hermes* spaceplane project. If Europe was not to have a manned means of supplying its module on the space station, it should at least have some, albeit unmanned means. *ATV* was finally approved by the ESA council meeting held in Toulouse in 1995. *ATV* is a can-shaped spacecraft. It can hold up to 869 kg of fuel and ferry up 5.5 tonnes of supplies in a variety of combinations, including individual modules. It is 4.57 m in diameter, 10 m long, up to 20 tonnes in weight, with a volume of 45 m^3 and designed to be launched by *Ariane 5*, bringing fuel and cargo up to the *ISS* and to remove waste. *ATV* has four solar wings, 18 m across. It can carry supplies that are the equivalent of three *Progress* freighters.

Construction soon began of preflight models. One was for structural and thermal tests in ESTEC (European Space Technology Centre), Noordwijk, the Netherlands; a second for electrical tests in Toulouse, France. The first of nine initial *ATV*s to fly to the *ISS* was rolled out for presentation to the press in April 2002. In honour of one of Europe's greatest science writers, ESA named it the *Jules Verne*. It is ESA's intention to launch one each year. The first is set for September 2004, to arrive a month ahead of Europe's *Columbus*. This was one of Europe's largest-ever space projects, its first large spacecraft able to manoeuvre independently in orbit, a combined tug, supplier and living space. Developing the *ATV* cost €600m and will cost €180m to launch each one.

Once in orbit at 300 km, *Jules Verne* will use its four manoeuvring engines to close in on the *ISS*. It will be guided by its GPS receivers and two star trackers toward the Russian *Zvezda* service module. Video will be used for the last 150 m, but in case things go horribly wrong there will be an airbag to cushion a collision. *ATV* will stay attached for up to six months for unloading and the transfer of fuel. It will be used to boost the *ISS* into a higher orbit to compensate for atmospheric drag, and for this an additional fuel load of 4,700 kg is available. Before leaving it will be filled with 6.5 tonnes of rubbish. Its job done, *Jules Verne* will undock and burn up in the upper atmosphere, over the Southern Ocean away from the shipping lanes. MAN Technologie of Augsburg, Germany, is building the fuel tank, thrusters and water tanks for *ATV*, Contraves the structure and Fokker the solar panels. The main contractors are Aérospatiale (prime contractor) with Contraves, Daimler Benz, Matra, Alenia and Alcatel.

7.8 FIRST EUROPEANS TO THE *ISS*

With the arrival of *Columbus* and the *ATV*, Europe should have a permanent presence on the *ISS*. The first Europeans have already reached the *ISS*, and the first presence there was from Italy. On 8 March 2001 *Discovery* carried aloft the first

Figure 7.6. *Jules Verne* heads for *ISS*.

of three Italian logistics modules on shuttle mission STS-102. The first was *Leonardo*. Pictures from the *ISS* showed cosmonaut Yuri Gizdenko floating in the middle of the relatively cavernous *Leonardo*, the walls filled with racks of equipment and experiments. Everything went smoothly with the berthing, unloading, reloading and unberthing of *Leonardo*, which made a real, if unheroic contribution to the smooth running of the station.

The second logistics module *Raffaello* was brought up to the station on 19 April 2001 on STS-100, this time accompanied by Italy's ESA astronaut Umberto Guidoni. His official job was loadmaster – responsible for ensuring that all the items on the module were properly taken out and stowed and that downcoming materials were appropriately brought back. He supervised the unloading of almost 2.7 tonnes of cargo. He also helped in berthing and unberthing with the shuttle robotic arm.

Leonardo returned to the *ISS* at the end of the summer, bringing up 3 tonnes of cargo: 12 racks of experiments, supplies, hardware, clothes and food. On its way back *Leonardo* transported 1.3 tonnes of experimental results, down cargo and waste. *Raffaello* made its second visit at the end of the year. By this time the procedure had become well established and the berthing process took less than an hour. Berthing and unberthing the logistics module was one of the first and last tasks undertaken by shuttle crews as they arrived at and left the space station. The third

S102E5095 2001/03/10 08:59:45

Figure 7.7. Italian module in shuttle bay.

logistics module *Donatello* arrived on a Beluga Airbus transport at the Kennedy Space Centre in early 2001 for outfitting and was set for launch on STS-130 in 2005.

These modules were a godsend for NASA, which hit fresh budget trouble in mid-2001: the *ISS* budget had overrun by over €1bn. Italy was the unlikely horseman riding to the rescue that summer offering €200m as a contribution to the construction of the habitation module. It became clear that NASA would not be in a position to increase the three-person *ISS* crew to its desired complement of six for some time.

STS-111 *Endeavour* brought the fifth permanent crew of Peggy Whitson, Valeri Korzun and Sergei Treschev to the space station in June 2002 with the *Leonardo* module. The shuttle included French astronaut Philippe Perrin who made three space walks with veteran shuttle astronaut Franklin Chang-Diaz to install new systems on the station's remote arm. A significant part of the *Leonardo* cargo was ESA's Microgravity Science Glovebox Facility, an experimental package to operate on the *ISS* for over 10 years providing a space for the testing of materials, fluids, combustion, biotechnology, electronics and superconductors. Perrin was the first European to make a space walk from the *ISS*. He was also responsible for the operation of the *Leonardo* module and experiments into the effects of radiation on the immune system. On his return he became involved in the *ATV* programme.

For shuttle mission STS-116, Crister Fuglesang at last received a crew assignment. He had been selected as far back as 1992 and had trained for over 10 years in Moscow and Houston.

Italian modules to the ISS			
8 March 2001	*Leonardo*	STS-102	
19 April 2001	*Raffaello*	STS-100	Umberto Guidoni
10 August 2001	*Leonardo*	STS-105	
5 December 2001	*Raffaello*	STS-108	
5 June 2002	*Leonardo*	STS-111	Philippe Perrin

7.9   ASSURED CREW RETURN VEHICLE

Although the Russians supplied a *Soyuz* spacecraft as an assured crew return vehicle (CRV) or lifeboat, NASA always planned to build a purpose-built CRV. NASA hoped to develop an assured CRV based on the X-38 spaceplane. ESA participated with NASA in a series of tests for an assured CRV and made proposals for a jointly developed CRV. Models of the X-38 were drop-tested in 2000, ESA providing the guidance, navigation and control software for the parafoil descent. The first model X-38 was dropped on 2 November 2000 from a B-52 bomber 11 km above Edwards Air Force Base in California. The model called the V-131R used an aerodynamic form designed by Dassault and control, navigation and guidance furnished by ESA. This was the first of several drop tests [180]. Germany spent €100m in developing a more advanced test version of the X-38 called the V-210, to be brought into orbit on the shuttle sometime later.

Although the tests went successfully, the X-38 became the victim of continuing budget overruns and cuts in the NASA budget. The X-38 was cancelled in 2001, although in NASA's view the need for an assured CRV remained. The cancellation annoyed the Europeans, since it was made without informing them in advance (ESA only found out when someone leaked an internal NASA memo).

7.10   AFTER *ARIANE 5 ...?*

Jules Verne will use *Ariane 5* to reach orbit. Even assuming the success of European occupation of the *ISS*, consideration will still need to be given to a successor rocket to *Ariane 5*. At the ESA council meeting held in Brussels on 23–24 June 1998 approval was given for the start of improvements in the *Ariane 5* programme. In June 2000 ESA adopted *A Proposal for a European Strategy in the Launch Sector*, outlining the future paths of development. *Ariane 5* improvements were divided into a number of stages, called *Perfo 2000*, *Ariane 5 Evolution* and *Ariane 5+*, with a budget of €700m. *Ariane 5 Evolution* involved a new *Vulcain* engine for the main stage, called the *Vulcain 2*, with over 20% more thrust. Several problems showed up during the lengthy 45,000 seconds of engine tests, the biggest being cracks in the turbine disks of the liquid oxygen turbopump. The new Vulcain 2 engine made its first extended test in Lampoldshausen in June 1999. These were the main lines of improvement:

- *Ariane 5 ESC V* (ESC stands for Etage Supérieure Cryogénique and V for versatile), with a 7.5-tonne payload (up from 6.2 tonnes) made possible by the reduction of weight and addition of more propellant to the solid rocket boosters and the main stage, with a restartable upper stage.
- *Ariane 5 ESC A* is an improved, non-reignitable HM7B cryogenic upper stage that is able to put 10.5 tonnes into geosynchronous obit. The HM7B is based on the HM7A of the *Ariane 4* upper stage. First flight was *Ariane 517*, the *Stentor* communications technology satellite.
- *Ariane 5 ESC B* is an improved, reignitable cryogenic hydrogen-fuelled upper stage called *Vinci* that is able to put 12 tonnes into geosynchronous orbit with improvements to the main stage and solid rocket boosters. First flight is due in 2006.

In 2000 ideas for a further programme of upgrades were mooted *Ariane 2010*. This programme would raise lift capacity to 20 tonnes to low Earth orbit, 15 tonnes to geostationary orbit and reduce production costs by 30%. The main means of doing so would be a new main engine, the *Vulcain 3*, with 20% more thrust, new turbo-pumps and gas generators.

The successful introduction of *Ariane 5* was the biggest challenge facing ESA in the final years of the 1990s. That having been done, despite the initial setbacks, Europe could still not afford to relax. International competition for the launching

Figure 7.8. *Vinci* engine.

of satellites was intense, especially with the arrival of the Russian *Proton M* and *Zenit 3SL* and, on the American side, the new *Delta IV* and *Atlas V*.

7.11   EUROPEAN REUSABLE LAUNCH VEHICLES (RLVs)

In 1993, long before *Ariane 5* had even flown, ESA first began to give consideration to the nature of rocketry in the post-*Ariane* period. Funding of €3.2m was put in place for feasibility studies of reusable rockets, spaceplanes, single- and two-stage ascent to orbit rockets under FESTIP (Future European Space Transportation Investigation Programme) [181]. These studies were led by DARA in Ottobrunn, Germany, which began to examine a number of possibilities for reusable launch vehicles and involved Aérospatiale, Matra, DaimlerChrysler and MAN Technologie. The resultant projects were, respectively, the *Ares/Themis* system, the *Hopper* plane and a liquid flyback booster, the last developed by MAN Technologie as a step toward a fully reuseable system.

In the next stage ESA approved the Future Launchers Technology Programme (FLTP) and €48m were approved for the preliminary research studies, funded by France and Belgium. A sum of €70m was allocated, but Germany, Britain and Italy would not join, leaving the balance to be raised. *FLTP 1* was intended to pave the way for *FLTP 2*, which would be a five-year, €600m programme to test selected designs and bring Europe to the stage where it could actually begin a proper RLV programme. FLTP 1 had four aims:

- identification of the next generation of launchers from 2007;
- development of reusable concepts and RLVs;
- identification of the technologies required for a new generation of cost-effective launchers; and
- elaboration of the demonstration and testing work required.

This was the beginning of European work on RLVs. RLVs are considered to be the next step after the space shuttle era. They differ from the shuttle in so far as they are *fully* reusable (all parts), can land at conventional airfields, have quick turnaround times (two days being optimum), use small ground crews and provide quick, low-cost access to space for crews and cargo. RLVs are single stage to orbit: the vehicle that is launched is the vehicle that comes back. In other words, RLVs will do what the shuttle was originally intended to do when first conceived in the 1960s. The problem with the shuttle is that it is large, only partly reusable (the external tank is discarded, the solid-fuel rocket boosters are retrieved from the ocean to be refurbished), has slow turnaround times (two or three months) and is at €500m a mission, expensive.

Developing the RLV has been the holy grail of space travel ever since the shuttle was introduced. The Americans invested in several RLV projects in the 1990s. The main project was the X-33, a fat spaceplane that would take off from Edwards Air

Figure 7.9. Next generation of spaceplanes.

Force Base, California and return to Malmstrom Air Force Base, Montana. Development was not easy, and it was daunting to combine a new, high-performance engine with heat-resistant materials, an efficient fuel tank and the level of operational ruggedness required. NASA had hoped that the X-33 would lead to an operational version called the *Venturestar* by 2007 to 2010 and the shuttle could then be phased out. X-33 got increasingly behind schedule, cost more and more and, in straightened budgetary circumstances, NASA pulled the plug. At the same time and at a fraction of the cost, an upright, bullet-shaped spaceplane called the DC-X managed to develop vertical take-off and landing techniques. At least it flew. By 2002 NASA was facing having to stretch out the lifetime of the shuttle to 50 years.

After the *Hermes* experience, Europe might have been expected to give RLVs a wide berth. Far from it. By 2001 several countries had developed their own RLV designs. France had *ANGEL*, Germany had two RLV designs (*Phoenix* and *Astra*) and even Austria had joined in. Italy had a programme called the Unmanned Launch Vehicle (ULV). The Dutch had a suborbital demonstrator called *Dart*. CNES (Centre Nationale d'Etudes Spatiales) was holding out for European leadership of the new programme, but the Italians were blocking this, still sore over French opposition to the *Vega* launcher. Well might the Americans call the European RLV effort 'sputtering and fragmented' [182].

FLTP aimed to enable Europe to catch up with the level of RLV research in Russia, Japan and the United States and position Europe to take firm decisions on RLVs by 2010 at the latest. A working group was set up in 2001 to try and bring some coherence to the various national efforts then under way. ESA began consideration of how it might cooperate with Russia, which had extensive spaceplane experience and test facilities (but no money) and Japan, which had carried out the *HOPE-X* study.

7.12 FUTURE SCIENCE MISSIONS

We have just described manned missions and the future launcher programme, but what about science – the core of European space efforts?

A successor programme for *Horizon 2000* was proposed in 1995, *Horizon 2000+*, to operate from 2006–17 . A call for mission concepts was made in October 1993 and 110 proposals were received. The survey committee reported in 1994 and approval was given at the ESA Council meeting in Toulouse in 1995. The budget set for the 2000–6 part was €1,869m. As originally proposed, this would involve three large-scale missions (planet Mercury, *Hipparcos* successor, gravitational observatory) and between two and four medium-size missions. The medium-size missions were limited to a cost of €176m. The first one approved was *Mars Express*. In October 2000 the ESA council finally approved five missions, with one in reserve:

- participation in the American Next Generation Space Telescope (NGST);
- *Gaia*, to analyse the composition, formation and evolution of our Galaxy (2012);
- *Bepi Colombo*, to orbit Mercury in 2009;
- *Lisa*, to observe gravitational waves in space;
- a solar orbiter, to replace *SOHO* and *Ulysses*;
- in reserve, *Eddington*, to find orbiting planets around stars.

The introduction of these missions also marked a change of personnel in ESA. In 2000 Roger Bonnet retired. He had guided Europe's science programmes from 1984. He was succeeded by David Southwood from Britain, formerly from Imperial College, the new Director of Scientific Programmes.

These missions were elaborately prepared. Long before they went to ESA for approval, each was submitted to years of scientific appraisal. A key phase was the system and technology study report. Here, a design consortium would go through the rationale for a mission, why it was important, what might be learned and why the target was of particular interest. In the case of the *Bepi Colombo* mission, the consortium comprised Alenia and Dornier (the orbiter) with Hunting in the UK (the lander). The consortium had a science advisory group, but would also consult with other experts, engineers and scientists on an *ad hoc* basis. ESA scientific personnel were involved in the whole process. The system and technology report would sketch the proposed spacecraft, the trajectories involved, the launcher requirements and the types of experiments that could most usefully be carried. But this was not a final design and might leave open a number of important decisions or options about the launcher, propellant or experiments. It was not the final word, but enough to make a decision. These were nicely produced and illustrated studies, all publicly available on request.

ESA's science programme entered some difficulty in late 2001. At the November 2001 ESA Council meeting in Edinburgh, there was agreement for only a very gradual rise in the ESA science budget, 2.5% instead of the 4% that programme promoters had hoped for [183]. The decision created a €500m shortfall. David Southwood, ESA science director, managed to save planned missions by stretching them out, combining resources, restructuring management and reducing slack, meaning that there were no spare resources should any of these programmes experience difficulty in their preparation. The revised programme was called *Cosmic Horizons*. The following are some of the scientific missions in the pipeline, drawn from old and new *Horizon* programmes and collaborative projects. Note that the dates are tentative and some may well slip.

7.13   *ROSETTA* (2003)

Rosetta was the third cornerstone mission of the *Horizon 2000* programme. It got its name as follows. The French scholar Jean-François Champollion found, near the town of Rosetta (Rashîd, Egypt) in 1799, the stone that explained the hieroglyphs of Egypt. It later became the famous Rosetta Stone. Similarly, *Rosetta* will hopefully unlock the secrets of our solar system.

The pre-studies went through a number of evolutions. Originally, the mission was to retrieve a sample from a comet and bring it back to Earth. This was later considered to be too ambitious, and it was decided instead to land a laboratory on the comet to make *in situ* measurements. Telespazio of Italy was made responsible for the technological development of this spacecraft in early 1991. The main experiments and instruments on the mission are being devised in Germany, France, Italy, Austria, Finland, Hungary, Britain and Ireland.

Rosetta is still an enormously ambitious scientific exploration mission, the most difficult since *Giotto*. The aim of the *Rosetta* project is to rendezvous with and then orbit comet Wirtanen from November 2011. Comets are the only bodies of primal matter left over from the early days of the solar system, and it is hoped that by examining one we may unlock the archive of early creation. Wirtanen is an elderly comet in the sense that it has visited the Sun many times and exhausted most of its energy. Wirtanen was discovered quite recently by Carl Wirtanen who was examining photographic plates of comets taken by Californian observatories. At that stage Wirtanen was circling the Sun every 6.65 years, but this was reduced to

Figure 7.10. *Rosetta.*

5.5 years after its orbit was perturbed by Jupiter in 1972 and again in 1984. The exact size of Wirtanen is not known, but the best estimates suggest it is 1.5 km in diameter [184]. *Rosetta* will also visit two asteroids: Otawara and Siwa.

Launch is set on *Ariane 5* for 2003, its first deep space mission. Intercepting Wirtanen will bring the spacecraft far outside the orbit of Mars for the chase. Tracking will be from the new dish being built in New Norcia in Western Australia, near the East River north of Perth. It is a demanding mission, one which requires the spacecraft to hibernate, act autonomously when out of radio contact with Earth and generate solar power at a long distance from the Sun.

Rosetta schedule

Day	Event	Distance (km)	Distance from Earth (AU)
1	Launch		
949	Mars fly-by	219	0.69
1,040	Earth fly-by	3,332	
1,267	Otawara	714	2.34
1,766	Earth fly-by	2,315	
2,011	Siwa fly-by	1,630	1.46
3,471	Wirtanen	1	2.6

Rosetta is box-shaped, 2 m by 2.8 m by 2.1 m, weighing 3 tonnes. It is a most demanding mission, by any standards. First, *Rosetta* must be equipped to last a long time – 10 years – a period requiring physical durability and the ability to go into hibernation for long periods. Second, it must be able to generate solar power a long distance from the Sun where the light is feeble. Its 14-m-wide solar wings have an area of 68 m^2. Even 5.25 AU out (790 million km) they should still generate 400 W of power. Third, it will need a high data transmission rate to cross long distances, so a 2.2-m dish is to be provided. Finally, station-keeping will require a substantial fuel reserve of 1,108 litres.

On its way to Wirtanen, *Rosetta* will swing by Mars once (2005) and Earth twice (2005, 2008) to build up the energy for the encounter. The trajectory will take it past two asteroids that it will visit en route – Otawara and Siwa – at a distance of 714 km and 1,630 km, respectively. For much of its mission, *Rosetta* will be a long way from Earth, at least 600 million km away.

Interception of the comet is set for November 2011, followed by entry into cometary orbit in April 2012. *Rosetta* will circle Wirtanen at a distance of 10 km to 40 km, observing Wirtanen closely as it begins to heat up approaching the Sun. The idea is that *Rosetta* will gradually match Wirtanen's orbit, shadowing it for the best part of 17 months and orbiting it at a distance of 10 km at effectively walking speed. The interception will take place a long way out, 400 million km from Earth and 500 million km from the Sun. Rosetta will map the surface carefully and analyse the dust and vapours given off by the comet.

A small lander *Champollion* (named after the discoverer of the Rosetta Stone) will be dropped to the surface in July 2012, to operate for several months. The 100-kg lander has nine experiments. One is COSAC which will drill 1 m into the comet's nucleus, analyse the icy and gaseous material on board the probe and attempt to determine the exact chemical composition of comets. The Open University is providing the laboratory to analyse the material: normally the size of a room, it has been reduced to that of a sugar bag. *Rosetta* is one of the first missions to involve the countries of eastern and central Europe. Hungary is participating in the command and data management system, the power supply of the lander and four experiments. *Rosetta* will leave the comet in July 2013.

In advance of its early 2003 launch, *Rosetta* was subjected to three weeks of thermal tests at the European Space Research and Technology Centre in Noordwijk. This aimed to reproduce, as accurately as possible, the kind of conditions that *Rosetta* would experience during its mission, with a hundred sensors measuring how it would respond. The thermal chamber was able to simulate conditions from $-100°C$ to $+250°C$.

7.14 *SMART* TO THE MOON (2003)

ESA studies of a possible Moon probe date to 1990 and the first formal proposals were made to the ESA council in 1995. By 1998 the concept had emerged called *SMART 1* (Small Missions for Advanced Research and Technology). This would be a technology demonstrator that would go to the Moon. *SMART 1* will test technologies to be used on later missions such as *Bepi Colombo* to Mercury.

SMART will be Europe's first Moon probe, following in the steps of three other countries that have already sent probes to the Moon: the USSR, USA and Japan. *SMART* is a 350-kg technology demonstrator to be launched on *Ariane 5*. The main purpose of the €86m mission is to test a solar electric propulsion system. *SMART* is to go into a 1,000-km by 10,000-km lunar orbit for six months. To reach such an orbit it must use solar electric propulsion from an elliptical Earth orbit of 7,000 km by 42,000 km. Its 20-kg payload includes a miniature high-resolution camera, infrared spectrometer and X-ray spectrometer to map the chemistry and mineralogy of the Moon. Also tested will be high-rate telemetry laser communications and a highly autonomous navigation system. The engine, developed by Alcatel and SNECMA, is a stationary plasma, xenon engine developing 70 mN thrust, fed by up to 82 kg of xenon. The mission will be managed by ESOC (European Space Operation Centre – mission control) Darmstadt and supported by the Perth and Redu tracking stations, with laser links to a station in the Canaries.

Even as *SMART 1* was in preparation, ESA in 1998 contracted Surrey Satellite Technologies Ltd (SSTL), the Technical University of Munich and the Swedish Institute of Space Physics to make a feasibility study of a 100-kg, lunar-orbiting, small subsatellite. However, to their disappointment, the project did not progress further and did not get beyond the ESA council meeting of autumn 1999. Later, probably around 2006, *SMART 2* will test the technologies of formation-flying, laser

Figure 7.11. *SMART 1.*

communications, sensors, spacecraft control, positioning, navigation and guidance, in advance of the *Lisa* and *Darwin* missions.

7.15   *MARS EXPRESS* (2003)

Mars Express is Europe's first Mars mission, following the USSR's *Mars 1* (1962), the United States' *Mariners 3* and *4* (1964) and Japan's *Nozomi* (1998). Several ideas had been circulated from the late 1970s for an unmanned European mission to Mars, and the project had even been baptized with the name *Kepler*. Britain had flown a pressure modulator infrared radiometer on the ill-fated American *Mars Observer* of 1992. The concept of *Mars Express* was a low-cost mission that would use some of the European experiments lost on the *Mars 96* mission. Twenty-five companies were involved from 15 countries.

Eventually, ESA approved in late 1998 a €180m Mars orbiter for a 2003 launch. Matra Marconi became the prime contractor and was awarded a contract to develop a Mars orbiter. The spacecraft is due to fly on a Russian *Starsem Soyuz Fregat* rocket in Summer 2003, to reach Mars' orbit for December 2003. The aim of the main 1,100-kg orbiting spacecraft is to compile a high-resolution, 10-m map of the surface topography and morphology, to make a 100-m mineralogical map, determine the composition of the planet's atmosphere and provide a radio relay

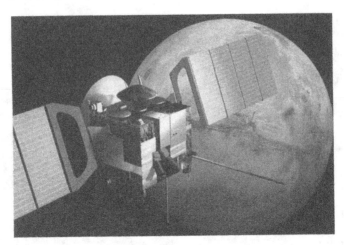

Figure 7.12. *Mars Express.*

for future missions. One of its main objectives will be to map subsurface water on the planet. The experimental package weighs 120 kg. The orbiter instruments are:

- subsurface sounding radar altimeter (Italy/USA);
- ultraviolet atmosphere spectrometer (France);
- radio science experiment (Germany);
- planetary spectrometer (Italy);
- high-resolution stereo colour imager (Germany);
- energetic neutral atoms analyser (Sweden/France);
- infrared mapping spectrometer (France).

A British consortium led by Open University Professor Colin Pillinger and involving the University College London, Matra Marconi and the University of Leicester proposed a small lander called *Beagle 2*, named after HMS *Beagle* in which Charles Darwin explored the Galapagos in the 19th century. *Beagle 2* is a lander designed to search for signs of life using a mole that would burrow underneath the surface. *Beagle 2* has a saucer shape, and as it lands the top half of the saucer opens up and divides into four petal-style solar cells. A weather station pops out on top. The lander is intended to add significantly to our knowledge of the weather, chemistry and mineralogy of the planet. It weighs 25 kg on the surface, about a tenth the size of the successful American *Pathfinder* mission of 1997. The lander will carry several cameras, a mass spectrometer, microscope, robotic arm and small rover. The mole is 12 cm long and will crawl 1 cm every six seconds, with a mass spectrometer in its nose. The automatic arm will be able to take 5-cm samples and analyse them. Landing is set for 25 December 2003. *Beagle*'s landing site has been determined as a sedimentary basin called Isidis Planitia, about 10°N, 270°W, as flat an area as possible to avoid a landing accident and then warming in the martian spring.

Figure 7.13. UK's *Beagle 2* descends.

ESA did not have funding for *Beagle 2*, but expressed a preparedness to carry the mission if the project sponsors could raise the necessary €45m themselves. The British government announced a contribution of €9m for the project in August 1999. In early 2001, just as the financial situation had become quite worrying, ESA decided to invest €24m in the project. ESA also decided to tighten its supervision of the project, especially to ensure that signals were correctly relayed to the orbiter. Assuming all went well with *Mars Express*, ESA aimed to develop a follow-on mission to its sister planet called *Venus Express*, using similar equipment in 2005–6.

7.16 *DOUBLESTAR* (2003)

In a follow-on to *Cluster*, ESA signed an agreement with China in July 2001 for the launching, between December 2002 and April 2003, of two satellites to study the Sun's relationship with the Earth's magnetosphere, their findings to be cross-referenced to those of *Cluster*. *Doublestar* will carry 18 instruments, 10 of which will duplicate those on *Cluster* and eight will be new, purpose-built Chinese instruments. Some are indeed spares or engineering models built during the preparation for the *Cluster* mission – meaning that they are known to be reliable. The use of existing equipment will keep costs down. ESA contributed a modest €8m to the mission. The equatorial satellite will fly in a 550–60,000-km orbit, 28.5°, while the polar satellite will fly from 350 km to 25,000 km at 90°, orbiting every 7.3 hours. The equatorial satellite will study the Earth's magnetic tail while the polar satellite will check what is going on over the magnetic poles and their aurorae.

7.17   *PLANCK* (2008)

Planck was approved in 1999. The overall purpose of the Planck mission is to measure the residual radiation left over from the Big Bang that started the universe billions of years ago. This residual or microwave background radiation was first discovered in 1965 and the initial exploration was done by the *COBE* satellite (COsmic Background Explorer). *Planck*'s sensors will be cooled to the lowest temperature yet achieved on a spacecraft −0.1°K, necessary because they will try and detect background radiation with a temperature as low as 3°K simultaneously in nine frequencies. *Planck* and *Herschel* share a common spacecraft bus, both will use the same *Ariane 5* as a launcher and have a common operating position L2, on the far side of the Earth from the Sun. There will be two main instruments – a high-frequency detector and a low-frequency detector. The first will be built by the Institute of Space Physics in Orsay, the second by the Institute for Technology and the Study of Extraterrestrial Radiation in Bologna. The mirror for *Planck* is being made at the Danish Space Research Institute.

7.18   *FIRST/HERSCHEL* (2008)

FIRST (Far InfraRed Submillimetre Telescope) was also approved in 1999. It will have the largest-ever infrared telescope, a record 3.5 m in diameter, and will also be located at L2. *FIRST* will try and pick up the coldest objects in the universe. It will look for planetary systems, gas and dust in and between galaxies. The telescope will be cooled down to +2°K. *FIRST* will have three instruments: a photoconductor array camera and spectrometer made in Germany, a spectral photometric imaging receiver (Britain) and a heterodyne instrument (Netherlands). In late 2000 *FIRST* was renamed *Herschel*, in honour of the Anglo-German astronomer Sir William Herschel, the discoverer of infrared light in the 18th century. *Herschel* is a 3.25-tonne telescope and its mirror will study previously unexplored wavelengths in the far-infrared and submillimeter spectrum (80 μm to 670 μm) to see how stars and galaxies were formed. In 2001 Alcatel won the €369m contract for the manufacture of *Planck* and *FIRST/Herschel*. Astrium and Alenia Spazio will be key contributors to the project and there will be scientific teams involved in all the 15 ESA countries,

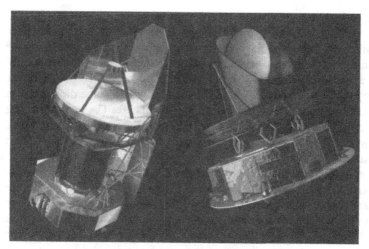

Figure 7.14. *FIRST/Herschel and Planck.*

including recently joined Portugal. One *Ariane 5* will be used to fly both *Planck* and *FIRST*, saving considerable costs, partly made possible because both satellites will be heading for very similar orbits.

7.19   *BEPI COLOMBO* (2009)

Bepi Colombo is named after Giuseppe Colombo (1921–1984). From Padua, he was an expert in celestial mechanics and space structures, dividing his life between Italy (Catania, Modena, Pisa, Padua, Genoa) and the United States (Jet Propulsion Laboratory [JPL]). As noted already, he was most interested in the idea of tethered satellites, but one of his greatest achievements was devising the trajectory for *Mariner 10* to Mercury, putting together a route with a fly-by of Venus and two visits to Mercury. By the new century, *Mariner 10* still remained the only earthly visitor to Mercury. Although relatively near to the Earth, Mercury remains much less well known than many of the other planets of the solar system and their moons. ESA made preliminary studies of a Mercury orbiter as far back as 1993. It was modified many times subsequently to take account of problems of complexity and cost, being formally agreed as an ESA mission in 2001.

The *Bepi Colombo* mission is designed to fly a payload of 1,100 kg to Mercury using a *Soyuz Fregat* launcher in 2009 for a 2012 arrival (a number of different trajectories are under consideration). Solar electric propulsion, pioneered by the *SMART* Moon probe, will be involved. The mission involves big design challenges as the probe must be able to survive extreme levels of heat and light.

The spacecraft is to consist of a polar orbiter, magnetospheric orbiter (provided by Japan) and lander. The 360-kg polar orbiter will be the main spacecraft, circling the cratered planet at an elliptical 400-km to 1,500-km orbit, and will carry 11 instruments, including a 20-m-resolution camera (so far, only 50% of the planet has been mapped), X-ray and gamma ray spectrometers. The 160-kg magnetospheric orbiter will be modelled on the *Cluster* spacecraft and will go into an elliptical orbit of 400 km to 15,000 km and study the interaction of Mercury's magnetosphere and the solar wind. It will be drum-shaped, with a deployable, rigid boom and a deployed wire, carrying 11 experiments. The 260-kg lander would come down near one of the Mercurian poles in darkness, in order to prolong its life. The type of braking engine, solid or liquid fuel, has still not been decided. Landing bags, pioneered by the *Pathfinder* on Mars in 1997, would be used so as to make the final 1,000 m of the descent survivable. The lander will carry seven experiments, including a camera, microrover and mole, X-ray spectrometer for surface analysis, seismometer and other instruments for assessing the nature of the surface. It will be expected to last seven (Earth) days. The lander looks like a flattened barrel: once it lands, the lid rises on the top to collect solar rays and a camera peeps out. The surface instruments will be ejected through the side. There is even a sunshade to try and cool some of the instruments.

7.20   EUROPEAN PARTICIPATION IN THE NGST *WEBB* TELESCOPE (2010)

The *Hubble* telescope has already revolutionalized our knowledge of space-based astronomy, more than fulfilling its designers' dreams, even after its very uncertain start. With a further servicing mission, *Hubble* is expected to operate until 2010. No-one is quite sure what to do with it then – either to send it into an orbit so far out that it won't come down for a very long time, or bring it back to the Air & Space Museum on the Mall in Washington, DC (an expensive museum delivery). Europe took a 15% share in *Hubble*, a level of commitment that ESA would like to see continued in a successor project. Collaboration on *Hubble* was considered a success story on both sides. The competitions for viewing time on *Hubble* in fact gave Europe more than 15%, so the threshold was never formally invoked.

Even as NASA considered life after *Hubble*, so too did Europe give thought to how it might work with NASA on a successor project. Accordingly, a study group was convened, one that included NASA representatives, working on the assumption of a 15% level of European participation in the new telescope, originally called the Next Generation Space Telescope (NGST), later named the *James Webb* telescope in honour of NASA's administrator during the Moon race. The 15% level has since been agreed, with ESA represented on all the project advisory bodies. Workshops were organized in Europe and the United States to consider the project in further detail and their outcomes were posted on the Internet.

The *Webb* telescope will be designed to operate at its greatest effectiveness in the near-infrared area, although it will also be able to see visible light. It will be an 8-m, 3.7-tonne telescope cooled to 30°K, located at the L2 point and as a result will not be serviceable by the space shuttle. Although it will be a general-purpose telescope, it will concentrate on the first light in the universe and the processes that started off the very first stars, a period of time that some astronomers call the dark ages. *Webb* is aimed to focus on the cosmology and structure of the universe, the origin and evolution of galaxies, the history of the Milky Way and its neighbours, the birth and formation of stars, and the origins and evolution of planetary systems. The three main instruments will be a wide-field camera, multi-object spectrograph and combined camera and spectrograph. *Webb* will also look different from *Hubble*, with an instrument box, large transmission dish and big, blue, vane-ear solar panels, all of them behind a huge Sun-facing sunshield to provide passive cooling for the telescope. A major engineering challenge is to find a way of folding the large telescope into the shroud.

ESA will be responsible for the service module and two instruments: the multi-object spectrometer and half of the camera spectrograph. *Webb* will be a smaller project than *Hubble* as NASA is still conscious of the enormous cost of *Hubble* and the high cost of its maintenance through the space shuttle. The launcher will either be the Russian-based *Atlas 5*, *Delta IV* or *Ariane 5* and is set for June 2010.

7.21   *SOLAR ORBITER*

Solar Orbiter is designed to be the first space probe to fly close to the Sun. With *Solar Orbiter*, ESA plans to build on the success of *Helios*, *Ulysses* and *SOHO* to achieve major breakthroughs in solar and heliospheric physics. The aim of the mission is:

- to explore the inner reaches of the solar system;
- fly by the Sun at a distance of 45 solar radii or 0.21 AU (the Earth is at a mean distance of 1 AU from the Sun);
- define and characterize the activities of the Sun from close range, including the polar regions;
- improve our understanding of the dynamics of solar activity.

Solar Orbiter will be a small probe, 1,308 kg in weight, with a payload of 130 kg, comprising 13 instruments. Five are designed for remote-sensing of the Sun from a distance: the two imagers, spectrometer, coronograph and radiometer. For the solar regions the probe will deploy a solar wind plasma analyser, radio and plasma wave analyser, magnetometer, energetic particles detector, dust detector, neutron detector, radio sounder and neutral particle detector. The imagers will provide a level of detail of activities on the Sun far greater than anything known so far. The spacecraft will build on the technologies developed for the *Bepi Colombo* mission to Mercury.

Solar Orbiter will be launched on a *Soyuz Fregat* from Baikonour. It will use solar electric propulsion and make several gravity assist manoeuvres with the help of Venus. Solar electric propulsion will be used for burns as long as 105 days at a time. *Solar Orbiter* will reach its closest point to the Sun after two years, after which it will orbit the Sun from 0.2 AU to 0.9 AU with a period of 149 days, less than half the Earth's period around the Sun. There will be two further Venus swing-bys, the purpose of which will be to raise the inclination of the orbit to 30°. This will take five years, after which the mission may be extended a further two years.

Solar Orbiter is in the shape of a tall box, 3 m tall, 1.6 m wide and 1.2 m in depth. It has two sets of solar arrays. The larger, 19-m-wide set will be used for 6.2-kW solar electric propulsion and this will be dumped when solar orbit reaches its closest distance to the Sun.

7.22 *LISA* (2011)

Lisa (Laser Interferometer Space Antenna) must be one of the most exotic scientific space missions planned for some time. The purpose of *Lisa* is to detect gravitational waves from black holes and thereby answer some of the fundamental questions about the nature of black holes, as well as to observe other stellar phenomena such as compact binary systems. Apparently, black holes emit low-frequency gravitational waves, ripples, across the time–space continuum that are undetectable on Earth because of our own loud gravitational waves. The concept of gravitational waves was the subject of a paper by Einstein in 1916 and they address some of the key issues of space, time, gravity and relativity.

The idea of the *Lisa* mission goes back to American studies in Colorado in the early 1980s. The concept has gone back and forth between Europe and the United States over the years and is currently seen as an ESA-led project with an American launcher (the trusty *Delta II*, able to lift all three of *Lisa*'s spacecraft together) and American collaboration (JPL, California). The feasibility study was carried out by Dornier, Alenia and Matra and is seen as a particular interest of the Max Planck Institute in Germany.

Lisa is designed to detect and analyse these low-frequency gravitational waves. The concept behind the mission is to place into solar orbit three small, 50-cm-high, 180-cm-diameter, drum-shaped spacecraft orbiting in a triangle about 5 million km or 20° behind Earth's orbit around the Sun. Solar electric propulsion, with 2-mN-thrust ion engines using between 14 kg and 19 kg of xenon propellant, will manoeuvre each 400-kg spacecraft into position about 13 months after their launch. Inside each spacecraft is a Y-shaped, 30-cm telescope to pick up gravitational waves in cooperation with the other two spacecraft of the trio. Each telescope transmits a laser beam to the other spacecraft, which rotate constantly around a centre of gravity in their solar orbit. Whenever a gravitational wave passes through the area of space of the three spacecraft, then that laser beam will be disrupted and the event will be recorded.

7.23 *GAIA* (2012)

Gaia is a 3-tonne spacecraft that will pinpoint the position of a billion stars in the sky using twin, 1.7-m-diameter telescopes. A by-product of the mission will be the discovery of extra-solar planetary systems, the location of 500,000 quasars and the mapping of small asteroid bodies in the solar system, from those that orbit near Jupiter (the Jovians) to others as far out as Pluto (the Plutonians) and beyond to the Oort cloud. In addition, Gaia is expected to provide detailed information on super-novae, variable stars, bursts, debris, dark matter, and brown and white dwarfs.

Gaia will be the logical successor to the troubled *Hipparcos* mission of 1989, but with enormously improved capabilities. *Gaia*'s stellar census will provide details of each star's location, brightness, temperature and composition down to 20 magni-tudes. Its telescopes will have the quality of being able to split a human hair at a distance of 1,000 km. The aim is to publish a complete stereoscopic map of the galaxy by 2024. *Gaia* would lead to a radical improvement in our knowledge of stellar populations, galaxy formation, the nature of star-forming regions, star histories and the spiral wings of galaxies. In studying the solar system, it could offer precise diameters for a thousand asteroids, improve 30-fold our knowledge of the orbits of asteroids and make a careful study of asteroidal objects that come close to Earth. To ensure proper accuracy, each object will be observed up to 100 times.

The suggested location for the mission is the L2 site, and it would be possible to place *Gaia* there using the *Ariane 5* launcher, the spacecraft taking about 260 days to reach the target orbit. The spacecraft would comprise a service module (subdivided into a main structure and a propulsion module) and a payload module, the latter carrying two identical telescopes. It would be tracked eight hours a day from the Perth, Australia station. This would be a larger spacecraft than many of the other scientific missions, weighing 3,137 kg, of which 1,010 kg would be fuel. In the middle of the spacecraft would be a rim solar array with an area of $24\,m^2$. The mission duration would be four to five years. Mission cost is estimated at €573m. *Gaia* is promoted as the ultimate observatory, promising to address many of the key questions in astronomy, with a broad-based appeal throughout the astronomical community and a mission that would demonstrate European excellence.

7.24   OTHER POSSIBLE FUTURE MISSIONS ...

Several other missions have been considered by ESA for *Horizon 2000 +*. These have not been approved so far, but there is the prospect that they or a version thereof will be in the future. They are briefly described here. They have been through and passed technological assessment, so approval is essentially a function of the resources available and making difficult choices about priorities.

7.25   *EDDINGTON*

In the *Star Trek* series *Enterprise*, first screened in 2002, the star ship *Enterprise* under the command of Captain Jonathan Archer sets out to visit habitable planets and to find life in that part of the galaxy closest to Earth. With *Eddington*, this mission moves a step closer. One of the scientific objectives of ESA is to detect and characterize habitable, Earth-like planets in other planetary systems, though there will be no room for Capt. Archer on board.

The spacecraft *Eddington* is named after Arthur Eddington, the modern theorist of modern stellar structure theory. *Eddington* is a 940-kg spacecraft designed to study the brightness of stars and thereby to test whether planets are passing before them. Each time a planet passes in front of a star, its brightness is reduced ever so slightly, just enough to be detected (this is called transient planet detection). Shaped like a box with a truncated, 1.2-m telescope and solar panels on each side, *Eddington* would be put into L2 orbit by a *Soyuz Fregat* from Baykonur. The emphasis of *Eddington* will not be on large planets (many of these are now suspected to exist), but on smaller, Earth-like, terrestrial planets orbiting in habitable zones around stars. The *Eddington* mission would have a number of other secondary scientific objectives, in such areas as quasars, the physical properties of stars, galactic halos and the light curves of supernovae. To keep costs down, *Eddington* could be based on the *Mars Express* design and would be, including the telescope, up to 4 m tall. The best observing point for *Eddington* is L2, located 1.5 million km away from the Earth on the outer side of the Earth's orbit (L1 is on the solar side): although the location takes some time to reach, it is a stable environment and offers the observer a complete view of the heavens every year.

7.26   DARWIN

Darwin also is a planet-finder, intended to find Earth-like worlds. Using the ever popular L2 location, it will study potential solar systems, blanking out the light of their Suns to look for encircling planets in habitable zones and telltale signs of life such as water, oxygen and ozone. Five 1-m, infrared telescopes cooled to 40°K are considered the best way to do this. Specifically, instruments will try to detect the existence, within these planetary atmospheres, of water and carbon dioxide so as to determine their capacity to sustain life as we know it. So far, over 40 planets have been discovered around other stars, but these 'discoveries' are qualified, for they are indirect and are based on inference, namely changes in a star's motion due to wobble or a change in light because of a planet passing in front. In all cases, they are large, gaseous planets, not suitable for life as we know it. A particular feature of the *Darwin* proposal is the use of seven separate, free-flying, interferometric spacecraft observatories, linked by a command and control spacecraft to relay data back. The proposers admit that it is a costly and ambitious project using new techniques and approaches – which may explain why the response so far has been cautious. Missions like *Darwin*, *Planck*, *Eddington* and *Herschel* meant that Europe would require much larger dishes than has hitherto been the case. Europe's largest dishes have been 15 m, but now dishes over 30 m were required. By 2001 the first of three new 35-m dishes had been approved (for Perth, Australia), with other sites under consideration in Madrid, Portugal and French Polynesia.

Figure 7.15. *Darwin.*

7.27 *HYPER*

Hyper (high precision cold atom interferometry in space) is a fundamental physics mission like *Lisa*, aimed at testing the nature of gravity, magnetism, matter and acceleration. A relatively small satellite is involved (770 kg), which could be launched by a *Rockot* rocket out of Plesetsk in northern Russia into a polar, 700-km, 98.2° orbit.

7.28 *STORMS*

Storms is a proposal for a three-spacecraft constellation to study the Earth's magnetic fields and its inner magnetosphere, and to improve our forecasting of solar storms. Although there have been many magnetospheric satellites, our knowledge of the interaction between the Sun and our magnetosphere is imperfect, the Van Allen radiation belts are still not well understood and forecasting is still poor. *Storms* will redress that by flying into orbits that have not been used before and focus on studying and analysing magnetospheric storms. Three drum-shaped spacecraft are proposed, each satellite weighing 1,873 kg. A large amount

(1,322 kg) is propellant in order get the satellites from the latitude of Baykonur (46°N) to equatorial latitudes (15° orbit), using a *Soyuz Fregat* launcher. *Storms'* orbits will be quite elliptical, at 700 km to 44,000 km, period 13.7 hr. The near-equatorial inclination, bringing the satellites into the radiation belts, also means that a certain amount of radiation shielding will be required.

7.29   *MASTER*

Master (Mars and Asteroid) is another ESA planetary project left at the blueprint stage, an assessment study having been concluded in 2000. It is a descendant of a projected Russian Mars and asteroid Vesta mission once scheduled for the 1990s. The ESA study proposes the use of a 1,500-kg *Mars Express* bus to fly by Mars, drop four landers and use Mars gravity assist to go on to the main asteroid belt to intercept the third largest asteroid, 520-km-diameter Vesta. Launch would be on *Soyuz Fregat* from Baikonour. The small *Netlander*-class Mars landers would investigate Mars' seismology, magnetic field, soil and rocks, meteorology, climate and atmosphere.

7.30   *XEUS*

XEUS (X-ray Evolving Universe Spectroscopy) is designed as a successor mission to *XMM/Newton*. The aim of *XEUS* is to focus on the origins of the universe in general and on super-massive black holes in particular. *XEUS* will have the most unusual design, comprising two spacecraft operating 50 m apart. The telescope mirror will fly separate from the detector in order to provide additional sensitivity, 250 times better than *XMM/Newton*. The main spacecraft will look conventional enough – a box with wings – but it will fly conjointly 50 m away from a rotating pipe-like assembly comprising the telescope mirror. Construction of the *XEUS* system may be complicated and some of it may be carried out from the *ISS*. Developing the technologies to make a radical design like XEUS work will take several years. Launch could be any time from 2012.

7.31   APPLICATIONS MISSIONS

We have just described the science missions, but what is Europe's future in space in the area of applications? The launching of *Envisat* was a major milestone for ESA. This was a huge mission with a long lead-in time, large cost, but a big dividend. Like many other space agencies, ESA has been moving away from these big missions toward smaller, faster missions that can be developed and flown more quickly and with lower risk. This redefinition has taken place in cooperation with the European Union under the GMES initiative (Global Monitoring for the Environment and Security). This has led ESA to articulate its approach to Earth observations under

Figure 7.16. *XEUS.*

what is called the Earth Observation Envelope Programme, which is worth €927m and will run until 2007. This has two main elements: *Earth Explorer* and *Earth Watch*. Four missions have been specified under *Earth Explorer*. *Earth Watch* will define missions for the long-term provision of regular environmental data from orbit, and various mission concepts are under study. Already €108m have been allocated to this process.

7.32 *EARTH EXPLORER*

Earth Explorer comprises core missions to address broad environmental concerns and opportunity missions to respond quickly to areas of immediate environmental interest and evolving situations. The latter are expected to be smaller satellites on shorter term programmes.

The first core mission is *GOCE* (Gravity Ocean Circulation Evolution), which is due to fly in 2005 on a two-year study of ocean circulation and gravity field. *GOCE* is a 770-kg satellite to go into a low, 260-km, Sun-synchronous orbit, with Italy's

Alenia as the prime contractor. The idea of *GOCE* is to improve our knowledge of oceanography, ice sheets and changes in sea level.

The second core mission is *ADM/Aeolus* (Atmosphere Dynamics Mission), to study global vertical wind profiles (2007). *Aeolus* was so named after the keeper of the winds in Greek mythology and is designed to use a lidar to make global observation of wind movements from a 400-km orbit.

The first opportunity mission is *Cryosat*, a radar altimetry mission to determine the thickness of the Earth's ice sheets and marine ice cover (2003). The second opportunity mission is SMOS (Soil Moisture and Ocean Salinity) to examine ocean saltness, the water cycle and soil moisture so as to better predict weather, climate change and extreme climatic events (2006). This is a €105m mission between ESA, CNES and Spain managed by the Centre for Biosphere Studies in Toulouse.

Three projects were chosen in summer 2002 for a second round of *Earth Explorer* missions, to fly after 2008:

- *ACE+* – an atmospheric and climate explorer satellite;
- *EGPM* (Earth Global Precipitation Mission) – a joint mission with NASA and NASDA to study rainfall; and
- *Swarm* – four small, polar-orbiting spacecraft to study the Earth's magnetic field.

7.33 PROJECT *AURORA*

In 2001 ESA began to define a long-term plan for the European space effort – *Aurora*. This was seen as a 30-year plan to develop new technologies and revitalize high-tech European industries by a focused European strategy for the exploration of near-Earth space, manned missions and journeys further afield into the solar system. For the preparatory period a sum of €14.1m was allocated, of which €3m came from France (Germany did not opt in). The strategy was seen as linked to the efforts of the European Commission to make Europe the world's leading industrial bloc for trade, science, engineering and industry. A call for ideas was made in 2001, leading to 300 *Aurora* concepts being received and sent for scrutiny to what was called the Experts Scientific Exploration Group. An *Aurora* coordinator was appointed – Franco Ongara. Contenders for leading *Aurora* projects were likely to be reusable launchers, Mars missions and the search for life in the Jovian system. *Aurora* did not intend to challenge the United States head-on, but did intend to develop specialized areas where Europe could take the lead (e.g., in smart technologies).

In October 2002 ESA announced approval by its Exploration Programme Advisory Committee of the first two *Aurora* flagship missions, intended to be major milestones to advance scientific and technical knowledge, and two less complex, less expensive precursor projects called *Arrow* missions. The flagships selected were *ExoMars* and *Mars Sample Return*. The aim of *ExoMars* was to characterize the biological environment of Mars before more advanced unmanned or even manned missions: *ExoMars* involved a Mars rover with a 40-kg experimental package, autonomous navigation system and life-detecting payload. The *Mars Sample Return* (2011–2017) was for a lander to drill for samples where it came down, sending a recoverable capsule into a 150-km-high Mars orbit where it would be picked up by an Earthbound return vehicle. The *Arrow* missions were for a return-from-Mars re-entry demonstrator and a Mars aerocapture demonstrator. Ideally, these would pave the way for European participation in an international expedition to Mars in around 2025–2030. These proposals were then sent to the ESA ministerial council for approval. A subtext of the *Mars Sample Return* mission was that the project was a possible refuge for the planned French–American Mars sample return mission that by this time was beginning to suffer from American doubts and indecision.

7.34 ESA AND THE EUROPEAN UNION (EU)

ESA developed independently from, and evolved on quite separate paths from the European Community, later called the EU. ESA included several countries that were not members of the EU (Norway and Switzerland being the most obvious examples).

During the Delors presidency of the EU, there was a strong emphasis on the need for Europe to catch up with the industrial performance of the United States and Japan. Central to the Commission's analysis of Europe's underperformance was that European countries had underinvested in cutting-edge and mainstream technologies. Funding programmes strategically could kick-start European companies to get ahead again.

In 1988 the Directorate General for Research and Technology in the European Commission published a communication, or policy paper, on spaceflight. Following a meeting between the Director General of ESA and President of the Commission Jacques Delors, five working groups were formed to explore areas of common interest. The EU made it clear that it was not challenging ESA and must have been conscious that several member states of ESA were not part of the EU. The definitive involvement of the EU in space activities may be traced to the policy communication issued 1992 called *The European Community and Space – Challenges, Opportunities and New Actions*. This outlined the principle that the EU should support the European space industry in its development and modernization. Environmental and observational programmes were singled out for special attention.

Under FP4 (Fourth framework programme for research), 1995–8, there was funding of €55m for telecoms, €20m for navigation and €275m for observation. The amount devoted to space-related projects was expected to have doubled by the end of FP5 (Fifth framework programme), 1999–2003. An example of this investment was the vegetation instrument on board *SPOT 4* and *5*, where the EU contribution was in the order of €35m. This was the first time that the EU paid for the development of space hardware.

This modest policy did not last for long. Several high-level experts began to look at the synergies, logic and benefits of a closer link between ESA and the EU. In November 2000 three wise men were delegated to produce a report. Those chosen were Carl Bildt (former Swedish prime minister), Jean Peyrelevade (President of Crédit Lyonnais) and Lothar Späth (Chief Executive Officer of Jenoptik) who issued a report *Toward a Space Agency for the European Union*. In effect, it called for ESA to take on a broader role and for its integration into the European political mainstream, including the common foreign security policy and other fields of action covered by the EU. A joint declaration called *A European Strategy for Space* was adopted by a special session of the ESA council and the EU's research council. In effect, there was pressure to align ESA with the EU, as the space agency of the EU. Antonio Radota, speaking to the ESA council meeting in Edinburgh in November 2001, said:

> Additional effort is now required to make space one the pillars of tomorrow's Europe and to make ESA the space agency of Europe.

In the late 1990s there were concrete signs of a closer organizational association between the two. In 2001 European Commissioners Loyola de Palacio and Philippe Busquet visited ESTEC in Noordwijk to meet ESA Director General Antonio Radota. This led ESA and the EU to agree the setting up of a Joint Strategic Space Advisory Group with a task force (sometimes called the joint task force) with a remit to further develop and implement the strategy. The task force was formally charged 'to put together their resources in order to establish a joint space programme with a United Europe.' Joint resolutions on a European strategy for space were signed between ESA and the EU in Brussels.

So far, formal cooperation between the two institutions has been limited to the *Galileo* satellite navigation project and the GMES programme. GMES was an initiative of the European Commission in 1998 to improve, develop and coordinate the information available in Europe from space in the areas of environment, natural and man-made disasters and climate change. Apart from the European Commission the other bodies involved are ESA, the national space agencies, Eumetsat, the European Association of Remote Sensing Companies, the Western European Union Observing Centre and Eurospace. According to Michael Praet, the ESA representative in Brussels:

> The question is whether to create a space agency within the Commission – which would not make any sense – or to use ESA as a space agency for Europe, which will also become an implementing agency serving EU policy needs [185].

The closer practical connexions between ESA and the EU were underlined when the ESA council met in Edinburgh, Scotland in November 2001. It was chaired by

German research minister Edelgard Bulmahn, but sitting beside her was the European President Romano Prodi. Informal meetings were set up between representatives of the ESA council and the European Council of Research Ministers to progress the integration of the two bodies, with a common European space policy and ESA functioning as the space agency of the EU. There were hints of an agreement for their institutional integration to begin in 2003. A unit responsible for space research was set up within the European Commission's directorate general for research in 2001.

7.35 THE *GALILEO* SATELLITE NAVIGATION PROJECT

Galileo is much the largest joint project between ESA and the European Union. The idea of a separate European navigation system was first explored by ESA in the mid-1990s. Up to that point European users were dependent on the American global positioning system (GPS), whose military accuracy was degraded for civilian use, or the Russian system (GLONASS), which was used by planes and ships in Russia (the Chinese also developed their own, more advanced system – Beidou). The idea of a European satellite navigation system had four attractions: it would provide experience in the construction of a cutting-edge technology, make Europe independent of the United States, develop ESA–EU collaboration and be profitable.

The first studies were carried out in 1994. The French company Alcatel won a €27m architecture definition phase in 1999 to work out the detail of the concept. The outcome was a baseline design review held in Frascati, Italy, on 3 August 2000 and attended by 150 satellite navigation experts. They came to the conclusion that the *Galileo* project would need about 30 satellites in medium Earth orbit and four in geosynchronous orbit. Each spacecraft would weigh 700 kg and orbit at 23,616 km, a little like the GLONASS system, at 56° in three orbital planes. The individual satellites could be launched by *Ariane 5* (eight at a time) or *Soyuz Fregat* (two at a time).

A European system would make the EU self-sufficient for navigation and provide a strong technological stimulus, so the argument went. It would cost €3bn, of which €750m would come from the EU, €500m from ESA and the rest was expected from private industry. The transport directorate-general of the Commission was confident that the investment costs could be recovered through navigation use and fees from the military, ships at sea, aircraft and cars. The scale of the project did make some countries nervous.

Approving the *Galileo* project proved to be a prolonged and difficult process. At the 2001 autumn ESA council meeting held in Edinburgh, Scotland, all the member states except Britain and Denmark supported the programme, indicative levels of support from the major countries being Germany 25%, Italy 22%, France 17% and Spain 11%. At EU level it was eventually approved at the meeting of the council of European transport ministers on 26 March 2002. The European Commission's formula for public–private investment was the sticking point, private companies

Figure 7.17. *Galileo* project team at ESTEC.

being slow to come forward with offers to invest in the project (though plenty of companies were queuing up for contracts). The ministers agreed to release €550m for the project, money originally assigned to trans-European networks and hitherto used to fund railways, roads and bridges. Some saw *Galileo* as a public resource, others as a commercially driven proposition. Defining the correct mixture of public and private funding proved difficult. Several sceptics questioned the value of the project, given the availability of the American GPS and Russian GLONASS, while others responded by pointing to, despite the complex start-up processes, the ultimate success of *Airbus* and *Ariane* [186].

The first satellite was expected to launch in 2005, with *Galileo* operating from 2008. Among the companies expected to be involved in *Galileo* are Thales and Alcatel in France, Telespaziale and Enav from Italy and Aena from Spain.

7.36   EUROPE'S PLACE IN THE WORLD

It is too early to judge what may be the outcome of the courtship between ESA and the EU. Much will depend on the carrying out of the *Galileo* project. At this stage it should be possible to assess the scale, nature and characteristics of the European space effort in the context of global spaceflight endeavour.

First, at a financial level, the following table situates European space spending in a world context:

World space budgets, 2001, estimated

Country	Budget (€m)
United States	35,888
Europe	5,865
ESA	2,835
National programmes (civilian)	2,297
Japan	2,033
China	1,640
Russia	750
India	580

Adapted from Sevig (2002) *European Space Directory*, 17th edn, Sevig Press, Paris.

These figures should be treated with some caution. Space budgets are rarely accounted for in an internationally comparative and transparent manner and are distorted by incomplete information, imprecise calculations of the levels of military spending and the sometimes volatile movement of exchange rates. China's budget would be higher, but for its low labour costs. Russia's formal space budget is very low, but it has a huge infrastructural base, not reflected in the figures, and generates substantial export revenues not shown here.

Having said that, the predominance of the United States in space-spending is apparent. The Russian space leadership made the point after July 1969 that ultimately they could never beat the Americans to the Moon while the latter had a budget that was many times bigger than theirs. The breadth of American space activities has always been overwhelming, reflecting its level of spending. Global expenditure on space flight is estimated at €70bn (if we include communications satellites and private spending). Of the public spending, 76% is spent in the United States, divided between the military and civilian space programme, 14% by Europe and 10% by the rest of the world. The director of ESA's science programmes David Southwood has often made the point that Europe is also trying to run a broad-based programme – but with about a sixth of the allocation of the United States. The proportion of research spending going to space flight in Europe has traditionally been lower than the United States, 2.5% compared with 7% [187].

Despite being overshadowed globally by the United States, Europe still has a big space programme. About 33,000 people are directly employed in the space industry in Europe and a further 250,000 indirectly [188]. The Astrium company (Matra, Marconi, British Aerospace and DASA) is the largest space company in Europe and one of the largest in the world (only Lockheed Martin and Boeing are bigger). ESA's space budget in 2001 was in the order of €2.835bn.

ESA contributions and proportions, 2001

Country	Amount (€m)	Proportion (%)	Mandatory (%)	Optional programmes (%)
Austria	25.6	1.2	2.48	1.13
Belgium	112.8	5.5	3.27	9.09
Britain	141.1	6.9	13.97	4.55
Denmark	24.1	1.2	1.9	0.83
Finland	10.4	0.5	1.34	0.53
France	614.5	29.9	17.05	30.67
Germany	531.4	25.9	25	26.04
Ireland	6.5	0.3	0.72	0.2
Italy	287.4	14	13.46	13.01
Netherlands	58.8	2.9	4.65	2.64
Norway	20	1	1.67	0.75
Spain	91.9	4.5	6.85	4.36
Sweden	48.8	2.4	2.65	2.13
Switzerland	61.4	3	3.68	3.15
Canada	11.4	0.6	3.01	0.61
Czech Republic	0.3	0.1	–	0.06
Portugal	–	–	1.31	0.06
Hungary	–	–	–	0.01

Source: ESA.

In terms of spending, the following are the main priorities (2000):

ESA spending under headings

	€m	%
Navigation	114.8	3.4
Science	359.2	10.6
Earth observation	630.2	18.6
Microgravity	107.4	3.2
Telecommunications	330.7	9.8
Manned	541.2	16
Launchers	670.4	19.8
PRODEX (Programmes de Developpement d'Expériences Scientifiques [Programme for the Development of Scientific Experiments])	146.6	4.3
Transformation programme	3.2	0.1
Cooperation (US, Russia, etc.)	183.3	5.4
General budget	205.9	6.1
Miscellaneous	92.6	2.7

Source: ESA.

From these two tables it is apparent that France and Germany remain the main funders of ESA (29.9% and 25.9%, respectively), followed by Italy (14%, but rising).

There is then a group of upper middle-league funders, like Britain and Belgium; a group of lower middle-league funders like Sweden and the Netherlands; and then, to continue the footballing analogy, the minnows or very small countries. In terms of budgets, launchers swallow up the largest single proportion, almost a fifth, followed by Earth observations (18.6%), manned flight (16%) and space science (10.6%). Administrative budgets are low, with only 1,820 staff in 1997 and offices in Paris, Washington, Brussels, Kourou, Moscow and Toulouse. So much for the ESA budgets and spending. What about the national programmes? As the table below indicates, the national programmes are almost as large as the ESA budget itself.

National space budgets, 2001

Country	€m
Austria	5.2
Belgium	30
Britain	115
Denmark	4
Finland	25
France	1,683
Germany	158
Italy	180
Netherlands	40
Norway	8
Portugal	0.5
Spain	30
Sweden	16
Switzerland	2.2
Total	2,297

Source: Sevig Press (2002) *European Space Directory*, 17th edn, Sevig Press, Paris.

This table applies only to those countries that have national space programmes. Several other countries have space budgets that are contributed entirely to ESA (e.g., Ireland). At one stage Germany had no national programme, all its spending being devoted to ESA. France's dominance of this table is startling, spending €1,683m. France actually comprises 73% of all *national* space-spending in Europe – on top of the fact that France is the largest contributor to the ESA budget in the first place. The European country with the second largest national space programme is Italy, at €180m, a long way behind France. Next comes Germany and Britain and then a group of countries with small national space programmes (Spain and Belgium €30m each, Sweden €16m, Finland €15m) and then some minnows: Norway, Austria, Switzerland and Portugal. However, as we have seen, some small countries have achieved impressive results with very small programmes.

7.37 EUROPE IN SPACE: AN ASSESSMENT

In October 2002 European rocketry was 60 years old. The circumstances that marked the launching of the first modern rocket, the A-4 or V-2, were not such that it was an anniversary to be marked with much enthusiasm or pride, despite the technical achievements involved.

Nevertheless, 60 years later there was much that could be celebrated. Despite the post-war difficulties Europe had pulled itself together to organize a modern space programme. Although the baton of leadership was first picked up in Britain, it rapidly passed to France which emerged as Europe's leading spacefaring nation, not only with a vibrant national programme, but also becoming the driving force of European space collaboration through good times and bad. Despite the role of nations in the building of European space efforts, it must be remembered that it was a group of European scientists who came spontaneously together to prompt the formation of ESRO (European Space Research Organization) and ELDO (European Launch Development Organization) over 1960–2. Whatever the quarrels between countries in the years that followed, European scientific collaboration has generally been of a high quality, transcending national, political, linguistic and cultural boundaries.

The 1960s were a difficult learning experience for European rocketry and science, with patience sorely tested by rockets crashing into the sea, poor programme management and heavy-handed political interventions. There was a real prospect that the European space effort would terminally collapse in 1972–3. During its darkest hour France managed to drive the faltering project forward once more with the obscurely named *L3S* programme, assisted by Belgian bureaucratic heroes and well-timed diplomatic interventions from Britain and Germany. They concocted what seemed at first a shaky political formula to rebuild the European space effort in the name of the European Space Agency. The formula proved durable and withstood its greatest test, the Kenneth Clark challenge of 1987, something that would certainly have overcome a weaker project.

The improbable *L3S* was, defying the classical gods, renamed *Ariane*. The rest is history and the ESA never looked back. From its triumphant first launch in the jungle of French Guiana at Christmas 1979 to the end of 2002, ESA completed over 150 launches and won 230 satellite orders, securing 56% of world commercial satellite launches and 27% of commercial communications satellites, a remarkable achievement in only 20 years. Not only was launching space satellites profitable but the technologies required drove European reindustrialization as well, especially in France. ESA's scientific programmes brought a huge haul of results, from the Infrared Space Observatory (ISO) to the giant *XMM/Newton* telescope. *Giotto* was shot into the heart of comet Halley. Although Europe's rockets did not always fly perfectly, no science mission ever failed. Europe's applications satellites were to the fore in developing meteorology (*Meteosat*), communications (*Olympus*, *Artemis*) and Earth observations (*ERS*, *Envisat*). Europe gradually built an experience in manned spaceflight, both through national and European programmes, sending visitors to the Russian space stations *Salyut* and *Mir* and on board

Europe's *Spacelab* and on the space shuttle. ESA's space efforts were supplemented by national space programmes, from the enormously extensive French programme, to small countries like Sweden, Belgium, and the Netherlands, which carved out distinctive projects and missions that completed a rich picture of scientific and engineering effort. Italy, the rising star of European space activities, built on its historical traditions of engineering excellence to build modules for the International Space Station (*ISS*).

With these achievements behind it, Europe's best years are probably still to come. Lined up are a series of applications missions to help us better understand our fragile planet, new meteorological programmes, an indigenous satellite navigation system, the first European probe to Mars, ambitious space telescopes and *Rosetta*, an adventure straight out of science fiction to land on a comet. Europe is building space modules for the *ISS* and is constructing an automated vehicle to fly supplies to the station for Europe's corps of astronauts. Called *Jules Verne*, this spaceship typifies a programme that has turned imagination into reality.

References and further reading

Baudry, Patrick (1994) *Ariane*. Flammarion Letters, Paris [in French].

Carlier, Claude & Gilli, Marcel (1994) *The First Thirty Years at CNES, 1962–1992*. La Documentation française, Paris [in French and English translation].

Chadeau, E. (1999) *Naissance d'Ariane Ambition Technologique*. Rive Droite, Paris [in French].

Caprara, Giovanni (2000) *Living in Space – from Science Fiction to the International Space Station*. Alenia & Firefly Books, Ontario, Canada.

Clarke, Arthur C. (ed.) (1967) *The Coming of the Space Age*. Victor Gollancz, London.

de Galiana, Thomas (1968) *Concise Encyclopaedia of Astronautics*. Collins, Glasgow and Follett, Chicago.

Gatland, Kenneth (1975) *Missiles and Rockets*. Blandford, London.

Gatland, Kenneth (ed.) (1989) *The Illustrated Encyclopaedia of Space Technology, 2nd edn*. Salamander, London.

Lewis, Richard (1983) *The Illustrated Encyclopaedia of Space Exploration*. Salamander, London.

Madders, Kevin (1997) *A New Force at a New Frontier. Europe's Development in the Space Field in the Light of its Main Actors, Policies, Law and Activities from its Beginnings up to the Present*. Cambridge University Press.

Massey, Harrie & Robbins, M. O. (1986) *History of British Space Science*. Cambridge University Press.

Millard, Douglas (2001) *The Black Arrow Rocket – A History of a Satellite Launch Vehicle and Its Engines*. Science Museum, London.

Morgan, Tom (1997–8) *Jane's Space Directory*, 14th edn. Jane's Information Group, Couldson, UK.

Myra, David (2002) *Saenger – Germany's Orbital Rocket Bomber in World War II*. Schiffer, Atglen, PA.

Neufeld, Michael (1997) *The Rocket and the Reich – Peenemünde and the Coming of the Ballistic Missile Era*. Harvard University Press, Boston.

Pilippe, J. P. (1993) *Ariane Horizon 2000*. J.P. Taillandie, Paris [in French].

Sevig Press (2002) *European Space Directory*, 17th edn. Sevig Press, Paris.

Shapland, David & Rycroft, Michael (1985) *Spacelab – Research in Earth Orbit*. Cambridge University Press.
Tavis, Robert (1998) *Operation Ariane 5*. Gerard de Villiers, Paris [in French].
Turnill, Reginald (1974) *The Observer's Book of Unmanned Spaceflight*. Frederick Warne & Co., London & New York.

PERIODICALS

- *Air & Cosmos*
- *Spaceflight*
- *Aviation Week & Space Technology*
- *ESA Bulletin*
- *Flight International*
- *CNES Magazine* (1997–) and, before that, *Lettre d'information* (1994–7) and *La Lettre de CNES* (–1992).

WEBSITES

Space agencies

- Austrian Space Agency   www.asaspace.at
- Hungarian Space Office   www.hso.hu
- British National Space Centre   www.bnsc.gov.uk
- European Space Agency   www.esa.int
- Swedish Space Corporation   www.ssc.se
- DLR Germany   www.dlr.de
- Netherlands National Aerospace Laboratory   www.nlr.nl
- Netherlands Agency for Aerospace   www.nivr.nl
  Programmes
- French Space Agency, CNES   www.cnes.fr

Companies

- EADS Launch Vehicles   www.launchers.eads.net
- Galileo Avionica   www.spazio.galileoavionica.com
- Contraves   www.contravesspace.com

List of European space launchings

The following is a listing of all European launchings to date under national and ESA programmes. First come national programmes, which apply to only three countries: France, Italy and Britain. Then there is a full listing of ESA launches – all on *Ariane* versions from French Guiana.

France's space launches

26 November 1965	*A-1 Asterix*	*Diamant A*	Hamaguir
17 February 1966	*D-1A Diapson*	*Diamant A*	Hamaguir
8 February 1967	*D-1C Diadème*	*Diamant A*	Hamaguir
15 February 1967	*D-1D Diadème 2*	*Diamant A*	Hamaguir
12 January 1970	*Péole*	*Diamant B*	Kourou
10 March 1970	*Wika/Mika*	*Diamant B*	Kourou (for Germany)
15 April 1971	*D-2A Tournesol 1*	*Diamant B*	Kourou
5 December 1971	*D-2B Polaire*	*Diamant B*	Kourou (fail)
21 May 1973	*Castor/Pollux*	*Diamant BP-4*	Kourou (fail)
6 February 1975	*Starlette*	*Diamant BP-4*	Kourou
17 May 1975	*Castor/Pollux*	*Diamant BP-4*	Kourou
27 September 1975	*D-2B Aura*	*Diamant BP-4*	Kourou

Italian launches

26 April 1964	*San Marco 2*	*Scout*	Formosa Bay, Kenya
24 April 1971	*San Marco 3*	*Scout*	Formosa Bay, Kenya
18 February 1974	*San Marco 4*	*Scout*	Formosa Bay, Kenya
25 March 1988	*San Marco 5*	*Scout*	Formosa Bay, Kenya

British launches

28 October 1971	*Prospero*	*Black Arrow*	Woomera

ESA launches

V1	*Ariane 1*	24 December 1979	Technical payload
V2	*Ariane 1*	23 May 1980	*Firewheel, Amsat 3A* (failure)
V3	*Ariane 1*	19 June 1981	*Meteosat 2, APPLE*
V4	*Ariane 1*	28 December 1981	*Marecs A*
V5	*Ariane 1*	10 September 1982	*Marecs B, Sirio 2* (failure)
V6	*Ariane 1*	16 June 1983	*ECS 1, Oscar 10*
V7	*Ariane 1*	18 October 1983	*Intelsat V F7*
V8	*Ariane 1*	5 March 1984	*Intelsat V F8*
V9	*Ariane 1*	22 May 1984	*Spacenet 1, Brazilsat 2*
V10	*Ariane 3*	4 August 1984	*ECS 2, Telecom 1A*
V11	*Ariane 3*	10 November 1984	*Spacenet 2, Marecs B-2*
V12	*Ariane 3*	8 February 1985	*Arabsat 1A, Brazilsat 1*
V13	*Ariane 3*	8 May 1985	*GStar 1, Telecom 1B*
V14	*Ariane 1*	2 July 1985	*Giotto*
V15	*Ariane 3*	12 September 1985	*ECS 3, Spacenet 3* (failure)
V16	*Ariane 1*	22 February 1986	*SPOT 1, Viking*
V17	*Ariane 3*	28 March 1986	*GStar 2, Geostar*
V18	*Ariane 2*	31 May 1986	*Intelsat 5A F14* (failure)
V19	*Ariane 3*	16 September 1987	*ECS 4, Aussat K-3*
V20	*Ariane 2*	20 November 1987	*TVSat 1*
V21	*Ariane 3*	11 March 1988	*Spacenet 3R, Telecom 1C*
V22	*Ariane 44LP*	15 June 1988	*Panamsat 1, Meteosat 3, Oscar 13*
V23	*Ariane 2*	17 May 1988	*Intelsat V F13*
V24	*Ariane 3*	21 July 1988	*Insat 1C, ECS 5*
V25	*Ariane 3*	8 September 1988	*GStar 3, SBS 5*
V26	*Ariane 2*	28 October 1988	*TDF 1*
V27	*Ariane 44LP*	11 December 1988	*Skynet 4B, Astra 1A*
V28	*Ariane 2*	27 January 1989	*Intelsat V F15*
V29	*Ariane 44LP*	4 March 1989	*JCSat 1, Meteosat 4*
V30	*Ariane 2*	2 April 1989	*Tele-X*
V31	*Ariane 44L*	5 June 1989	*Superbird 1, DFS 1*
V32	*Ariane 3*	11 July 1989	*Olympus 1*
V33	*Ariane 44LP*	8 August 1989	*Hipparcos, TVSat 2*
V34	*Ariane 44L*	27 October 1989	*Intelsat 602*
V35	*Ariane 40*	22 January 1990	*SPOT 2, UoSat 3, 4, Microsats 1–4*
V36	*Ariane 44L*	22 February 1990	*Superbird B, BS-2X* (fail)
V37	*Ariane 44L*	24 July 1990	*TDF 2, DFS 2*
V38	*Ariane 44LP*	30 August 1990	*Skynet 4C, Eutelsat 2F1*
V39	*Ariane 44L*	12 October 1990	*SBS 6, Galaxy 6*
V40	*Ariane 42P*	20 November 1990	*Satcom C1, GStar 4*
V41	*Ariane 44L*	15 January 1991	*Italsat 1, Eutelsat 2F2*
V42	*Ariane 44LP*	2 March 1991	*MOP 2, Astra 1B*
V43	*Ariane 44P*	4 April 1991	*Anik E2*
V44	*Ariane 40*	17 July 1991	*ERS 1, UoSat 5, SARA, Orbcom X, Tubsat A*
V45	*Ariane 44L*	14 August 1991	*Intelsat 605*
V46	*Ariane 44P*	26 September 1991	*Anik E1*
V47	*Ariane 44L*	29 October 1991	*Intelsat 601*

V48	*Ariane 44L*	16 December 1991	*Inmarsat 2F3, Telecom 2A*
V49	*Ariane 44L*	26 February 1992	*Arabsat 1C, Superbird B1*
V50	*Ariane 44L*	15 April 1992	*Telecom 2B, Inmarsat 2F4*
V51	*Ariane 44L*	9 July 1992	*Insat 2A, Eutelsat 2F4*
V52	*Ariane 42P*	11 August 1992	*Topex Poseidon, S80T, Kitsat 1*
V53	*Ariane 44LP*	10 September 1992	*Hispasat 1A, Satcom C3*
V54	*Ariane 42P*	28 October 1992	*Galaxy 7*
V55	*Ariane 42P*	1 December 1992	*Superbird A1*
V56	*Ariane 42L*	12 May 1993	*Astra 1C, Arsène*
V57	*Ariane 42P*	25 June 1993	*Galaxy 4*
V58	*Ariane 44L*	22 July 1993	*Hispasat 1Bm, Insat 2B*
V59	*Ariane 40*	26 September 1993	*SPOT 3, Stella, Kitsat 2, Itamsat, PoSat, Healthsat 2, Eyesat*
V60	*Ariane 44LP*	22 October 1993	*Intelsat 701*
V61	*Ariane 44LP*	20 November 1993	*Solidaridad 1, Meteosat 6*
V62	*Ariane 44L*	18 December 1993	*DBS 1, Thaicom 1*
V63	*Ariane 44LP*	24 January 1993	*Turksat 1, Eutelsat 2FS* (fail)
V64	*Ariane 44LP*	17 June 1994	*Intelsat 702, STRV 1A, STRV 1B*
V65	*Ariane 44L*	8 July 1994	*PAS 2, BS 3N*
V66	*Ariane 44LP*	10 August 1993	*Brasilsat B1, Turksat 1B*
V67	*Ariane 42L*	9 September 1994	*Telsat 402*
V68	*Ariane 44L*	8 October 1994	*Solidaridad 2, Thaicom 2*
V69	*Ariane 42P*	1 November 1994	*Astra 1D*
V70	*Ariane 42P*	1 December 1994	*PAS 3* (failure)
V71	*Ariane 44LP*	28 March 1995	*Hot Bird 1, Brasilsat B2*
V72	*Ariane 40*	21 April 1995	*ERS 2*
V73	*Ariane 44LP*	17 May 1995	*Intelsat 706*
V74	*Ariane 42P*	10 June 1995	*DBS 3* (failure)
V75	*Ariane 40*	7 July 1995	*Helios 1A, Cerise, UPMSat 1*
V76	*Ariane 42L*	3 August 1995	*PAS 4*
V77	*Ariane 44P*	29 August 1995	*NStar A*
V78	*Ariane 42L*	24 September 1995	*Telstar 402R*
V79	*Ariane 42L*	19 October 1995	*Astra 1E*
V80	*Ariane 44P*	17 November 1995	*ISO*
V81	*Ariane 44L*	6 December 1995	*Telecom 2C, Insat 2C*
V82	*Ariane 44L*	12 January 1996	*PAS 3R, Measat 1*
V83	*Ariane 44P*	5 February 1996	*NStar B*
V84	*Ariane 44LP*	14 March 1996	*Intelsat 707*
V85	*Ariane 42P*	20 April 1996	*MSAT 1*
V86	*Ariane 44L*	16 May 1996	*Palapa C2, Amos*
V87	*Ariane 44P*	13 June 1996	*Intelsat 708*
V88	*Ariane 501*	4 June 1996	*Cluster* (failure)
V89	*Ariane 44L*	9 July 1996	*Arabsat 2A, Turksat 2C*
V90	*Ariane 44L*	8 August 1996	*Italsat 2, Telecom 2D*
V91	*Ariane 42P*	11 September 1996	*Echostar 2*
V92	*Ariane 44L*	13 November 1996	*Arabsat 2A, Measat 2*
V93	*Ariane 44L*	30 January 1997	*GE 2, Nahuel 1A*
V94	*Ariane 44P*	1 March 1997	*Intelsat 801*

(*continued*)

ESA launches (*cont.*)

V95	*Ariane 44L*	16 April 1997	*Thaicom 3, BSat 1A*
V96	*Ariane 44P*	25 June 1997	*Intelsat 802*
V97	*Ariane 44L*	3 June 1997	*Inmarsat 3F4, Insat 2D*
V98	*Ariane 44P*	8 August 1887	*Panamsat 6*
V99	*Ariane 44LP*	2 September 1997	*Hot Bird 3, Meteosat 7*
V100	*Ariane 42L*	23 September 1997	*Intelsat 803*
V101	*Ariane 502*	30 October 1997	*Maqsat H, B*
V102	*Ariane 44L*	12 November 1997	*Sirius 2, Cakrawarta 1*
V103	*Ariane 44P*	2 December 1997	*JCSAT 5, Equator S*
V104	*Ariane 42L*	22 December 1997	*Intelsat 804*
V105	*Ariane 44LP*	4 February 1998	*Brasilsat B3, Inmarsat 3F5*
V106	*Ariane 4*	27 February 1998	*Hot Bird 4*
V107	*Ariane 40*	24 March 1998	*SPOT 4*
V108	*Ariane 44P*	29 April 1998	*Nilesat 101, BSB 1B*
V109	*Ariane 4*	25 August 1998	*ST 1*
V110	*Ariane 44LP*	16 September 1998	*Panamsat 7*
V111	*Ariane 44L*	5 October 1998	*Eutelsat W2, Sirius 3*
V112	*Ariane 503*	21 October 1998	*ARD*
V113	*Ariane 44L*	28 October 1998	*Afristar, GE-5*
V114	*Ariane 42L*	6 December 1998	*Satmex 5*
V115	*Ariane 42L*	21 December 1998	*Panamsat 6B*
V116	*Ariane 44L*	26 February 1999	*Arabsat 3A, Skynet 4E*
V117	*Ariane 42P*	2 April 1999	*Insat 2E*
V118	*Ariane 42P*	12 August 1999	*Telkom 1*
V119	*Ariane 504*	10 December 1999	*XMM*
V120	*Ariane 42P*	4 September 1999	*Koreasat 3*
V121	*Ariane 44P*	25 September 1999	*Telstar 7*
V122	*Ariane 44LP*	19 October 1999	*Orion 2*
V123	*Ariane 4*	13 November 1999	*GE 4*
V124	*Ariane 40*	3 December 1999	*Helios 1B, Clementine*
V125	*Ariane 44L*	21 December 1999	*Galaxy 11*
V126	*Ariane 42L*	25 January 2000	*Galaxy 10R*
V127	*Ariane 44LP*	18 February 2000	*Superbird 4*
V128	*Ariane 505*	21 March 2000	*Asiasat, Insat 3B*
V129	*Ariane 42L*	19 April 2000	*Galaxy 4R*
V130	*Ariane 506*	14 September 2000	*Astra 2B, GE 7*
V131	*Ariane 44LP*	17 August 2000	*Brasilsat B4, Nilesat 102*
V132	*Ariane 44P*	6 September 2000	*Eutelsat W1*
V133	*Ariane 42L*	6 October 2000	*N Sat 110*
V134	*Ariane 44LP*	29 October 2000	*Europestar 1*
V135	*Ariane 507*	16 November 2000	*PAS 1R, Amsat P3D, STRV 1C, 1D*
V136	*Ariane 44L*	21 November 2000	*Anik F1*
V138	*Ariane 508*	20 December 2000	*Astra 2D, GE8, LDREX*
V137	*Ariane 44P*	10 January 2001	*Turksat 2, Eurasiasat 1*
V139	*Ariane 44L*	7 February 2001	*Sicral 1, Skynet 4F*
V140	*Ariane 509*	8 March 2001	*Eurobird 1, BSAT 2a*
V142	*Ariane 510*	12 July 2001	*Artemis, BSAT 2b*

V144	*Ariane 44P*	25 September 2001	*Atlantic Bird TM 2*
V141	*Ariane 4*	9 June 2001	*Intelsat 901*
V143	*Ariane 44L*	30 August 2001	*Intelsat 902*
V146	*Ariane 44LP*	27 November 2001	*DirecTV 4S*
V147	*Ariane 42L*	23 January 2002	*Insat 3C*
V145	*Ariane 511*	28 February 2002	*Envisat*
V150	*Ariane 44L*	16 April 2002	*New Skies Satellite 7*
V148	*Ariane 44L*	23 February 2002	*Intelsat 904*
V149	*Ariane 44L*	28 March 2002	*Astra 3A, JC Sat 8*
V151	*Ariane 42P*	3 May 2002	*SPOT 5, Idefix*
V152	*Ariane 44L*	5 June 2002	*Intelsat 905*
V153	*Ariane 512*	5 July 2002	*Stellat 5, N-Star C*
V155	*Ariane 513*	28 August 2002	*MSG 1, Atlantic Bird 1*
V154	*Ariane 44L*	5 September 2002	*Intelsat 906*
V157	*Ariane 517*	11 December 2002	*Stentor, Hot Bird 6* (failure)
V156	*Ariane 44L*	17 December 2002	*New Skies Satellite 6*

Notes

1 Sir Harry Massey & M. O. Robbins (1986) *History of British Space Science.* Cambridge University Press.

2 Sir William E. Congreve, 1772–1828. *Spaceflight,* **23**, March 1981.

3 Carl E. Franklin (2001) The early rockets of William Congreve in British service. *Journal of the British Interplanetary Society,* **54**(9/10), October–November.

4 For a history of the VfR, see Frank Winter (1977) Birth of the VfR – the start of modern astronautics. *Spaceflight,* **19**(7/8), July–August.

5 Henry Matthews (1995) Max Valier, 1895–1930. *Spaceflight,* **37**(5), May.

6 Frank Winter (1979) 1928–30 forerunners of the shuttle – the von Opel flights. *Spaceflight,* **21**(2), February.

7 For an appreciation of von Braun, see A. V. Cleaver (1997) Wernher von Braun – a special tribute. *Spaceflight,* **19**(9), September.

8 Tom Huntington (1993) V-2 – the long shadow. *Air & Space,* February–March.

9 Jürgen P. Esders (1998) History centre of war and rockets. *Spaceflight,* **40**(2), February.

10 Nicolae-Florin Zaganescu, George Popa, Rodcia Zaganescu & Lucia Popa (1999) *Dr Irene Saenger-Bredt – A Life for Astronautics.* 50th Astronautical Congress, the Netherlands, 4–8 October.

11 David Myra (2002) *Sänger – Germany's Orbital Rocket Bomber in World War II.* Schiffer, Atglen, PA.

12 Roland S. Speth (1998) Visiting Peenmünde. *Spaceflight,* **40**(12), December. See also Roland S. Speth (2000) Visiting the Mittelwerk – past and present. *Spaceflight,* **42**(3), March.

13 Mitchell R. Sharpe (1977) The day the British army launched the V-2s. *Spaceflight,* **19**(5), May.

14 L. J. Carter (1979) Visit to PERME. *Spaceflight,* **21**(6), June.

15 Michael Neufeld (1997) *The Rocket and the Reich – Peenemünde and the Coming of the Ballistic Missile Age*. Harvard University Press, Boston.

16 Readers are recommended the following extensive, informative and well-illustrated history of CNES: Claude Carlier & Marcel Gilli (1994) *The First Thirty Years at CNES*. Paris, La Documentation française.

17 G. V. E Thompson (1979) The British Interplanetary Society. *Spaceflight*, **21**(10), October. See also Frank H. Winter (1983) The formative years of the BIS – from Liverpool to London. *Spaceflight*, **25**(11), November.

18 Charles Tarratt (1985) In retrospect. *Spaceflight*, **27**(5), May.

19 Redstones for Australia. *Flight International*, 30 December 1965.

20 C. N. Hill (2001) *A Vertical Empire – A History of the UK Rocket and Space Programme, 1950–71*. Imperial College Press, London.

21 Roy Sherwood (1991) Britain in space. *Spaceflight*, **33**(5), May.

22 For a history of Woomera, see Isaac Boxx: Woomera. *Spaceflight*, **37**(6–7), June, July 1995 (two parts).

23 Rob Baker (2000) The Wilson government policy toward ELDO. *Journal of the British Interplanetary Society*, **53**.

24 ELDO's perigee/apogee system and Comsat application of the ELDO vehicles. *Flight International*, 28 July 1966.

25 ELDO analysed. *Flight International*, 29 June 1972.

26 Europa II: ELDO takes action. *Flight International*, 2 March 1973.

27 Europe's advanced engine. *Flight International*, 26 October 1973.

28 Europe agrees at last – new hopes for European cooperation as space agency is agreed. *Flight International*, 4 January 1975.

29 Alan Lawrie (1985) Space at East Fortune. *Spaceflight*, **21**(11), November. See also Cooped up. *Spaceflight*, **27**(4), April.

30 Sir Harry Massey & M. O. Robbins (1986) *History of British Space Science*. Cambridge University Press.

31 ESRO discord. *Flight International*, 10 December 1970.

32 Simon E. Dinwiddy (1984) ESA's telecommunications programme. *Spaceflight*, **26**(1), January.

33 The significance of Diamant. *Flight International*, 16 December 1965.

34 Robert Bramscher (1980) A survey of launch vehicle failures. *Spaceflight*, **22**, (11–12), November–December.

35 William McLaughlin (1997) Oceanography from space. *Spaceflight*, **29**(6), June.

36 Heading for Mars. CNES, 1999.

37 Christian Lardier (2000) Accord Franco – Americain sur Mars. *Air & Cosmos*, No. 1769, 3 November. See also Michael A. Taverna (2000) France, US target 2001 kickoff for joint sample return missions. *Aviation Week & Space Technology*, 13 November.

38 Douglas Millard (2001) *The Black Arrow Project – A History of a Satellite Launch Vehicle and its Engines*. Science Museum, London.

39 John Harlow (1998) Woomera, October. *Spaceflight*, **41**(1), January 1999.

40 A. R. Thompson (1998) British space rockets. *Spaceflight*, **40**(3), March.

41 D. J. Brown (1980) Skylark sounding rocket facilities for materials science experiments. *Spaceflight*, **22**, May.

42 A. R. Thompson (1998) British space rockets. *Spaceflight*, **40**(3), March.

43 The case for a national space agency. *Spaceflight*, **27**(1), January.

44 Roy Gibson (1984) A time for greatness. *Spaceflight*, **26**(4), April.

45 Clive Simpson (1986) New British space initiative. *Spaceflight*, **28**(1), January.

46 UK – Soviet space deal. *Spaceflight*, **28**(11), November 1986.

47 Space plan flounders. *Spaceflight*, **29**(9), September 1987.

48 William E. Smith (1987) A proud Ariane again takes flight. *Time*, 5 October.

49 Roy Gibson (1988) Britain's space future. *Spaceflight*, **30**(2), February.

50 Arthur Pryor (1989) British space must pay its way, in Reginald Turnill (ed), *Space technology International Annual*.

51 Andrew Wilson (1979) Scout – NASA's small satellite launcher. *Spaceflight*, **20**(11), November.

52 Michael Coe (1978) Ariel 5 – a British triumph. *Spaceflight*, **20**(2), February.

53 A. R. Thompson (1999) Britain's first interplanetary science mission. *Spaceflight*, **41**(2), February.

54 R. L. Harris (1995) Military satellite communications in the UK. *Spaceflight*, **37**(1), October.

55 Ian Hayes (1978) Britain's military satellite. *Spaceflight*, **20**(2), February.

56 Neil Pattie (1989) Skynet 4 – the unknown soldier. *Spaceflight*, **31**(5), May.

57 Frank Morrings (2001) At Surrey Satellite Technology, small and simple is beautiful. *Aviation Week & Space Technology*, 18 June.

58 Tim Furniss (1992) Italy aims high. *Flight International*, 25 December 1991– 7 January 1992.

59 Michael A. Taverna (2001) Twin milsats enhance Europe's telecom net. *Aviation Week & Space Technology*, 12 February.

60 Theo Pirard (1998) European modules for the International Space Station. *Spaceflight*, **40**(6), June.

61 Suzane Parch (1980) Rolf Engel – fifty years in rocketry and spaceflight. *Spaceflight*, **22**(6), June.

62 Richard Taylor (1986) X-ray comet. *Spaceflight*, **38**(6), June.

63 Michael A. Taverna (2001) Germany earmarks $3.5bn for space. *Aviation Week & Space Technology*, 28 May.

64 Marius Werner (1997) Germany's No.1 spaceflight hardware producer. *Spaceflight*, **39**(2), February.

65 W. I. McLaughlin & W. H. de Leeuw (1978) Infrared astronomical satellite. *Spaceflight*, **20**(5), May.

66 William I. McLaughlin (1987) IRAS asteroid survey. *Spaceflight*, **29**(1), January.

67 IRAS and galaxies. *Spaceflight*, **28**(12), December.

68 Stefan Zenker (1997) *Space Is Our Place – A Personal Memoir on the Occasion of the 25th Anniversary of the Swedish Space Corporation*. Stockholm, Swedish Space Corporation.

69 Alan D. Farmer (1983) Sweden in space. *Spaceflight*, **25**(3), March. See also L. J. Carter (1986) Vikings on the space path. *Spaceflight*, **28**(2), February.

70 Theo Pirard (1995) The Astra constellation. *Spaceflight*, **37**(10), October. See also Theo Pirard (1998) The evolving Astra system. *Spaceflight*, **40**(8), August.

71 Norway in space. *Spaceflight*, **31**(12), December 1989. See also Michael Brady (2002) Putting space to work in Norway. *Spaceflight*, **44**(3), March.

72 Europe to try again next week. *Flight International*, 26 July 1973.

73 For unity in Europe. *Flight International*, 5 July 1973.

74 Bruno Gire (1982) The Ariane launcher. *Spaceflight*, **24**(9–10), September–October.

75 David Velupillai (1979) Europe's equatorial launch site. *Flight International*, 17 February.

76 David Velupillai (1979) Ariane: First flight and after. *Flight International*, 8 December.

77 Tim Furniss (1986) Ariane's big fix. *Flight International*, 20 September. See also Gilbert Sedbon (1987) Tension mounts in Kourou. *Flight International*, 12 September.

78 Bruno Gire & Jacques Schibler (1984) The Ariane 3 launcher. *Spaceflight*, **26**(12), December.

79 David Velupillai (1983) Ariane 3 uprated. *Flight International*, 30 April.

80 Louisa Wright (1984) Competitor in the cosmos – Europe's Arianespace is giving NASA a run for its money. *Newsweek*, 26 November.

81 Julian Moxon (1985) Ariane 4 – Europe's launcher grows. *Flight International*, 4 May.

82 David Velupillai (1979) Ariane – first flight and after. *Flight International*, 8 December.

83 Gilbert Sedbon (1989) Vulcain forges ahead. *Flight International*, 18 February; See also Gilbert Sedbon (1990) Vulcain burning to soar. *Flight International*, 17–23 October.

84 Michael A. Taverna (2001) New Ariane user facility opens for business. *Aviation Week & Space Technology*, 18 June.

85 Ariane 5 – Europe's new heavy payload launcher. *Spaceflight*, **38**(5), May 1996.

86 Two-comet mission. *Spaceflight*, **22**(6), June 1980.

87 William I. McLoughlin (1984) Through the tail of a comet. *Spaceflight*, **26**(11), November.

88 L. J. Carter (1986) Comet flyby – first results. *Spaceflight*, **28**(3), March.

89 Earth gets a halo. *Spaceflight*, **21**(5), May 1979.

90 R. M. Bonnet (2002) History of the Giotto mission. *Journal of the British Interplanetary Society*, **22**(1), Space Chronicle series.

91 The Giotto spacecraft – why was Giotto so special? *Journal of the British Interplanetary Society*, **22**(1), 2002, Space Chronicle series.

92 Darren Burnham (1992) The second coming of Giotto. *Spaceflight*, **34**(6–7), June–July (in two parts). See also Journey to Grigg – Skellerup. *Spaceflight*, **35**(4), April 1992.

93 Richard Flower (1992) Health check on Giotto. *Spaceflight*, **35**(4), April.

94 Neville Kidger (1982) Hipparcos – Europe's astronmetric satellite. *Spaceflight*, **24**(4), April.

95 For an end of mission report, see Norman Longdon (1993) Astrometry reaps a rich harvest of data. *Spaceflight*, **35**(11), November.

96 Europe's new space telescope in action. *Spaceflight*, **38**(5), May 1996.

97 Clive Simpson (1985) Space science in Europe. *Spaceflight*, **27**(4) April. See also Gordon Whitcomb (1985) The ESA science programme. *Spaceflight*, **27**(5), May.

98 Ulysses probes the Sun's south pole. *Spaceflight*, **36**(12), December 1994.

99 Richard Marsden (2000) Four dimensional view of Sun – Ulysses anniversary marked by solar pass. *Spaceflight*, **42**(12), December.

100 Europe's Titan probe. *Spaceflight*, **38**(7), July 1996.

101 Bernhard Fleck & Stein Haugan (2000) SOHO uncovers solar secrets. *Spaceflight*, **42**(11), November.

102 Jacques van Oene (2000) Ariane 504 delivers XMM telescope into space. *Spaceflight*, **42**(2), February.

103 Europe's OTS satellite. *Flight International*, 4 June 1977.

104 Telecommunications: the OTS legacy. *Spaceflight*, **31**(1), January 1989.

105 Europe's community satellite. *Flight International*, 17 December 1977.

106 D. E. B. Wilkins (1989) Olympus in orbit. See also Deborah Smith (1989) Olympus: a giant among satellites. *Spaceflight*, **31**(11), November.

107 Dave Wilkins (1981) The Olympus recovery – mission impossible? *Spaceflight*, **33**(11), November.

108 Theo Pirard (1998) Artemis – next European technological satellite. *Spaceflight*, **40**(9), September. See also Jacques van Oene (2001) Artemis ready for launch this summer. *Spaceflight*, **43**(7), July.

109 Meteosat – Europe's geostationary weather satellite. *Spaceflight*, **20**(4), April 1978. See also Meteosat – a success story with a long tradition. *Spaceflight*, **35**(10), October 1993.

110 Norman Langdon (1991) Bright eyes on Earth. *Spaceflight*, **33**(5), May. See also **33**(9), September 1991 (in two parts).

111 Norman Langdon (1993) From oil slicks to green shoots – all in a day's work for ERS 1. *Spaceflight*, **35**(4), April.

112 For a history of its evolution, see Mark Hempsell (1993) Early history of the polar platform, 1984–7. *Spaceflight*, **35**(8), August.

113 Simon Chalkley & Judith Simpson (1993) The polar platform – what it is and what it provides. *Spaceflight*, **35**(8), August.

114 Jacques Louet (2001) Europe's Envisat to provide wealth of data on Earth. *Spaceflight*, **43**(10), October.

115 Joel W. Powell (1996) Biopan and Biobox – ESA's recoverable experiment carriers. *Spaceflight*, **38**(2), February.

116 Joel W. Powell (1998) German reentry vehicle flies on Foton 11. *Spaceflight*, **40**(6), June.

117 Joel W. Powell (2001) Microgravity upgrade. *Spaceflight*, **43**(8), August.

118 David Buchan (1994) Spies on high. *Irish Times*, 7 November.

119 Michael A Taverna (2001) France readies sigint satellite system. *Aviation Week & Space Technology*, 1 October.

120 Philippe Ninane (1999) Europeans in space. *Spaceflight*, **41**(10), October; **41**(11), November; **41**(12), December (in three parts).

121 Roland S. Speth (1999) The endless fall of Soyuz 29. *Spaceflight*, **41**(4), April.

122 Erik Seedhouse (1998) Reinhold Ewald – the fourth German on Mir. *Spaceflight*, **40**(3), March.

123 Bert Vis (1997) Launch to Mir. *Spaceflight*, **39**(5), May.

124 Philip Corneille (2001) Austrian becomes father on launch day. *Spaceflight*, **43**(11), November.

125 Francis French (2002) First British ride into space. *Spaceflight*, **44**(5), May.

126 Neville Kidger (1995) ESA studies effects of long – term flights. *Spaceflight*, **37**(2), November.

127 Darren Burnham (1996) Astronaut tells both sides of the story. *Spaceflight*, **38**(6), June. See also Tim Furniss (1994) A month on Mir. *Flight International*, 14–20 September.

128 Neville Kidger (1995) EuroMir 95. *Spaceflight*, **37**(12), December. See also **38**(4), April 1996.

129 UK stirs on post – Apollo. *Flight International*, 27 April 1972.

130 Post Apollo – a new quandary. *Flight International*, 20 July 1972.

131 The continuing conflict. *Flight International*, 2 November 1972.

132 Europe gets shuttle deadline. *Flight International*, 2 March 1973.

133 Congress at Vienna. *Flight International*, 2 November 1972.

134 David Baker (1978) Space shuttle – a user's guide. *Flight International*, 10 May.

135 D. J. Shapland & F. Rossitto (1994) European manned space programmes crew assignments. *Spaceflight*, **36**(2), February. See also Gordon R. Bolton & Wernher Riesseknabb (1998) Spacelab era ends. *Spaceflight*, **40**(8), August.

136 David Baker & Michael Wilson (1976) Space shuttle debut. *Flight International*, 25 September.

137 David Velupillai (1982) Europe creates a workplace in space. *Flight International*, 15 May.

138 Europe signs for Spacelab. *Flight International*, 4 October 1973.

139 I am grateful to Rex Hall for supplying a copy of *The Story of the French Astronaut Selection*, CAP Espace paper, undated, with details of this and subsequent selections of French astronauts.

140 For a full description, see John A. Pfannerstill (1984) The first flight of Spacelab, *Spaceflight*, **26**(5), April.

141 David Velupillai (1984) Spacelab scrutinized. *Flight International*, 25 February. See also Sharon Begley (1983) Launching an orbiting lab. *Newsweek*, 5 December; Not for astronauts only. *Newsweek*, 12 December; and Those balky computers again. *Time*, 19 December 1983.

142 Erik Seedhouse (1998) Heike Walpot – to become the first German woman in space. *Spaceflight*, **40**(5), May.

143. Daniel Fisher (1990) Invitation to Mir brings new challenges to German

astronaut corps. *Spaceflight*, **21**(6), June. See also Germany books third Spacelab. *Flight International*, 5 August 1989.

144 Steven Young (1992) International Spacelab mission an outstanding success. *Spaceflight*, **34**(3), March.

145 Phillip Corneille (2002) First of a new generation. *Spaceflight*, **44**(4), April.

146 Roelof Schuiling (1996) Columbia breaks duration record. *Spaceflight*, **38**(10), October.

147 Ben Evans (1999) SRTM – mapping the world in 3D. *Spaceflight*, **41**(9), September.

148 Bert Vis (2001) The ESA parabolic flights programme. *Spaceflight*, **43**(5), May.

149 European astronaut centre celebrates its tenth anniversary. *Spaceflight*, **42**(8), August 2000.

150 Norman Longdon (1992) Eureca and Europe can cry Eureka! *Spaceflight*, **34**(7,8), July, August (in two parts); All the makings of a great success – Eureca – first report on early results. *Spaceflight*, **34**(10), October.

151 Steven Young (1992) STS-46 – tethered satellite mission ends in disappointment. *Spaceflight*, **34**(9), September.

152 Rodica Ionasecu & Paul A Penzo (1988) Innovative use of space tethers. *Spaceflight*, **30**(5), May.

153 Tim Furniss (1992) Tethered trouble. *Flight International*, 15–21 July.

154 Martin Sénéchal (1986) Hermes – the French shuttle. *Spaceflight*, **28**(1), January.

155 Tim Furniss (1988) Ariane 5 – T minus seven years and counting. *Flight International*, 16 January.

156 Tim Furniss (1989) Changing times for Hermes. *Flight International*, 5 August.

157 Tim Furniss (1990) Hermes holds firm. *Flight International*, 7–13 March.

158 Gilbert Sedbon (1990) Heat is on Hermes. *Flight International*, 7–13 March.

159 Hermes design nearly finished. Hermes in doubt. *Spaceflight*, **32**(11), November 1990.

160 A new look at Hermes. *Spaceflight*, **31**(10), October 1989.

161 Hermes costs soar as capabilities are lost. *Spaceflight*, **33**(8), August.

162 Steven Young (1992) Euro plans hit by costs. *Spaceflight*, **34**(1), January.

163 Jacques Chirac veut rétablir Hermes. *Air & Cosmos*, No. 1411, 1–7 February 1993.

164 *La lettre de CNES*. No. 140, November 1992.

165 For an appreciation of *Saenger* in comparison to *Hermes* and *HOTOL*, see Saenger joins Hermes and HOTOL. *Flight International*, 13 September.

166 Doug Millard (1993) Spaceplanes – back to the future. *Spaceflight*, **35**(3), March.

167 Mark Hempsell (1995) HOTOL's secret engines revealed. *Spaceflight*, **35**(5), May.

168 For a sketch of the *HOTOL* mission, see Clive Simpson (1987) Flying into the future. *Spaceflight*, **29**(1), January.

169 Robert C. Parkinson (1988) HOTOL – status report. Spaceflight, **30**(9), September.

170 Richard Varvill & Alan Bond (1993) Skylon – a new element of a future space transportation system. *Spaceflight*, **35**(5), May.

171 Roy Gibson (1984) Europe and the space station. *Spaceflight*, **26**(1), January.

172 NASA and ESA 'agree' on space station. *Flight International*, 16 August.

173 Britain rethinks Columbus. *Flight International*, 23 April 1988.

174 Space station faces EVA, weight and power problems. *Spaceflight*, **30**(9), September 1990.

175 ESA ready to quit space station project. *Spaceflight*, **32**(12), December 1990.

176 Phase out shuttle, scale down space station. *Spaceflight*, **33**(2), February 1991.

177 NASA plans a smaller, cheaper space station Freedom. *Spaceflight*, **33**(5), May 1991.

178 ESA seeks space cooperation with Russia. *Spaceflight*, **35**(10), October 1993.

179 Jacques van Oene (2001) Dutch space station contribution. *Spaceflight*, **43**(7), July.

180 Christian Lardier (2000) Vol mouvementé pour le X-38. *Air & Cosmos*, No. 1170, 10 November.

181 Norman Longdon (1993) FESTIP – a long distance look into the future. *Spaceflight*, **35**(5), May.

182 Michael A. Taverna (2001) Europe strives to put RLV effort back on track. *Aviation Week & Space Technology*, 9 April.

183 Michael A. Taverna (2002) Europe revamps science programme. *Aviation Week & Space Technology*, 3 June.

184 Rosetta science. *Spaceflight*, **39**(7), July 1997.

185 Three questions for Michael Praet. RTD info, No. 30, June 2001.

186 Mike Healy (2001) Galileo – the countdown continues. *Spaceflight*, **43**(11), November.

187 A third community in space. *Spaceflight*, **26**(9–10), September–October 1983.

188 The direct estimate comes for Sevig Press, *European Space Directory 2002*, 17th edition, Paris.

Index